Building Sustainability Through Environmental Education

Lynn A. Wilson
SeaTrust Institute, USA

Carolyn N. Stevenson
Purdue University Global, USA

A volume in the Advances in Environmental Engineering and Green Technologies (AEEGT) Book Series

IGI Global
DISSEMINATOR OF KNOWLEDGE

Published in the United States of America by
IGI Global
Engineering Science Reference (an imprint of IGI Global)
701 E. Chocolate Avenue
Hershey PA, USA 17033
Tel: 717-533-8845
Fax: 717-533-8661
E-mail: cust@igi-global.com
Web site: http://www.igi-global.com

Library of Congress Cataloging-in-Publication Data

Names: Wilson, Lynn A., 1954- editor. | Stevenson, Carolyn, 1968- editor.
Title: Building sustainability through environmental education / Lynn A. Wilson and Carolyn Stevenson, editors.
Description: Hershey, PA : Engineering Science Reference, [2019]
Identifiers: LCCN 2018035051| ISBN 9781522577270 (hardcover) | ISBN 9781522577287 (ebook)
Subjects: LCSH: Environmental education. | Sustainable development. | Water security. | Climate change.
Classification: LCC GE70 .B84 2019 | DDC 333.7071--dc23 LC record available at https://lccn.loc.gov/2018035051

This book is published in the IGI Global book series Advances in Environmental Engineering and Green Technologies (AEEGT) (ISSN: 2326-9162; eISSN: 2326-9170)

British Cataloguing in Publication Data
A Cataloguing in Publication record for this book is available from the British Library.

Advances in Environmental Engineering and Green Technologies (AEEGT) Book Series

ISSN:2326-9162
EISSN:2326-9170

MISSION

Growing awareness and an increased focus on environmental issues such as climate change, energy use, and loss of non-renewable resources have brought about a greater need for research that provides potential solutions to these problems. Research in environmental science and engineering continues to play a vital role in uncovering new opportunities for a "green" future.

The **Advances in Environmental Engineering and Green Technologies (AEEGT)** book series is a mouthpiece for research in all aspects of environmental science, earth science, and green initiatives. This series supports the ongoing research in this field through publishing books that discuss topics within environmental engineering or that deal with the interdisciplinary field of green technologies.

COVERAGE

- Cleantech
- Air Quality
- Green Technology
- Contaminated Site Remediation
- Pollution Management
- Radioactive Waste Treatment
- Electric Vehicles
- Green Transportation
- Policies Involving Green Technologies and Environmental Engineering
- Renewable energy

IGI Global is currently accepting manuscripts for publication within this series. To submit a proposal for a volume in this series, please contact our Acquisition Editors at Acquisitions@igi-global.com or visit: http://www.igi-global.com/publish/.

The Advances in Environmental Engineering and Green Technologies (AEEGT) Book Series (ISSN 2326-9162) is published by IGI Global, 701 E. Chocolate Avenue, Hershey, PA 17033-1240, USA, www.igi-global.com. This series is composed of titles available for purchase individually; each title is edited to be contextually exclusive from any other title within the series. For pricing and ordering information please visit http://www.igi-global.com/book-series/advances-environmental-engineering-green-technologies/73679. Postmaster: Send all address changes to above address.

Titles in this Series

For a list of additional titles in this series, please visit:
https://www.igi-global.com/book-series/advances-environmental-engineering-green-technologies/73679

Harnessing Marine Macroalgae for Industrial Purposes in an Australian Context ...
Göran Roos (Economic Development Board of South Australia, Australia) Anthony Cheshire (Balance Carbon Pty Ltd, Australia) Sasi Nayar (South Australian Aquatic Sciences Centre, Australia) Steven M. Clarke (South Australian Research and Development Institute, Australia) and Wei Zhang (Flinders University, Australia)
Engineering Science Reference • ©2019 • 291pp • H/C (ISBN: 9781522555773) • US $195.00

Advanced Methodologies and Technologies in Engineering and Environmental Science
Mehdi Khosrow-Pour, D.B.A. (Information Resources Management Association, USA)
Engineering Science Reference • ©2019 • 506pp • H/C (ISBN: 9781522573593) • US $295.00

Transitioning Island Nations Into Sustainable Energy Hubs Emerging Research ...
Catalina Spataru (UCL Energy Institute, UK)
Engineering Science Reference • ©2019 • 172pp • H/C (ISBN: 9781522560029) • US $135.00

Recent Challenges and Advances in Geotechnical Earthquake Engineering
T.G. Sitharam (Indian Institute of Science, India) and J. S. Vinod (University of Wollongong, Australia)
Engineering Science Reference • ©2019 • 315pp • H/C (ISBN: 9781522569480) • US $195.00

Agrifood Economics and Sustainable Development in Contemporary Society
Gabriel Popescu (Bucharest Academy of Economic Studies, Romania)
Engineering Science Reference • ©2019 • 401pp • H/C (ISBN: 9781522557395) • US $195.00

Smart Farming Technologies for Sustainable Agricultural Development
Ramesh C. Poonia (Amity University Jaipur, India) Xiao-Zhi Gao (University of Eastern Finland, Finland) Linesh Raja (Amity University Jaipur, India) Sugam Sharma (Iowa State University, USA) and Sonali Vyas (Amity University Jaipur, India)
Engineering Science Reference • ©2019 • 308pp • H/C (ISBN: 9781522559092) • US $195.00

For an entire list of titles in this series, please visit:
https://www.igi-global.com/book-series/advances-environmental-engineering-green-technologies/73679

IGI Global
DISSEMINATOR OF KNOWLEDGE

701 East Chocolate Avenue, Hershey, PA 17033, USA
Tel: 717-533-8845 x100 • Fax: 717-533-8661
E-Mail: cust@igi-global.com • www.igi-global.com

Editorial Advisory Board

Table of Contents

Detailed Table of Contents

formal, and informal education. Vulnerable young people are often excluded from educational systems. Inclusive polices are needed to ensure access to education for poorest youth in cities and remote areas, youth affected with HIV, refugee youth, and migrant youth. This chapter explores education and sustainable development.

The international community is increasingly recognizing the importance of youth as key stakeholders in achieving the Sustainable Development Goals and Agenda 2030. Young people need to be continuously empowered and provided with opportunities to enhance their competences and networks. There are multiple capacity-building and networking events that seek to engage youth, yet the learning and impact generated in such spaces remains less understood. This chapter explores the value of youth-targeted and youth-led conferences and events centered on sustainability themes in advancing the learning experiences of youth participants. Testimonials from young people provide key insights into their engagement. The findings highlight the need for such platforms to be more empowering through focusing more directly on fostering collaborative actions amongst youth and organizations rather than merely on capacity-building on the spot; and through garnering financial or technical support for advancing youth action on the SDGs, especially at the local community level.

The United Nations Sustainable Development Goal 4 Quality Education has increased awareness in pre-university environmental education efforts. Environmental education is important to not only creating awareness of world environmental issues, but taking action towards fostering positive change. Environmental education programs such as SeaTrust Institute's AWARE (Action Within a Resilient Environment) assist teens in learning about issues that directly impact their communities and their world. AWARE combines environmental education with hands-on experiential learning projects that help promote environmental awareness in their communities. Through education and experience with active scholars and professional practitioners, students gain an increased understanding of environmental challenges and ways to make a positive impact – both domestically and globally. This is especially critical to developing countries which lack the educational programs and resources to address the impact environmental changes have on their nations and communities.

This chapter describes and explains the role of science communication to advance environmental education on climate change with a special reference on Africa being one of the regions suffering from the effects of climate-induced disasters and risks in the increased anthropogenic effects of modern development. It is argued that scientists are poor communicators on what they do, and hence, are often misunderstood by the media and society. This then calls for attention to be paid with regards to science communication, which has to be packaged in ways that make it easier for the generality of citizens to interpret and understand. Efforts in this regard are made possible through environmental education, which has proved to be useful in the discourse of disaster risk management in different parts of the world. The buttressing methodological philosophy to this chapter is applied systems approach. Critical areas of reference are health and diseases, resilient communities, coastal adaptation, and farming practices and technologies.

Well before island nations began to consider rising ocean levels, a feature of global climate change, they have been concerned with the allocation of water resources. The purpose of this chapter is to revisit the efforts of Zanzibar's academic, as well as private and public institutions, as they promote environmentally responsible entrepreneurial projects, while advancing women's economic empowerment. Analysis will examine the history of seaweed production and consider how Zanzibar's seaweed farmers have recently responded to the dislocations associated with global climate change. This discussion will also consider to what extent Zanzibari seaweed production reflects the norms enshrined in the United Nation's Rio + 20 platform, and the language of the UN's 2030 sustainable development goals.

Informed action by the leaders of the future is critical for creating resilient communities. Preparing these future leaders through formal and informal education, research, and environmental/climate change programs that interweave local knowledge with the most current global science positions them to becomes the catalysts that propel community leaders to engage a wider range of possible futures. This chapter integrates findings from a SeaTrust Institute research project with the sustainable development goals in an analysis supporting dynamic and reconfigurable combinations of agents that promote the attributes of elasticity, future orientation, and motivation to address the high stakes choices for resilience to climate and environmental/social change. Author objectives in this chapter are to illustrate the optimum roles of youth in the process and what preparations and conditions are needed to instill and support youth in their ability to flip a process at the point of catastrophe to restore equilibrium and promote resilience.

The fate of empires depends on the education of the youth. – Aristotle

Foreword

I met the senior editor of this volume in 1995 at an environmental writing workshop in Vermont. At that time Dr. Lynn Wilson was passionately writing about saving the world's oceans. Now 23 years later, she along with her colleague and coeditor Dr. Carolyn Stevenson and the other esteemed contributors to this book are writing passionately and profoundly about what we now know to be the greatest threat to not only to our seas, but to humans and the entire planet. Specifically, this book focuses attention on the critical role that environmental education can, and should play in addressing the enormous and ubiquitous climate change challenges.

I believe that one of the first things that readers will notice when they pick up this book is that it is truly global in scope. The list of contributors includes authors from such far ranging locations as Zimbabwe, The Netherlands, and Indonesia. This of course is appropriate as climate change is a global problem. However, when one digs deeper into the individual chapters the reader find that the focus is on local initiatives the collectively address the global climate change crisis.

Conservationist and author Wendell Berry (1991) wrote eloquently about this need to focus locally to address global problems.

Unless one is willing to be destructive on a very large scale, one cannot do something except locally, in a small place. ... You can't act locally by thinking globally. Look at one of those photographs of half the earth taken from outer space, and see if you recognize your neighborhood. In order to make ecological good sense for the planet, you must make ecological good sense locally. If you want to keep your local acts from destroying the globe, you must think locally.

This book artfully strikes the balance of presenting local activities and ideas that are globally distributed and have inherent global implications.

It allows us to hear the voices of youth. It is through the ideas and energy of youth that resiliency will come. In fact a common thread throughout the book is the role that youth in tackling current challenges within the shadowy prospect of an uncertain future. For example, this book presents a description the *SeaTrust Institute's*

AWARE program, a promising effort aimed at engaging youth in environmental stewardship. AWARE is a community service program for high school and middle school students. As with the other contributions that characterize this book, the AWARE program blends the local with the global. Students are involved in local community-based projects, but they also have the opportunity to collaborate with students around the world who share similar interests and environmental issues. Moreover, *AWARE* helps students of all ages build a life-long global professional peer network. Students explore science, engineering, social and policy processes that make their community more resilient to environmental change.

This book makes a significant contribution to informing and inspiring the reader to attend to the needs of our communities and ecosystems in responding to future climate changes. It has been noted that the greatest threat to conservation is the loss of hope. Environmentalist and author Scott Russell Sanders (1998) wrote that, "To live in hope we needn't believe that everything will turn out well. We only need to believe we are on the right path." This book gives reasons for hope. It encourages readers to choose and travel down a path of resiliency and prosperity in an age of climate change.

Ted Cable
Kansas State University, USA

REFERENCES

Berry, W. (1991, February). Out of your car, off your horse: Twenty-seven propositions about global thinking and the sustainability of cities. *The Atlantic.* Retrieved from https://www.theatlantic.com/magazine/archive/1991/02/out-your-car-your-horse/309159/

Sander, S. (1998). *Hunting for hope: A father's journey*. Boston: Beacon Press.

Preface

As the interconnection of the physical and social aspects of sustainability are increasingly well recognized by nations and cultures around the world, discussions and decisions about what is best for humankind in the face of extreme change and upheaval have moved from a primary, almost singular focus on mitigation, to one that elevates adaptation and capacity building for resilience. Sustainability activities, tied to building resilience, have largely coalesced around the Sustainable Development Goals, known as the SDGs. The SDGs are 17 globally agreed-upon goals with 169 targets that encapsulate the major social issues in the world today. These goals range from poverty and hunger, health, education, gender and inequality, to water and oceans, cities, infrastructure, energy, climate change, natural resources, peace, jobs and partnerships. The impetus for this book is the urgent need for the education community, at all levels, to purposefully address the SDGs in actionable ways that contribute directly to helping individuals, communities, nations, and the world build resilience to extreme and often violent change.

Educators not only convey knowledge and information; they also conduct research and help define what counts as knowledge. This opportunity to create the kind of global awareness across generations that catalyzes meaningful action in response to the most critical issues impacting the environment and society also carries somber responsibilities. In 2017 at the United Nations High Level Political Forum on Sustainability, UN General Assembly President and Higher Education Sustainability Initiative (HESI) keynote speaker Peter Thompson identified education as a key driver of sustainable development asserting that "no enabler more powerful than inclusive, equitable education for all" (United Nations Department of Social and Economic Affairs, 2017).

Education is specifically addressed in SDG #4, but it has a much more far reaching role as a building block for peace, resilience, and sustainable development. Educators are charged with scaling up and changing educational delivery on the ground and virtually in ways that change the conversation about sustainability and resilience. Knowledge empowers people to take ownership of the SDGs and to take positive steps towards resilience. Alongside Mr. Thompson, UN Deputy

Secretary-General Amina Mohammed labeled education as "the cornerstone of sustainable development…We know when we deliver education to a young person, we're not only delivering the knowledge and skills they will need to chart their own future — we're preparing them to lend their hands, their mind, and their heart to shaping a more peaceful, prosperous future for their society, and indeed, for the world" (UN Department of Social and Economic Affairs, 2017). Towards that end, Mr. Thompson explicitly asked for universities to incorporate the SDGs into their curricula and learning processes.

In May 2018, the University Network for Climate Capacity (UNCC) was launched by a group of educational and research institutions that came together though their work with the Research Constituency of the United Nations Framework Convention on Climate Change (UNFCCC). The UNCC is charged with sharing curricula, developing research partnerships across the Global South and Global North, and helping to provide research to countries working to deliver their promised contributions to the Paris Agreement and the Sustainable Development Goals. Implicit in its mission is building global youth capacity to lead SDG implementation and long-term stewardship of the legacy of the SDGs. Developing the next generation of leaders for building resilience under extreme social and physical change represents one of the most poignant areas in which the youth experience with science and environmental issues is not only desirable, but crucial for infusing thoughtful actions into increasingly difficult and complex sustainability decision processes. The escalating importance accorded to education in enabling achievement of the SDGs guides the consideration of education's responsibility to sustainability as addressed by the authors in this book.

A BRIEF HISTORY OF EDUCATION AND THE SDGS

Multilateral process documents leading up to the SDGs unambiguously endorse the value of youth education in addressing the crises of global sustainability and resilience development. In 2002, the United Nations General Assembly proclaimed the United Nations Decade of Education for Sustainable Development (2005–2014), emphasizing the indispensable role of education in achieving sustainable development. The Rio Declaration on Environment and Development of 1992 (RIO) calls youth an indispensable component in achieving sustainable development, a statement that was echoed in official transcripts from RIO+10 in 2002, RIO+20 in 2012, and in Agenda 21, in Chapter 25 *Children, Youth and Sustainable Development* and in Chapter 36 *Promoting Education, Public Awareness and Training* (Agenda 21, 1993). The UNFCCC position on the value of education in sustainability (Article 6) and the Kyoto Protocol (Article 10) promoting awareness and education were both reaffirmed through the 2014 Lima Ministerial Declaration on Education and Awareness-Raising at COP20 held in Lima, Peru, in December 2014 and as part of the 2015 Paris Agreement. Authorized at the Rio Earth Summit on sustainability in 1992, and as the parent treaty to the Kyoto Protocol from which the SDGs evolved, the UNFCCC is the convention through which the Sustainable Development Goals (SDGs) became linked with the 2015 Paris Agreement (UNFCCC, 2015). The UNFCCC also oversees the framing, implementation, monitoring, and evaluation of the SDGs, and countries' progress towards achieving the goals.

Contained in Agenda 2030 (UN General Assembly, 2015), the SDGS were conceived at the Rio+20 Conference in 2012 which concluded with the agreement by Member States to launch a process to develop a set of Sustainable Development Goals (SDGs) as a successor framework to the Millennium Development Goals (MDGs). The SDGs came into effect in September 2015, just prior to the authorization and first signatory countries pledges in support of the Paris Agreement in November at the meeting of the UNFCCC. Several landmark changes to multilateral paradigms were inaugurated through Agenda 2030 including the integration of the three fundamental pillars of development: economic, social, and environmental. As an accepted, universally applicable approach, countries from both the Global North and Global South agreed to implement it. And, it includes issues that had previously been outside the scope of development, especially peace and climate change (Independent Commission on Multilateralism, & International Peace Institute, 2016).

SDG Goal #4 is Quality Education, which is designed to "ensure inclusive and equitable quality education and promote lifelong learning opportunities for all" (UN General Assembly, 2015). Each SDG has an associated set of established targets and global indicators associated with those targets. (See the Appendix for a list of education goal targets and indicators.) But addressing targets does not always equate with sustainability. For example, while progress has been reported for access to education for both boys and girls, particularly at the primary school level, millions of children in over 40 countries remain without pre-primary education. Access does not necessarily mean quality of education or even completion of primary school leaving over 100 million youth worldwide without basic literacy skills. More than 60 percent of those lacking basic skills are women which puts Target 1 of Goal #4 (to ensure that, by 2030, all girls and boys complete free, equitable, and quality primary and secondary education) in jeopardy. Indicators for SDG Goal #4 targets for learning outcomes (target 1), early childhood education (target 2), and effective learning environments rely on incomplete and aggregate data, making it difficult to analyze and identify the children at greatest risk of being left behind. This one example highlights some of the challenges with a global evaluation process that depends upon local data and local implementation under voluntary reporting and in the absence of context-sensitive global oversight. In recognition of this potential vulnerability in achieving the SDGs, during 2014 the UN called for gathering more information from countries about how non-state actors can contribute to all the SDGs, including quality education.

In July 2017, the High Level Political Forum on Sustainable Development (HLPF)was held in New York City under the auspices of the Economic and Social Council, ECOCOC. One session featured the Higher Education Sustainability Initiative (HESI), a partnership between United Nations Department of Economic and Social Affairs, UNESCO, United Nations Environment, UN Global Compact's Principles for Responsible Management Education (PRME) initiative, United Nations University (UNU), UN-HABITAT and UNCTAD. HESI was created in 2012 in the run-up to the United Nations Conference on Sustainable Development (Rio+20). With commitments from over 300 universities from around the world, HESI accounted for more than one-third of all the voluntary commitments that were launched at Rio+20. Through its strong association with the United Nations, HESI provides higher education institutions with a unique interface between higher education, science, and policy making.

The HESI sessions targeted government officials, academia, university administrators, UN representatives, sustainability professionals, major groups and other relevant stakeholders.

The 2017 meeting of HESI at the HLPF showcased how the 2030 Agenda for Sustainable Development, including the SDGs, is being integrating into sustainability strategies, research, teaching, pedagogy, and campus practices, and to position higher education institutions as key drivers for achieving the SDGs. Throughout, it was stressed that partnerships will be the only way to achieve the central vision of SDGs. Four categories were defined by HESI:

1. Teach sustainable development across all disciplines of study
2. Encourage research and dissemination of sustainable development knowledge
3. Green campuses and support local sustainability efforts
4. Engage and share information with international networks.

While international and national level declarations are important attestations of high-level support, outcomes from meaningful environmental education in creating preferable outcomes to climate and other environmental change play out at local levels illustrating how everyone's "global" is someone else's "local." While there is no endpoint to environmental and social change, research points to behaviors with the largest potential benefits having a requirement of political engagement that is developed through experiences associated with young people's interests in public issues (Chawla, 2007). Often beginning with personal actions like those in the home or in public arena like schools and communities, these experiences require a personal sense of competence and a sense of collective competence, or confidence in one's ability to achieve goals by working with a group.

Youth Development Within the Multilateral Arena in Parallel to the SDGs

Against the backdrop of the dialogues about youth and sustainable development leading up to the 2015 release of the SDGs, youth had been developing an increasingly sophisticated international climate and sustainability policy voice through their formally recognized constituency at the UNFCCC. In 2009, youth were denied access to the COP15 premises in Copenhagen and responded by stripping down to their underwear and jumping up and down in the snow chanting "don't leave youth out in the cold!" Five years later at the 2014 COP20 Youth Day, the official Youth Constituency presented "Intergenerational Inquiry: Youth as Agents of Change," inside the negotiating venue, pointing to how youth are self-organizing to take on climate change governance and actions such as in the 2014 New York City HLPF climate meetings, and calling on global youth to participate in community government events. Youth are exuberant and even dramatic. These presenters were no exception, exhibiting emotional despair at the state of the environment and what that means to their future. The Brazilian representative railed at the people in the room that "the men of paper are the new men of war," while the Danish Youth Council representative pointed to the professional audience proclaiming that "You have condemned us to an uncertain future" (Wilson, 2014).

Alongside the admonitions to older generations, other youth participants offered suggestions. World Scout representative Nathan Nguyen reminded the audience that among the issues youth face is the issue of meaningful participation and that to have that kind of participation, youth need to work with older people. Others chimed in that while youth from the Global North had well prepared statements for governments, those from the Global South were ill-prepared and the differences in global ideologies were threatening the opportunities for dialogue. Youth colleagues called for increasing global awareness among all youth regarding the science and the social implications of inaction or maladaptation so that the youth perspective can explicitly contribute to solutions whether at the UN or in their local communities. At COP23 hosted by Fiji, youth leader Timothy Damon presented as part of a panel in the host country's Pavilion about the transformational abilities and unique contributions of youth in building resilience when provided with the knowledge, training and opportunities from older generations (Wilson, 2017).

Outside the UN, educators had been laying the groundwork for the youth platform far before COP20, claiming that inaction on climate change is a dangerous and improper action that is a detriment to resilience and sustainability. Environmental education is a recently evolved academic field; climate change education is newer still and fraught with the expected controversy surrounding a new far-reaching discipline (Schreiner, 2005). In 1968 the United Nations Educational, Scientific and Cultural Organization (UNESCO) called for curriculum development in environmental education at all grade levels and the stimulation of global awareness to environmental problems (Palmer 2001). As the Millennium Development Goals (MDGs) morphed into the Sustainable Development Goals (SDGs) in 2015, climate change as a focal point for youth action increasingly targeted initiatives such as UNEP and UNESCO's *youthXchange* guidebooks that enable educators and young people to personalize learning and climate change resilience actions (Anderson 2012). Those actions range widely from public health to ocean acidification, multilateral processes, collaboration and conflict, international capacity development, climate equity, green building, food security, insurance, finance and business, sustainable development and disaster risk reduction.

A roundtable which was overseen by the U.S. Board on Science Education, the Board on Environmental Change and Society, and the Division on Earth and Life Studies, met during a recent workshop to focus on the teaching and learning of climate change and climate science in formal education settings (National Resources Council 2012). Participants linked formal education to the necessary links between scientific and technical analysis with public deliberation and decision making as stated in America's Climate Choices (National Resources Council 2011) as critical to effective responses to sustainability that results in resilience in which citizens understand the risks of both action and inaction and to engage in effective deliberation about all available choices.

This short chronicle demonstrates how the evolving relationship between education and the SDGs clarifies the roles and responsibilities for education in achieving the Sustainable Development Goals. Through experience, the editors of this volume know that the most effective sustainability work is, as Sea Trust Institute says, "Local to Global and Back Again (SM)." Engaging youth with the science, engineering, social and policy processes that make their community more resilient to environmental change creates an opportunity for global dialogue and focuses this text on developing a deeper understanding of social learning in the transition from awareness to action. Opportunities for youth in engaging with the SDGs may be as change-makers who act and catalyze others, critical thinkers who question and challenge power structures and barriers to change, communicators who take the messages of agreement and impasse to peers and communities, innovators who bring fresh ideas and alternative approaches, and as leaders who are prepared to take their place at the head of discussions and actions at all levels.

EXPLORING HOW ENVIRONMENTAL EDUCATION IMPACTS LOCAL AND GLOBAL RESILIENCE DEVELOPMENT

At the highest risk for climate change disasters are the world's poorest people. Most of these are youth, under age 24, who make up nearly half of the world's population with most living in developing countries. These young people have a significant role to play in creating a sustainable and resilient future. Education makes regional and international interdisciplinary scientific research, education and policy accessible for building resilience to climate change at local and global levels. This book, *Building Sustainability Through Environmental Education,* demonstrates ways in which education can be used to impact both sustainability and resilience to environmental and social change.

Both sustainability and resilience require reconciliation of environmental, social equity and economic demands - often referred to as the "three pillars" of sustainability. Together, these comprise the definition of the general subject area. Diverse contributions to the book allow readers to understand new theoretical and practical approaches emerging from current research and identify ways to adapt or transfer them across sectors, geography, ecosystems and cultures. Authors utilize an interdisciplinary approach, blending environmental and climate science with social, political and economic knowledge impacting local conditions on topics ranging from adaptation and resilience in developing countries to the role of experiential and designed learning spaces, science communication, vulnerability, roles of different players in promoting sustainability, and a thought experiment about environmental education in the postnormal future. Chapters highlight case studies, analyses and descriptions of scientific, use-inspired research projects of interest to policy-makers, scientists, environmental and civic leaders and the public to demonstrate what it means to promote climate change awareness for a new generation of environmental leaders. Each author places their work within the context of the SDGs.

The first chapters offer a high level view of the research, principles, governance and policy issues at the nexus of education and sustainability. With emphasis on how actors across scales and sectors can positively impact resilience development in communities and government, authors explore how education impacts water security, changing environmental conditions, and community resilience in a complex future. In the next section, authors provide more in-depth analyses of conferences, educational programs and non-traditional learning spaces. Chapter authors then move into broad discussions about disaster risk learning strategies, communication, potential limits to the public sector's ability to address levels of complexity in a postnormal future, before concluding with an examination of women's empowerment through a specialized entrepreneurial approach that directly responds to the SDGs and shared global norms, and the roles of youth as change agents and catalysts for creating and sustaining resilience.

BUILDING RESILIENCE THROUGH ENVIRONMENTAL EDUCATION: A CONTRIBUTION TO THE LITERATURE

What are scholars of sustainability, policy, and environmental education discovering about the links, wise approaches, long lasting effects and pivotal changes in approaches under evolving conditions of change and uncertainty? This book is designed to address some of the gaps in that knowledge including perspectives that illuminate and resonate with messages from governments, communities, and especially youth who bring powerful new sources of information for expanding knowledge about how environmental education may best promote sustainability and resilience.

1. *Need for empirical evidence – showing examples that engage both youth and adults through environmental education* to equip them to lead and participate in resilience development action projects. Authors explore new ways to purposefully infuse education into local and global environmental decisions and actions in the face of escalating changes and challenges. Informal as well as formal learning approaches tailored to today's media culture and learning mechanisms in different settings and cultures must simultaneously address capacity building, organizational structure, and changing technologies to support resilience in a changeable and uncertain future.

Capacity—attitudes and competencies leaders need to guide, support or facilitate environmental and sustainability education to catalyze and sustain resilient communities. Examples of what constitutes a successful sustainability and environmental education project are measured through their effects on awareness, resilience and future action.

Organization—educational systems and structures that contribute to dynamic learning environments while respecting different physical, ecological, cultural, political, economic and social realities and combining perspectives to create unique outcomes suited to specific communities and situations. Potential new or augmented roles for educators, NGOs, scientists, students, businesses, government leaders, and environmental researchers to address both present crises and unknown future effects through education are considered.

Technology—effective technologies that embrace changing conditions. Flexible approaches need to be used effectively to enhance environmental educational collaborations and interactions between researchers, communities, governments, academics, and youth.

2. *Need for new perspectives on sustainability, combined environmental/ social awareness, and modes of education from academic and professional*

research for academics, practitioners, educators and governmental leaders. Different modalities of environmental education in action suggest new ways to purposefully engage education with local and global environmental decisions and actions in the face of escalating changes and challenges. Educators will use it as a tool for their own professional work and research; policy makers will use the examples as a guide to incorporating education more purposefully into environmental, climate change, and other sustainability decisions; and non-governmental organizations (NGOs) will find linkages between their mission and education that will enhance both sectors.

3. *Need for evaluating scholarly findings through the voices of young professionals and through a futures perspective* to evaluate effectiveness of theories and approaches to building resilience.

This book is meant to be thought- provoking and practically useful. By examining academic theories and studies through the lenses of resilience development and the Sustainable Development Goals, it is hoped that this book can open new channels and opportunities for cross-disciplinary exchange leading to new, innovative discoveries that empower and energize new types of leaders equipped with the elasticity and flexibility along with the knowledge required for future environmental and social resilience development. Through sharing research and experiences based around the theme of education as capability building in sustainability strategies and innovations as spelled out by the SDGs, readers of *Building Sustainability Through Environmental Education* will expand the community of scholars, practitioners and citizens to aid in our quest. We invite you to generate new dialogue, study, and collaborative practices that promote resilience to the current and coming changes, and to assist young people in achieving their full potentials as leaders in this important academic and social pursuit.

Lynn A. Wilson
SeaTrust Institute, USA

Carolyn N. Stevenson
Purdue University Global, USA

Figure 1.

REFERENCES

Agenda 21: programme of action for sustainable development; Rio Declaration on Environment and Development; Statement of Forest Principles: The final text of agreements negotiated by governments at the United Nations Conference on Environment and Development (UNCED). (1992). Rio New York, NY: United Nations Dept. of Public Information.

Anderson, A., & Strecker, M. (2012). Sustainable development: A case for education. *Environment*, *54*(6), 3–16. doi:10.1080/00139157.2012.730010

Chawla, L., & Cushing, D. F. (2007). Education for strategic environmental behavior. *Environmental Education Research*, *13*(4), 437–452. doi:10.1080/13504620701581539

Independent Commission on Multilateralism & International Peace Institute. (2016). *The 2030 Agenda for Sustainable Development and Addressing Climate Change.* Discussion Paper. Retrieved from https://www.icm2016.org/the-2030-agenda-for-sustainable-development-and-addressing-climate-change

National Research Council. (2011). *America's climate choices. Committee on America's Climate Choices. Board on Atmospheric Sciences and Climate, Division on Earth and Life Sciences.* Washington, DC: The National Academies Press.

National Research Council. (2012). Climate change education in formal settings, K-14: A workshop summary. A. Beatty, Rapporteur. Steering Committee on Climate Change Education Goals and Objectives. Board on Science Education, Division of Behavioral and Social Sciences and Education. Washington, DC: The National Academies Press.

Palmer, J. (2002). *Environmental education in the 21st century: Theory, practice, progress and promise*. Routledge.

Schreiner, C., Henriksen, E. K., & Kirkeby Hansen, P. J. (2005). *Climate education: Empowering today's youth to meet tomorrow's challenges*. Academic Press.

UN Department of Economic and Social Affairs. (2017, June 28). *'Inclusive, equitable and quality education' at the heart of high-level UN event.* Retrieved from https://www.un.org/development/desa/en/news/sustainable/high-level-event-education.html

UN General Assembly. (2015, October 21). *Transforming our world: The 2030 Agenda for Sustainable Development.* A/RES/70/1. Retrieved from http://www.refworld.org/docid/57b6e3e44.html

UNESCO & UNEP. (2011). *YouthXChange guidebook series: Climate change and lifestyles*. Paris: United Nations Educational, Scientific and Cultural Organization (UNESCO). Retrieved from www.youthxchange.net

UNFCCC. (2015). *Paris agreement. FCCC/CP/2015/L.9/Rev1*. Retrieved from http://unfccc.int/resource/docs/2015/cop21/eng/l09r01.pdf

Wilson, L. (2014, December). *Notes taken at the United Nations Framework Convention on Climate Change, COP20/CMP10 Special Session*. Intergenerational Inquiry: Youth as Agents of Change, Lima, Peru.

Wilson, L. (2017, November). The AWARE(SM) Pacific Islands/Global Climate Ambassadors Program. Presentation at *United Nations Framework Convention on Climate Change COP23*, Bonn, Germany.

Acknowledgment

To every person affected by environmental and climate change, and especially to the younger generations: may you inherit a world that free from boundaries that inhibit the fulfillment of your highest and best dreams.

Our sincere gratitude goes to the authors who contributed their time and expertise to this book, and to the editors and reviewers who vastly improved the quality, coherence, and content.

Lynn A. Wilson
SeaTrust Institute, USA

Carolyn N. Stevenson
Purdue University Global, USA

Chapter 1
Building Community Resilience Through Environmental Education:
A Local Response to Climate Change

Mphemelang Joseph Ketlhoilwe
University of Botswana, Botswana

ABSTRACT

The call for collaborative efforts to respond to climate change is heeded through bilateral and multilateral agreements. The UN Sustainable Development Goals bears testimony to the call. Environmental education is one of the vehicles to raise awareness, understanding, and assessment of sustainable development goals at a community level to build resilience for sustainability. Environmental education enhances the accomplishment of the key competencies for sustainability. Climate change is a complex environmental problem that is not only naturally induced, but made more stressful by anthropocentric capabilities in the quest for a better lifestyle. Although climate change causes and impacts are known, its mitigation strategies are compounded by human wants at the expense of their own sustainable survival. This chapter explores ways of building sustainable development in communities. Environmental education is a core development strategy in local communities against the adverse impact of climate change, especially in vulnerable areas.

DOI: 10.4018/978-1-5225-7727-0.ch001

INTRODUCTION

Climate change is recognized by international and regional organisations such as the United Nations and its organs as a global problem dominating social, economic and environmental discourses. Its adaptation and mitigation strategies are a matter of local and global discourses. The impacts of climate change are calling for international attention and collaborative efforts through multilateral and bilateral agreements. It is a complex environmental problem that is not only naturally induced but made more stressful by anthropocentric capabilities in the quest for better lifestyles. Although climate change causes and impacts are known, its mitigation strategies are compounded by anthropogenic drive at the expense of their own sustainable survival. Human beings continue to explore different new knowledge and technologies that has negative impact on the resources and adding substances to the atmosphere causing increased temperatures and varied climatic conditions. Although mitigation and adaptation approaches and strategies are attempted, there are far exceeded by the damages caused and human induced constraints to succeed in their implementation. If climate change adaptation and mitigation strategies are compromised, it could affect the implementation and achievement of sustainable development goals. Climate change education, could be one of the vital tools through environmental education to enhance both adaptation and mitigation capabilities. Environmental education could be viewed as a vital tool for building local community resilience to climate change. Education for Sustainable Development Goals (UNESCO, 2017), goal 4 in particular, could enhance resilience to the impact of climate change at different levels including local community. Quality and relevant education is capable of fostering the creation of capabilities and appropriate knowledge and skills to enable local communities to adapt to climate change. It can help communities re-consider their contribution to damages associated with climate change regardless of their geographical position or wealth. Education is capable of enhancing the knowledge, skills to adapt, and to establish a greater responsibility especially in more vulnerable location such as those in Africa. Climate change awareness could be promoted through action-oriented policy implementation at local levels. This could be at local community, or school levels.

The focus of this chapter would be at local community level. Insights are drawn from different parts of the world where empirical evidence demonstrated the role of education to promote community resilience. The purpose of this chapter is to explore how environmental education could enhance community climate change awareness to achieve sustainability. It draws from education and climate change literature to make suggestions for local community-based environmental education programme that would promote climate change awareness. It also makes reference to SDG

13 on climate action learning objectives (UNESCO, 2017). It deliberates on how education impacts and is impacted by these goals. The chapter emphasizes the role of education at local community level in responding to an environmental issue such as climate change. Education provides a place for communities to gain knowledge and practical experience in conducting environmentally oriented adaptation and mitigation strategies. The chapter recommends the recognition of the role of local knowledge to enhance community's capabilities and resilience to climate change. Resilience to climate change would enhance the reaching of sustainable development goals at local community level.

CONTEXTUAL BACKGROUND

Public education needs drastic re-orientation to mainstream climate change and variability to promote learning for sustainable development. Community environmental education programme has a role to play in understanding, mitigating and adapting to climate change. A study conducted by Ardoin and Heimlich (2013, p. 106) indicated that

During the interviews, several of the decision makers commented on the role that education plays, particularly as a tool for awareness. One said, "The benefit to education is that it helps people develop a deeper understanding of an issue. . . . If [my organization] wants to encourage people to take a specific action around an issue, if people are already up-to-speed on it—they're knowledgeable about it, they care about it—it makes [my organization's] work easier.

Climate change is one of the key action themes of the UN Decade of Education for Sustainable Development, as was reaffirmed at the World Conferences on Education for Sustainable Development held in Bonn in April 2009 (UNESCO, 2009) and in Aichi-Nagoya in November 2014. Education is meant to encourage the changes in attitudes and behaviours needed to put our world on a more sustainable development path, and build a new generation of climate change-aware citizen (UNESCO, 2010). This is in line with the objectives of environmental education and education for sustainable development, both of which are viable tools for climate change adaptation and mitigation. Critical assessment of SDGs can enable the community to deal with accelerating climate change, increasing complexity, contested knowledge claims and inevitable uncertainty (Kopnina, 2017; Lotz-Sisitka et al. 2015). Recognition and implementation of SDGs 4 and 13 learning objectives (UNESCO 2017), as a package at local community level would enhance resilience. The SDGs learning objectives are described in the cognitive, socio-emotional and behavioural domains. Community quality education on climate change could be based on its impact or possible impact on livelihoods so as to encourage the community to develop adaptive strategies or resilience.

Promoting Adaptation and Mitigation

Adaptation is so complex and contextual. Research in Ethiopia has revealed that people do not "experience the same features of climate change within all localities. There is a considerable variation in climate change impacts, which challenges people in different ways and demands variations in strategies of adaptation." (Toger, 2011, p. 6). In addition to well laid out coping strategies, new and innovative strategies based on experiences should be allowed as adaptive capacities within the affected communities. Adaptive capabilities could be promoted by appropriate climate change knowledge and skills acquisition by both the community and their leaders.

Knowledge on local drought frequency and extent as one of the common indicators of climate change could be an integral part of public quality environmental education programme. Climate change awareness programme should encompass the use and recognition of local ecological knowledge especially the local climate variability coping strategies. Strategies need to be put in place to enable communities to cope with the effects and impact of drought. These could be in the form of local conservation strategies and provision of sufficient storage capacity for water demands during periods of less rainfall due to climate variability. Water managers need to be capacitated to implement a careful water education and management including groundwater resources. A careful management of water resource would have to be preceded by public education to reveal the degree of threat of prolonged or frequent drought and how that may affect related ecological systems such as vegetation and other sources of local livelihoods for the community.

Education could be used as a social response to climate change especially in areas that are vulnerable or prone to flooding. Floods are reported to

... have the greatest damage potential of all natural disasters worldwide and affect the greatest number of people. On a global basis, there is evidence that the number of people affected, and the economic damages resulting from flooding, are rising at an alarming rate. Society must move from the current paradigm of a post-disaster response: the current event–disaster cycle must be broken. More than ever, there is a need for decision-makers to adopt holistic approaches for flood disaster management. Interventions before, during and after floods can reduce short- and long-term health impacts." (Leonardi, 2010, p. 6).

Places at risk include crop growing areas, homes and schools. Flooding is an issue especially among vulnerable communities such as the poor, women and children. It requires pro-active action backed by well-designed education programme and policies to facilitate coordinated planning and action, and accompanied by high levels of cooperation in disaster preparedness. Lionardi (2010, p. 6) posits that early warning systems including "processing and disseminating understandable warnings to political authorities and the population" are necessary.

Climate change may affect sustainable development in several ways. It may lead to worsening rural vulnerability, indigenous survival as well as food security. The socio-economic impacts of climate change may also lead to poverty, food insecurity, environmental degradation as well as water shortages due to erratic rains and changing pattern of droughts. Local public education is necessary to promote mitigation and adaptation to climate change. Table 1 illustrates the distinction between mitigation and adaptation.

CLIMATE CHANGE EDUCATION

Education is one of the four core programmes of climate change initiatives and which is also supporting the UNFCCC article 6, on public awareness, education and training. This core programme address the provision of climate change education (CCE) for sustainable development in both formal and non-formal settings through improved education policy, analysis, research and planning; educator education and training. It encourages and enhances innovative teaching approaches through

Table 1. Distinction between mitigation and adaptation

	Mitigation	Adaptation
Timing	Costs now, benefits delayed	Costs whenever, benefits may be later or relatively soon
Temporal incidence	Costs now, benefits to inter generations	Benefits more or less appropriable by those bearing costs
Geographical incidence	Local costs, global benefits	Local costs, often relatively local benefits
Sectoral incidence	Focus on emissions from energy consumption	Very heterogeneous
Sectoral and regional sensitivity	Reduces impacts on all sectors and regions	More feasible for certain sectors and regions than others
Reaction to uncertainty	Must act earlier despite greater uncertainty	May act later after reducing uncertainty, although anticipatory adaptation may require earlier action

Sources: (Wilbanks, Kane, Leiby et al., 2003)

interdisciplinary approaches (UNESCO: 2010). Climate change education may be understood as a process dealing with complex weather systems that bring many challenges to the environment, social and economic life. It challenges the core values of the skeptics, curious and has profound qualities of dealing with uncertainties, poorly defined situations, diverging views, interests and reality constructions in the face of climate variations and its impact. To achieve climate change knowledge and understanding would require policy orientation and pedagogies that address community attitudes and values. Environmental education for sustainability could ensure that local communities have access to knowledge and opportunities for value addition and sustainable development. It could facilitate the implementation of resilient agricultural practices that would increase productivity and production, that help maintain ecosystems, and that strengthen the capacity for adaptation to climate change (UN, 2017). Environmental education programme within the community would encourage reflexive and social learning through transformative engagement. Environmental education at community level would enhance the achievement of key competencies for sustainability through sustainable development goals learning objectives. According to UNESCO (2017) the key sustainability competencies includes, but not limited to, systems thinking, critical thinking, problem solving, collaborative, strategic, anticipatory, and normative competencies.

Environmental Policy Making and Climate Change Education

Policy making process could be facilitated to address climate change effects that may occur in future, manifest at critical places calling for international support or that may be already affecting the local socio-ecosystems. Environmental education policy initiatives may be a demonstration of responsibility to deal with a global or local phenomenon that directly and indirectly threaten humanity, the environment, and socio-economic activities. It would also promote good environmental awareness and governance. Environmentally aware citizens can influence changes in different ways as citizens, consumers and as producers. World citizens can influence changes in different ways as citizens, consumers and as producers. van Dam-Mieras, (2006, p. 13) states that "As a citizen, that influence runs on the one hand via democratic institutions in the home country and on the other hand via civil society, as a consumer that influence runs via consumer behaviour. As a producer, that influence runs via governance structures and employees participation". These statements demonstrate that environmental education for climate change awareness requires a holistic approach to traverse other sectors outside formal education. Therefore, environmental education policy change could go beyond formal education to local community. It could further enhance the achievement of the majority of the Education for Sustainable Development Goals objectives (UNESCO, 2017).

Existing policies and structures could be modified to create space for climate change awareness through environmental education. The government ministry responsible for education should "develop and integrate climate change into education policies and also to include the education response to climate change into climate policies and action plans" (UNESCO, 2010, p. 6). The re-orientation of education should take into account the already overstretched nature of the national curriculum and use multi-disciplinary and trans-disciplinary approaches such as environmental education in local communities. Climate change awareness should be part of local environmental knowledge, skills, values and action-oriented approaches for adaptation and mitigation to achieve sustainable development.

Themes such as sustainable consumption, disaster preparedness, environmental protection, recycling, water, desertification and renewable energies should be discussed, taking into account their relevance in the specific national and local context. Finally, concepts such as ethics, human rights, gender equality, uncertainty, risk management, social justice, and decision making are crucial to prepare learners for a changing environment. (UNESCO, 2010, p. 9)

The relevant policy instruments in place should be reviewed and adapted to be able to deal with current and future negative impacts of climate change. They should be modified to ensure that adaptation is truly apparent in concrete decision. Integration of climate change adaptation through reviewing and modifying existing instruments shall not be restricted to the environmental sector or to the public authority; it also refers to economic sectors and private organisations such as businesses. Clearly assigned responsibilities (e.g. for coordination, information provision, taking actions), community leaders and staff-training and a sufficient financial budget will enhance the success of integrating climate change education. However, there may be cases where the modification of existing policies, structures and processes are insufficient to handle the climate change adaptation needs and the development of new instruments may be required.

Awareness on climate change adaptation or mitigation and practices to address disaster risk reduction, emergency preparedness and locally relevant sustainable development options could be promoted through public education policy. A suggestion was made in Kenya for policy support to integrate climate change in the curricular

The decision to integrate climate change into Kenya's education system would have to be formulated into a national policy. Once approved the subsequent stages will include; identification of climate change experts, curriculum design, climate change content development, curriculum support materials design and development, validation and approval of the curriculum, teacher preparation/orientation, piloting of the curriculum and then implementation of the curriculum on national scale. Integrating climate change into Kenya's education system is achievable and can be infused within the framework of the Medium Term Plan (MTP). (Republic of Kenya, 2012, p. 9)

The Kenyan case may be applied to local community by integrating climate change awareness in public environmental education. In terms of education policy, it is vital to bring about policy synergies in climate change. This would also call for capacity building and sharing good practices that would bring about minimizing causes of climate change and promote community resilience through quality education.

There are existing policy frameworks at regional and international levels that may guide the development of local policies to promote climate change awareness. For instance, The Rio Declaration on Environment and Development of 1992 has confirmed the indispensability of the community in achieving sustainable development. The United Nations Framework Convention on Climate Change (Article 6) and the Kyoto Protocol (Article 10) both encourage governments to educate, empower and engage all stakeholders and major groups on climate change policies. In 2002, the United Nations General Assembly proclaimed the United Nations Decade of Education for Sustainable Development (2005–2014), underscoring the indispensable role of education in achieving sustainable development. Facilitating access to information on climate change is critical to winning public support for climate related policies. UNICEF, 2014, pp 4. The product of the UN Decade of Education for Sustainable Development (ESD) is the post 2015 Global Action Programme (GAP), a Road Map on ESD. The GAP, had five priority areas on policy, curriculum and instruction, educators and trainers, youth and community leadership on ESD. This chapter focus on education at local community level that is linked to GAP priority area 5 on accelerating sustainable solutions at local level and the UN Sustainable Development Goal 4 on quality education and Education Agenda 2030 (UNESCO, 2017). Environmental education enrichment could be infusion of climate change awareness to achieve SDG 4 shown in Table 2. This is complemented by education for sustainable development goals learning objectives (UNESCO, 2017).

The information in Table 2, shows a synopsis of how education is reflected in other SDGs. However, it should be noted that education and awareness transgress other SDGs as well. Limited by the focus of this chapter, SDG, 6 and 13 have been used to show the synergistic and complementary relationship between the goals. Climate change awareness and knowledge in local communities could be part of SDG 13 on climate action to build community resilience.

Building community resilience for sustainability means "intentionally guiding the system's process of adaptation in an attempt to preserve some qualities and allow

Table 2. SDG 4 and other education related SDG targets and indicators

Sustainable Development Goal	Target	Indicator
4 Ensure inclusive and equitable quality education and promote lifelong learning opportunities for all	4.7 By 2030, ensure that all learners acquire the knowledge and skills needed to promote sustainable development, including, among others, through education for sustainable development and sustainable lifestyles, human rights, gender equality, promotion of a culture of peace and non-violence, global citizenship and appreciation of cultural diversity and of culture's contribution to sustainable development.	4.7.1 Extent to which (i) global citizenship education and (ii) education for sustainable development, including gender equality and human rights, are mainstreamed at all levels in: (a) national education policies, (b) curricula, (c) teacher education and (d) student assessment
6 Ensure availability and sustainable management of water and sanitation for all	6.b Support and strengthen the participation of local communities in improving water and sanitation management	6.b.1 Proportion of local administrative units with established and operational policies and procedures for participation of local communities in water and sanitation management
13 Take urgent action to combat climate change and its impacts	13.3 Improve education, awareness-raising and human and institutional capacity on climate change mitigation, adaptation, impact reduction and early warning	13.3.1 Number of countries that have integrated mitigation, adaptation, impact reduction and early warning into primary, secondary and tertiary curricula 13.3.2 Number of countries that have communicated the strengthening of institutional, systemic and individual capacity-building to implement adaptation, mitigation and technology transfer, and development actions

others to fade away, all while retaining the essence—or "identity"—of the system. In a human community, identity is essentially determined by what people value about where they live" (Lerch, 2015, p. i). Lerch (2015, p. 10) defines community resilience "as the ability of a community to maintain and evolve its identity in the face of both short-term and long-term changes while cultivating environmental, social, and economic sustainability". Building community resilience entails building regenerative capacity, sensing emerging risks, responding to disruption, learning and transformation linked to resilience (Lerch, 2015).

Building resilience relies upon flexible, adaptive and responsive approaches. Quality environmental education could be capable of facilitating such approaches within the local community. It would enable the community to recover, respond and adapt to climate change. It would empower local communities to take ownership of their own climate change adaptation and mitigation agendas fostering a shift towards greater environmental sustainability (Cinderby et al., 2015). Climate change knowledge would increase community's ability to reduce their vulnerability to risk and uncertainty. Education would encourage a shared vision and action of the future. Building community resilience in the face of climate change could be promoted through the development of resilience building action projects and reskilling with clear strategies for change within the community. Altering the way a community manages its environment requires resilience strategies such as cutting carbon emissions; activities encouraging environmentally friendly agricultural production; building positive attitudes and values and strengthening local identity. (Kais & Islam, 2016). However, resilience alone, cannot be a solution to intricate, multifaceted and contradictory problems such as climate change which no one single action can solve, and solutions themselves might have severe side effects (Kopnina, 2017; Lotz-Sisitka et al., 2015; Pokrant & Stocker, 2011).

ENVIRONMENTAL EDUCATION AND LEARNING FOR CLIMATE CHANGE

The processes of industrialization and technological advancement are some of the driving forces for climate change. These affect the ecological systems and challenges education to be more relevant and contextual. It has also contributed to global discourses on climate change. Environmental education is a viable tool for climate change awareness and knowledge building. Education is the primary agent of transformation towards sustainable development, increasing people's capacities to transform their visions for society into reality. Educated citizens who have learned to make decisions that consider the principles, values and practices of sustainable development (education for sustainable development) will create a more sustainable future in terms of environmental integrity, economic viability and a just society for present and future generations.

van dam-Mieras (2006, p. 14) describes learning as

... the result of the process of continuous interaction of an individual or group with its physical and social environment. The learning environment can be formal (the educational system), non-formal (e.g. training on the job) and informal (family life, leisure time, visit to a museum or a zoo), and the learning process continues life long, it can be calculated that on the average only about 5 – 10 per cent of the lifelong learning process of an individual takes place in the formal learning environment.

Climate change education must prepare individuals for functioning in professional and societal settings in which they will be confronted with complex problems and will have to work together with experts from different disciplinary domains. Acquiring climate change knowledge is an important for the local community and knowledge recipients must apply it within context. Emphasis should be on community active learning process. In addition, the acquisition of climate change knowledge and competences should be oriented towards giving more attention to individual role to solve environmental problems. It should emphasise both behaviouristic and cognitivist-oriented outcome to promote a shift from community educator oriented to learner pedagogy in its approach promoting interest in development of climate change adaptation and mitigation strategies for sustainable development and resilience.

Environmental education recognises that "…there can be no learning without re-learning, without the revision that must be undertaken when we realise the weakness of what we thought we knew…" (Granados, 2014, p. 2). The origin of climate change causes is advancement in knowledge that caters less for environmental degradation and hence GHG effects. In the words of Orr (2004, p. 17):

Education, in other words, can be a dangerous thing (...). It is time, I believe, for an educational 'perestroika', by which I mean a general rethinking of the process and substance of education at all levels, beginning with the admission that much of what has gone wrong with the world is the result of education that alienates us from life in the name of human domination, fragments instead of unifies, overemphasizes success and careers, separates feeling from intellect and the practical from the theoretical, and unleashes on the world minds ignorant of their own ignorance.

The United Nations Decade on Education for Sustainable Development brought in a paradigm shift in environmental education, the reconstruction approach to education shifting from just individual integrity but emphasizing continuity. Education is challenged to seek for new answers, imagining new ways of living and acting, assessing and bringing about social change for sustainable development. Climate change could be embedded in the four pillars of education set out in the Delors Report (1996): learning to know, learning to do, learning to live together and learning to be. In the context of environmental education and education for sustainable development, UNESCO (2008) suggested the inclusion of a fifth pillar: learning to transform oneself and society. Environmental education would promote climate change awareness leading to individual empowerment through the acquisition of the capacity to make decisions and act effectively in accordance with those decisions. This type of education would develop personal and social qualities, social conscience and a sense of the personal agency which allow action. This agency may restrict our interventions to increase the causes of climate change and make it necessary to decide our personal degree of action (Goldberg, 2009).

Imbedding Climate Change Awareness Through Environmental Education

Enhancing local community climate change awareness calls for strategic programmes including, inter and multidisciplinary approaches as well as innovations in pedagogical approaches. It has been noted that

The scale and complexity of the underlying challenges of adaptation, vulnerability, mitigation and sustainability is such that the traditional disciplinary epistemological frameworks may appear to be inadequate to develop the conceptual, methodological and analytical tools necessary to understand developments in a range of interdependent knowledge domains. (Kotecha, 2010, p. 9)

Multidisciplinary Approach to Climate Change Education

The traditional discipline-based approaches are comfort zones for the majority of educators. It may not be easy to facilitate multi and transdisciplinary approaches in educational programmes for climate change. The filtering of multi-pronged approaches to education to the local communities may be a viable option. Institutions charged with in-depth knowledge generation would have to proactively scale up their efforts and practices to promote climate change knowledge and understanding in the community. This could be facilitated by carefully crafting policy change to mainstream climate change awareness through environmental education to avoid a situation where intended innovation remains nobody's responsibility, while local communities continue practices that are not climate change adaptation compliant. From time-to-time, ideas and new knowledge gain momentum through research and debate. This new knowledge should be shared with local communities to improve on their local knowledge to enhance their resilience and copping strategies. For climate change to permeate the education system it should be weaved throughout curricular including local community levels.

Access to climate change information could be incorporated into environmental education programmes at public awareness forums to build knowledge and awareness. As noted by Prutsch, Grothmann, Schauser, Otto, and McCallum (2010, p. 11).

Due to the complexity of climate change and its potential impacts, particular attention needs to be given to knowledge transfer and awareness raising. Decision makers, stakeholders and the larger public should have access to reliable information about

potential climate change impacts, vulnerabilities, adaptation options and tools, good practice examples and advantages of adaptation, potential trade-offs, synergies and conflicts with mitigation and existing policies.

Environmental education programme would enhance environmental sustainability and community resilience to climate change. Community resilience through environmental education would entails coping with and respond to climate change and variability. It would encompass the ability to integrate environmental stewardship contributing to the ability to respond to future disturbances, which is adaptation. It could open up new niches to post-disturbance solutions or mitigation strategies. Environmental education programme could promote reactive and proactive community resilience by enhancing the community's ability to be stable during climate disruption and the capability to adapt to new change. Through climate awareness knowledge, resilient communities would have improved capabilities to live with change and uncertainty; to nurture diversity for reorganization and renewal; to combine different kinds of knowledge; and to create an opportunity for self-organization. It would be able to adapt to constant changes. In addition, a resilient local community would be knowledgeable and skillful in assessing, managing, and learn new skills building on past experiences. Different ways could be deployed to promote local community climate change awareness.

Use of Media and Collaboration

These can increase climate change awareness, for instance, through use of radio, television slots, conferences and print material resources such as pamphlets, leaflets, brochures, newspaper articles, books and many more including proper use of social media by the local community. Public consultations through public talks, small group discussions may build trust, allow learning from each other and effectively stimulate action. The use of social media through the internet where available and made accessible would facilitate knowledge transfer and raising awareness by reaching the general public. This could be enhanced further by researchers, policy makers and practitioners jointly bundling understandable usable information on potential impacts and adaptation to fill knowledge gaps especially for local levels (Prutsch, et al 2010). One of the most devastating effects of climate change, especially in arid, drought-prone areas, is the threat to water security. Both quantitative and qualitative evidence from Botswana indicate that between 2012 and 2014, there was rationing of portable water in the capital city, Gaborone, due to successive years of low rainfall in the south eastern part of the country. The dam that feeds the city with water was at its lowest, 7% in 2014. This 7% was estimated to supply the city with a population of over 300 000 inhabitants with water only for two months. The water authorities resorted to use of media to educate people on sustainable methods of using water at home, in offices, institutions, industries and in maintaining their gardens. The "save water" project was public awareness focused hoping to conserve water in the face of climate change. As a result, individuals and building contractors started obtaining borehole water from the outskirts of the city for gardens and constructions.

Project Based Learning

climate change awareness could be promoted through the creation of action projects that are protective and hands-on. In designing the action projects, risks and disaster preparedness measures must be the leading frameworks. The projects must be people-centered and inclusive and create messages that are environmentally friendly. Project methods or approaches may help people gain true experiences and practices.

Creation and establishing community-based projects concept has been tried in several parts of the world under natural resources management. In schools, different labels such as 'green schools' or 'school competitions' or 'green flags' has been used. The community based natural resources management has been successful in some areas such as Botswana. However, it has been with some challenges where accountability was a concern. Its success could be replicated and applied as an adaptation strategy for climate change. The community project-based approach could be linked to problem-based learning methods to enhance quality education in the face of climate change. For instance, depending on how climate change is manifesting, a project may be chosen. An empirical evidence from Tswapong Hills in Botswana where women dependent on natural resources for their livelihoods has demonstrated that given proper knowledge and skills, people could diversify their economy to avoid the impact of climate change. After experiencing low agricultural yields and low supply of natural resources women had formed a Trust through which environmental awareness was created and they started small business from the micro finance loan scheme created to assist them divert from using natural resources which were becoming scares due to continued climate variability (Ketlhoilwe and Jeremiah, 2013). This project demonstrated the role of education supported by the provision of financial support.

A project approach in the Gambia, known as *All schools Tree Nursery Competition* research has shown that schools and learners could be a viable approach to incorporating concepts such as environmental education in the Gambia. Schools were selected to provide pre-existing structures, allowing an easier means to facilitate communication (Paulete, Orr, and Samateh, 2008). Learning institutions and learners are an easier means of communicating climate change information to communities. Effective climate change awareness through environmental education could facilitate collaboration with different stakeholders. Collaboration may be supported through educator training on climate change and awareness programme training could be facilitated by institutions in local communities such as research centres, local authorities and environmental civic societies.

Community Educators' Education and Training

Retraining and training of educators and community leaders may enhance the necessary skills to promote climate change awareness and knowledge. An effective training programme may require a stronger interdisciplinary links to key stakeholders such as researchers, teachers, NGOs and public officers to eliminate possible obstacles. Community leaders and educators' education "must be locally relevant and culturally appropriate, and should include due reference to equity issues. It must balance local needs with global concerns." (UNICEF, 2014, p. 20). Educators' training would increase their understanding of the environment, climate and sustainability issues and help them develop the necessary skills (UNICEF, 2014). Training programmes for climate change awareness should recognize and appreciate the already existing local knowledge on climate change. Local ecological knowledge may need to be enhanced to promote adaptive capacities and capabilities of the relevant communities. There is need to cultivate an attitude of a caring society among locals. It is essential to integrate climate change aspects into existing traditional knowledge.

Pedagogy in Climate Change Education

Environmental education requires pedagogical resources to enhance knowledge and understanding, skills acquisitions, to positive values and attitudes as well as to encourage participation in the utilisation of the resources. It espouses learning methods that are mainly interactive (or experiential) and activity-based encouraging participation and cooperation. The aim of the teaching and learning outcomes in climate change awareness through environmental education programme should be to "include sustainable livelihood skills; abilities to cope with local environmental hazards and health issues; and positive attitudes and behaviours regarding environmental stewardship and sustainable consumption." (UNICEF, 2014, pp 11). People-centred pedagogy should be the main focus for climate change to promote adaptation and resilience among the communities. People should be at the centre of learning in environmental education for climate change. This would empower the learning community and make them feel like there are parts of the solutions. It would promote the spirit of stewardship thereby enhancing sustainable development. Active project-based and problem-based pedagogies would enhance climate change awareness among the local communities. Table 3 shows the examples of possible approaches or methods of education for sustainable development goals.

Table 3. Examples of learning approaches and methods selected SDG

Examples of Learning Approaches and Methods for SDG 4 Quality Education	Examples of Learning Approaches and Methods for SDG 13 Climate Action
Develop partnerships between schools, universities and other institutions offering education in different regions of the world (South and North, South and South)	Perform a role-play to estimate and feel the impact of climate change related phenomena from different perspectives
Plan and run a quality education awareness campaign	Analyse different climate change scenarios with regard to their assumptions, consequences and their preceding development paths
Conduct a case study on the education system and access to education (e.g. enrolment in primary education) in selected communities or countries	Develop and run an action project or campaign related to climate protection
Plan and run an ESD project at a school or university, or for the local community	Develop a web page or blog for group contributions related to climate change issues
Celebrate UN World Youth Skills Day (15 July), International Literacy Day (8 September) or World Teachers' Day (5 October); or take part in Global Action Week for Education	Develop climate friendly biographies
Organize ESD days at local, regional and national level	Undertake a case study about how climate change could increase the risk of disasters in a local community
Develop an enquiry-based project: "What is a sustainable school?"	Develop an enquiry-based project investigating the statement "Those who caused the most damage to the atmosphere should pay for it"

Source: Adapted from UNESCO (2017) *Education for Sustainable Development Goals: Learning objectives* for SDG 4 and 13.

FUTURE RESEARCH DIRECTIONS

Environmental education research for climate change would involve profound transformative knowledge and understanding that are relevant in the twenty first century when humanity is experiencing climate variation of significant magnitude. Aligned to Education 2030 (UNESCO, 2017) Agenda, the SDGs 4 and 13 targets and indicators, environmental education research could assist in reorienting implementation strategies to achieve sustainability. Knowledge about climate changes must be developed and shared to increase awareness and provide rationale for decision making. Research for climate change may not be separated from other related research such as research for sustainable development in environmental education. Scientific and social research and publishing would enhance knowledge transfer to the local community for application. Researched information would enhance community action on climate change and create general knowledge that may promote resilience. Action research approach with the community on climate related changes and impact is likely to promote understanding of the complexities of climate change and individual's role to contribute to its adaptation as well as build resilience where possible and necessary. Researched knowledge should not be for intellectual discourse only but should be accessed and debated by the people experiencing climate change impact. Researched information on climate change could be used to frame policy direction leading to both mitigation and adaptation strategies. Communities need help to regulate their energy use, production, purchasing building material and energy efficiency. Publication of researched information is likely to promote climate literacy and the understanding of the processes involved. These processes could be presented in a simplified form of climate science for the local community to promote its adaptive capacity.

Research institutes and regional bodies should work together to drastically reduce industrial emissions of greenhouse gases contributing to global warming. The challenge to environmental education is to explore relevant ways of creating high level of awareness to find direct values in mitigating the effects of climate change and variability. It has already been appreciated in most regions, through "incidence of high temperature events, floods and droughts, with resultant consequences for fires, pest outbreaks, and ecosystem changes" (SADC, 2008, p. 55), that climate change is a reality. For climate change awareness to be taken to greater heights, research should provide evidence-based guidance to policy makers. In addition, policy makers need to be educated so that they appreciate precautionary measures to take at policy level to mitigate greenhouse gas emissions and enhance "...the resilience of vulnerable systems by means of adaptation." (SADC, 2008, p. 55)

Research may focus on lack of awareness of the role of education in achieving climate change adaptation and mitigation. The research may extend to investigating policy initiatives and impacts at different community levels as well as on resources such as funding in some contexts and situations. As learning institutions are usually characterized by the prevalence of traditional disciplinary curriculum frameworks that makes transdisciplinary attempts arduous at times, local ecological knowledge could be explored and promoted to avoid delays in prioritizing climate mainstreaming in community education.

CONCLUSION

Climate change is now a reality and societies around the world are grappling with how to respond, to reduce or avoid further or new damages. Communities' sensibility to climate change can reduce the consequences and increase resilience to adverse effects. It is important to continue discussions on how best to respond to both natural and anthropogenic climate change to disrupt further impacts and avoid threats. Attempts had been made through the 1992 Earth Summit and 1997 Kyoto Protocol to reduce net greenhouse gas emissions to below their levels with little success. What is disturbing is that despite all attempts anthropogenic drives add to "a variety of phenomena such as glacial retreat, thawing of permafrost and shifts of plants and animal ranges". (Wilbanks, Kane, Leiby, et al, 2003).

It is important to consider education as a response to climate change impacts. This must be accompanied by research to monitor climate change complexities as well as to promote conceptual and empirical understanding. It should be noted that both adaptation and mitigation approaches without quality education may have some limits in reducing the impact of climate change. Local communities' use of traditional knowledge system such as using weather patterns to time planting and harvesting are adaptive strategy to climate change. However, abrupt climate events may work against this local practice. Climate change has a negative impact on the maintenance of environmental security (Scholtz, 2011) particularly among rural communities dependent on natural resources for their livelihoods. To mitigate this, knowledge sharing would be an important and sustainable strategy to promote adaptation in to climate variability in the local community.

Based on the insights from this chapter, it is recommended that policies aimed at reducing disaster and climate change risks be put in place and effectively implemented. It is equally important that policy makers and academic communities of practitioners consistently work together effectively to promote policy discourses and implementation outcomes that promote adaptation and mitigation to climate change. Policy discourses should focus on improving public awareness on climate change through education.

Local communities could be educated "through scholarship and scientific research, with the ideas and imagination as well as the social, technical and managerial capabilities for creating the conditions for long-term sustainability" (Kotecha, 2010, pp 9). Public educational awareness through environmental education should be stepped up through a variety of means. The use of social media, print media, radio and television programmes as well as through public gatherings. The use of local ecological knowledge promoting adaptation to climate change should be appreciated and encouraged. Where appropriate the acquisition and use of new technologies aimed at reducing the impact of climate change should be recognized and promoted especially in communities highly dependent on natural resources as their source of livelihoods. This could contribute to resilience building and adaptive strategies. Educators, in both formal and non- formal sectors need training and re-training on adaptation and mitigation strategies for climate change awareness.

Finally, both interdisciplinary and transdisciplinary approaches could be implemented and non-scientific forms of knowledge could be explored to promote climate change awareness among local communities. This could be characterized by new teaching and learning approaches that enable the development of critical and creative thinking. The competencies common to all learners could be determined and the corresponding expectations would be defined to facilitate social transformation to adapt to variable climate changes. Teaching and learning process must be more active, connected to real life, and designed with communities and their unique contexts in mind.

REFERENCES

Ardoin, N. M., & Heimlich, H. E. (2013). Views from the field: Conservation Educators' and practitioners' perceptions of education as a strategy for achieving conservation outcomes. *The Journal of Environmental Education*, *44*(2), 97–115. doi:10.1080/00958964.2012.700963

Cinderby, S., Haq, G., Cambridge, H. & Lock, K. (2015). Building community resilience: can everyone enjoy a good life? *Local Environment, International journal of justice and sustainability, 21*(10), 1252-1270. doi:10.1080/13549839.2015.1100597

Delors, J. (1996). *Learning: The treasure within – report to UNESCO of the international commission on education on education for the twenty-first century.* Paris: UNESCVO.

Goldberg, M. (2009). Social conscience. The ability to reflect on deeply-held opinions about social justice and sustainability. In A. Stibbe, (Ed.), The handbook of sustainability literacy: skills for a changing world (pp. 105-110). Devon: Green Books.

Granados, J. (n.d.). The challenges of higher education in the 21st century. *Global University Network for Innovation*. Retrieved from http://www.guninetwork.org/resources/he-articles/the-challenges-of-higher-education-in-the-21st-century

Graugaard, J.D. (2012). A tool for building community resilience? A case study of the Lewes Pound, Local Environment, *International journal of justice and sustainability,* 17(2), 243-260. doi:10.1080/13549839.2012.660908

Kais, S. M. and Saidul Islam, (2016). Community Capitals as Community Resilience to Climate Change: Conceptual Connections. *International Journal of Environment, Research and Public Health, 13*(12). Retrieved from https://www.ncbi.nlm.nih.gov/pmc/articles/PMC5201352/

Kopnina, H. (2017). *Teaching Sustainable Development Goals in The Netherlands: a critical approach. Environmental education research*, 1-16. doi:10.1080/13504622.2017.1303819

Kotecha, P. (Ed.). (2010). Climate change, adaptation and higher education: Securing our future. SARUA Leadership Dialogue Series, 2(4).

Leonardi, G. (2010). Adaptation to climate change.

Lerch, D. (2015). *Six foundations for building community resilience. Post Carbon Institute.* Retrieved from http://www.postcarbon.org/wp-content/uploads/2015/11/Six-foundations-for-building-community-resilience.pdf

Lotz-Sisitka, H., Wals, A. E. J., Kronlid, D., & McGarry, D. (2015). Transformative, transgressive social learning: Rethinking higher education pedagogy in times of systemic global dysfunction. *Current Opinion in Environmental Sustainability, 16*, 73–80. doi:10.1016/j.cosust.2015.07.018

Orr, D. (2004). *Earth in Mind: on education, environment and the human prospect* (2nd ed.). Chicago: Island Press.

Paulete, E. F., Orr, B., & Samteh, S. (2008). Incorporating EE into grassroots reformation efforts: The Gambia All School Tree Nursery Competetion. In A.B. Temu, S.A.O. Chamshama, J. Kung'u et al. (Eds.), *New perspectives in forestry education* (pp. 319 – 331). Republic of Kenya: Nairobi.

Pokrant, R. J., & Stocker, L. (2011). Anthropology, climate change and coastal planning. In H. Kopnina & E. Shoreman-Ouimet (Eds.), *Environmental anthropology today* (pp. 179–194). New York: Routledge.

Prutsch, A., Grothmann, T., Schauser, I., Otto, S., & McCallum, S. (2010). Guiding principles for adaptation to climate change in Europe (ETC/ACC Technical Paper 2010/6).

Scholtz, W. (2011). Climate change. USAID disaster reduction training course for southern Africa (USAID/OFDA. Project ACDS1).

Troger, S. (2011). Failing Seasons, ailing societies: Climate change and the meaning of adaptation. *Horn of Africa REC*. Retrieved from www.hoarec.org

UNESCO. (2008). Education and the search for a sustainable future, policy dialogue 1: ESD and Development Policy. Retrieved from http://unesdoc.unesco.org/images/0017/001791/179121e.pdf

UNESCO. (2009). Trends in global higher education: Tracking an academic revolution. Retrieved from http://unesdoc.unesco.org/images/0018/001831/183168e.pdf

UNESCO. (2010). The UNESCO climate change initiative. Climate change education for sustainable development. Paris: UNESCO. Retrieved from http://unesdoc.unesco.org/imageS/0019/001901/190101e.pdf)

UNESCO. (2017). Education for Sustainable Development Goals: Learning objectives. Retrieved from http://unesdoc.unesco.org/images/0024/002474/247444e.pdf

UNICEF. (2014). Climate change and environmental education: a companion to the child friendly schools. Retrieved from www.unicef.org/publications/CFS-climate-e-pdf

United Nations. (2017). Progress towards the Sustainable Development Goals. Retrieved from https://sustainabledevelopment.un.org/sdg2

Van Dam-Mieras, R. (2006). Learning for sustainable development: Is it possible within the established higher education structures? In J. Holmberg & B. E. Samuelsson (Eds.), *Drivers and barriers for implementing sustainable development in higher education. Education for sustainable development in Action* (pp. 13–18). Paris: UNESCO Education Sector.

Wilbanks, T. J., Kane, S. M., Leiby, P. N., Perlack, R. D., Settle, C., Shogren, J. F., & Smith, J. B. (2003). Possible Responses to global climate change: Integrating mitigation and adaptation. *Environment*, *45*(5), 28–38. doi:10.1080/00139150309604547

Chapter 2
When Good Waters Go Bad:
Sustainability and Education in a Postnormal Future

Lynn A. Wilson
SeaTrust Institute, USA & Walden University, USA

ABSTRACT

This chapter offers commentary on adaptation and resilience to stresses on water systems in a potentially catastrophic future. While considering futures studies as an integral part of science education is not new, reorganizing knowledge and its deployment to equip future leaders to address the complexity, paradox and unpredictability of problem requires new educational paradigms. Youth are poised as agents of change in a collaborative, networked, and complexity-embracing future. Through exploring the changes in waters due to climate change and human activity, and what those changes may mean for developing and maintaining resilience in the postnormal future, a complex adaptive systems (CAS) framework guides new alternatives for education and water policy action in these changing times and within the broad goals of sustainability.

INTRODUCTION

When April 21, 2018 was designated as "Day Zero" for Cape Town, South Africa, the day the water taps would be turned off in the first major city in the world ("Cape Town on the Verge," 2018), the world was put on notice. Prolonged drought broke with a single day of rain in February. Due to that event and agricultural water usage decline, the shutoff date was pushed back to May 11 and further delayed so the

DOI: 10.4018/978-1-5225-7727-0.ch002

water was not shut off entirely. However, during the interim period, daily water allocation drops from 23 gallons to 13 gallons per person, compared to the average 100 gallons per person daily use in the U.S. It was a wake-up call for adapting to a future in a water-scarce world.

Water is an essential element in the grand challenge humanity faces to provide for the basic needs of people in an equitable manner while attending to appropriate ecosystem functioning for the needs of future generations. While water systems' restoration, conservation, and ecosystem maintenance are needed now to preserve ways of life, livelihoods and even human survival, looking towards a future based on current environmental and social trajectories may require new policies, concepts, and even new priorities for our oceans, coastal waters, and freshwater systems. This new way of thinking needs to be reflected and expanded upon within sustainability education; educators at all levels need to address these issues in considering both the content and delivery of knowledge to different audiences at different scales. This requires innovative understandings of the interrelated environmental and social issues, a willingness to challenge traditional ways of thinking about delivery of environmental education, and a futures orientation. Learners at all levels also need to become knowledgeable about how these issues and outcomes are measured and evaluated so that they understand the ramifications of different ways of measuring environmental and social complexities, and so that their own work and decisions lead to chosen, preferable futures.

This chapter offers commentary on adaptation and resilience to stresses on water systems in a potentially catastrophic future. While considering futures studies as an integral part of science education is not a new idea (Lloyd & Wallace, 2004), reorganizing knowledge and its deployment to equip future leaders to address the complexity, paradox and unpredictability of modern problems requires new educational paradigms. In the WEF Outlook 2015, the World Economic Forum listed lack of leadership as the third most critical challenge facing humanity (World Economic Forum, 2015), because even with talented young people on the horizon, it will take time to redefine leadership models from their hierarchical present into its collaborative, networked, and complexity-embracing future. As part of the thought experiment "*When Things Fall Apart*" conceived by Christopher Jones and explained in his companion chapter in this book, this chapter explores the changes in waters due to climate change and human activity, and what those changes may mean for developing and maintaining resilience in the postnormal future.

Using secondary research together with personal accounts and observations, this reflection draws from the tradition of hermeneutics and relies heavily on a complex adaptive systems (CAS) framework (Gell-Mann, 1995; Folke, 2007) in considering water policy and action in these changing times and within the broad goals of sustainability. CAS has been used in fields such as health care to integrate interdisciplinary and transdisciplinary knowledge and theories to help make sense of natural phenomena, including human responses to problem solving (Ellis, 2011). For Cape Town, that consideration became reality in 2018; although they survived the crisis, the experience offers a cautionary tale about the practical relationship between sustainability and resilience.

BACKGROUND

When the United Nations World Commission on Environment and Development authors of the 1987 report 'Our Common Future' coined the term "sustainable development," it was designed to incite urgent action to protect the global environment (Bruntland, 1987). The report held a message of actionable hope that helped catalyze a renewal of environmental and social awareness, sowing seeds for the development of sustainability science and policy at local, national, and global levels. Fast forward to January 2018, and the World Economic Forum's annual meetings in Davos, Switzerland, where world leaders meet to craft global economic strategies. Environmental issues used to play a relatively minor role in these talks, but that has changed as climate change and the human toll from environmental destruction has taken an increasingly larger bite out of global economies. The 2015 Sustainable Development Goals (SDGs) that are part of Agenda 2030 and attached to the Paris Agreement played a major part in the WEF theme, *Creating a Shared Future in a Fractured World*, which looked to a future that reflected the interconnectedness of social and environmental goals in the global economy (Cassar, 2018) in the urgency to adapt to a changing climate.

Adaptation is often associated with the notion of resistance to perturbation and to rates of recovery from the effects of stress (Redman, 2015), therefore tending towards a risk assessment approach that measures adaptation success in terms of the speed and degree of a return to a system's original state rather than transformation to a potentially new state. This common assumption about adaptation is reflected in the short-hand rhetoric of multilateral climate change policy that historically has categorized climate change work as either mitigation or adaptation. According to a United Nations study, between 1995 and 2015 an average of 335 weather-related disasters occurred per year, nearly twice the level recorded from 1985 to 1995 (Centre for Research on the Epidemiology of Disasters, & United Nations Office for Disaster Risk Reduction, 2015). As disasters have intensified in frequency, strength and negative effects, adaptation has gained more air time in global policy circles, leading to a deeper consideration of the linkages between, and separate characteristics of, adaptation and resilience.

Resilience comes from the Latin word *resilíre*, meaning to spring back quickly or recover from difficulties (Oxford English Dictionaries, 2009). Contemporary theorists, researchers, and practitioners usually consider resilience within the constraints of a disciplinary or professional focus; while engineers focus on efficiency and make assumptions about systems as constant and predictable, ecologists tend to focus on persistence of a system or organism under significant environmental changes and under unpredictable circumstances. However, theories of adaptive capacity and resilience of complex systems such as the coupling of physical climate systems and society necessarily concentrate on the management of dynamic, unstable interdisciplinary systems that co-evolve, negotiate, and self-organize in cyclical patterns. These variations have led to debates regarding the optimum relationship of traditional resilience theory and sustainability science. Debates continue about whether to separate or combine resilience and sustainability in attempts to prioritize outcomes, reflect system dynamics, rely on community input, or employ other innovations more suited to complex systems (Redman, 2014). The definition of resilience used in this chapter is to consider resilience at a higher level that is inseparable from sustainability. Here, resilience is defined as maintaining core principles and integrity – of all kinds - under dramatically changed circumstances. Maintaining that core integrity in making decisions may challenge traditional adaptation perspectives about waters when resilience development is part of an adaptive response to catastrophic change.

Freshwater

Waters permeate the SDGs and discussions of human and planetary futures. As freshwater becomes increasingly scarce it affects other policy considerations including security, political stability, trade, development and human health. Only 1% of the world's freshwater is easily accessible (U.S. Geological Survey, 2016). Most exists within the world's glaciers and is inequitably distributed around the globe. The long-standing concept that freshwater is a renewable resource is compromised by the ongoing deterioration in water quality, leading to the degradation of aquatic ecosystems upon which human health, livelihoods, and development depend. The quality of available water is being compromised in some surprising ways with significant implications for human health and ecosystems worldwide. For example, permafrost, which occurs in approximately 24% of the Northern Hemisphere land surface surrounding the Arctic ocean, has recently been found to contain massive amounts of natural mercury. Climate models predict a 30 to 90% reduction in permafrost by 2100, depending on greenhouse gas emissions (Schuster, et al, 2018), leading to increased mercury concentration in available water.

Globally, upwards of 4 billion people face severe water shortages with 3 billion of those lacking sufficient water from 4 to 6 months a year (Mesfin, 2016).

Increasingly, traditional water sources from rivers and lakes, such as Cape Town's Voëlvlei Dam, are running dry. When surface water becomes scarce or inaccessible, communities, industry, and agriculture all turn to ground water which can drain reserves. Groundwater depletion, often caused by sustained or excessive groundwater pumping, exacerbates water shortages; when more water is taken from the system than is being naturally replenished, water quality suffers, costs increase, groundwater inflow into lakes and streams declines, and water tables decline. Social inequalities deepen as water becomes scarce; Cape Town's wealthy, government and industry are drilling bore holes into the aquifer with potentially dire future consequences for providing municipal water.

In the United States, groundwater is the source of drinking water for about half the total population and nearly all the rural population. Where water supply and availability has always been under stress in the desert Southwest, overstressed aquifers now extend through entire systems. The High Plains aquifer (including the Ogallala aquifer) flowing beneath eight states is being depleted at a rate that, if continued, will result 35% of the southern High Plains being unable to support irrigation within the next 30 years (Scanlon, Faunt, Longuevergne, Reedy, Alley, McGuire, & McMahon, 2012). In west central Florida, groundwater depletion has created sinkholes that consume infrastructure and allow saltwater intrusion into freshwater systems. Even in the rain-soaked Pacific Northwest, groundwater extraction in the Columbia River basin has caused significant declines in available groundwater and summer water scarcity is a new normal. Mexico City is another extreme case of groundwater depletion. The city is built on an old lakebed and surrounded by mountains, so no nearby surface water source is available for the city's 15.6 million people who must rely on the underground aquifer, supplemented by water pumped across the mountains at a high cost (Wilson, 2014).

IISD reports from the February 2018 28th UN-Water meeting that UN-Water is developing a baseline for developing indicators in preparation for the in-depth review of implementation of SDG #6 (water and sanitation) during the July 2018 meeting of the High-level Political Forum on Sustainable Development (HLPF) (Wagner & Pasini, 2018). Meeting participants emphasized the absolute dependence of agriculture on water, the interrelationship of water with food, energy and biodiversity, and stressed the need to purposefully link the work of UN-Water with other multilateral goals, such as the Aichi Biodiversity Targets. Water security policies and indicators that include both climate change adaptation and disaster risk reduction must integrate and mainstream water, sanitation and hygiene, climate change adaptation and finance and disaster risk reduction as part of mutually reinforcing the relationship between SDG# 6 targets and virtually all other SDGs (Csala, 2017).

The availability and quality of water impacts human health, economies, political stability and ultimately all human decisions. Without adequate clean water, social systems and even life shuts down.

As we begin the official UN International Decade for Action: Water for Sustainable Development 2018-2028, scientific monitoring and evaluation of water as part of the Sustainable Development Goal on clean water and sanitation (SDG #6) is being purposefully paired with education (SDG #4) in an effort to raise awareness of resource managers and policy makers to the importance of monitoring the quality of water and aquatic ecosystems (Paul, 2018). One of the global indicator targets for water, SDG #6.3.2, requires reporting on the proportion of bodies of water with good ambient water quality. Debbie Chapman of the UN Environment Global Environment Monitoring System for Water (GEMS/Water) emphasizes the role for education in raising awareness and developing the necessary technical and policy skills to meet this and other targets for SDG 6 (International Standards Organization, 2018).

Oceans

Freshwater systems are intimately linked with the ocean. Ocean and freshwater systems meet most dramatically at the coasts where approximately 40 percent of all people on earth reside. Humans face direct and indirect threats from changes in the ocean. Oceans create and drive natural disasters. Natural hazards that inflict heavy losses in countries globally such as heat waves, cyclones (include hurricanes, tropical storms, tropical depressions and typhoons), cold waves, drought, floods, tsunamis and storm surges are highly influenced by oceans. In 2016, the number of global loss events from ocean-related disasters was recorded by Munich RE was 750 compared to 730 loss events in 2015 (Munich RE, 2018).

As sea levels rise, some communities retreat from coastlines, often to already overcrowded cities, while planners in threatened cities like Miami and New Orleans continue to ignore dire warnings from the community of climate scientists (Pilkey, Pilkey-Jarvis & Pilkey, 2017). Salt water intrusion into fresh water systems makes food more difficult or impossible to grow, increases in the extent and nature of vector borne diseases, and spreads pollution that affects people well beyond coastal communities.

Ocean systems provide critical ecosystem services upon which humans depend including food, climate regulation, protection from storm surges, oxygen, pest control, recreation, and cultural heritage. Healthy oceans clean the air, absorbing greenhouse gases from sources such as industrial smokestacks and automobile emissions. Global ocean processes include primary production and nutrient cycling. Healthy oceans provide food and other resources like medicines and minerals, recreation, economic benefits, transportation, and resources for products like cosmetics, and materials for manufacturing and industry. Oceans provide oxygen. About 100,000 small organisms covered with tiny shells, called coccolithophores, are found in a single teaspoon of seawater. In some areas, they congregate together in such large numbers that they can be seen from space. Together with cyanobacter and diatoms, coccolithophores contribute between 50% and 70% of the world's oxygen (University of Leicester, 2015). These powerful tiny organisms are under attack from an increasingly acidic ocean that dissolves their calciferous shells and promotes viruses that kill phytoplankton. No matter where you are, you depend upon the ocean every day for the water you drink and the breaths you take.

The ocean is critical not only in its roles in breathable air and feeding or fouling freshwater sources, but the ocean impacts myriad issues related to the global mission to ensure a preferable future for present and upcoming generations. Warmer global surface temperatures and warmer sea surface temperatures are making extreme weather events more frequent and/or intense for a variety of reasons including ENSO, the El Nino Southern Oscillation (Taylor, 2012). Oceans create and drive the weather all over the world. A storm in the middle of the United States, in Europe or in the African Savannah began somewhere over an ocean. The best known ocean-driven weather process, El Niño, originates in the South Pacific and affects floods, droughts and storms worldwide. Droughts and floods can cause famine as crops fail; both can bring or escalate diseases. Damage from these events and from storms makes communities more vulnerable and less able to cope with additional stresses. It is therefore imperative that we do not base climate change adaptation efforts on the improving atmospheric conditions alone, but on the ocean as well.

Coral reefs, one of the most vulnerable marine ecosystems, provide not only nurseries that feed fish stocks, but protect coastal areas from erosion, storms, and subsiding coastal lands. Under attack from an increasingly acidic ocean that dissolves their calciferous structure and bleaching from increased ocean temperatures, those reefs are also facing degradation and potential destruction from deadly infections carried by plastic flotsam that has proliferated far beyond the famed Pacific Garbage Patch. A recent scientific study conducted by global marine scientists including Cornell University's Drew Harvell and Joleah Lamb reports that reefs with plastics snags show a rise in the likelihood of disease rises by 4% to 89% (Radford, 2018) depending upon other environmental conditions. A second conglomeration of microplastics the size of Mexico that cannot be seen at the surface was recently discovered by the Algalita Research Foundation in the southern Pacific off the coasts of Chile and Peru (Montanari, 2017). Tiny plastics are now found in over 90% of seabirds as plastic makes its way through the diet of the marine environmental food chain into human diets and into terrestrial fresh water systems.

Understanding the direct and indirect roles the ocean plays in human health and wellbeing is vital in making good decisions about the future for sustainability and resilience to environmental changes (Bowen, 2014).

Unhealthy oceans impact us in striking ways. While healthy oceans clean the air, unhealthy oceans add toxins, sometimes deadly ones, to the air we breathe. Oceans transport vectors and diseases like cholera to new places. As the oceans become more acidic at more than a 30% increase since the industrial revolution (Ocean Acidification, 2018), jellies are taking over niches of other marine organisms and depriving some fish and marine mammals of primary food sources. Acidic-loving viruses are flourishing and multiplying. Warmer oceans cause some surprising disruptions to the marine web such as the recent discovery that sea turtle hatchlings along the northern part of the Great Barrier Reef are now 99% female due to warmer coastal conditions (Hogge, 2018) because ocean temperature at the time of egg hatch is the sexual determinant for these animals.

Results of a study conducted by scientist Thomas Meissner (2012) show that a combination of ocean acidification, which results in reduced aragonite saturation levels and increasing sea-surface temperatures, will expose coral reefs to more severe thermal stress, resulting in bleaching. Coral reefs occupy about 10 percent of the tropical oceans globally and provide habitat for over one million species. They flourish in a relatively narrow range of temperature tolerance and are therefore highly vulnerable to sea-surface temperature increases. The effects of both ocean warming and ocean acidification are projected to cause major damages to coral reef systems and lead to losses in fish production, at least regionally. These outcomes also threaten human livelihoods and marine-tourism dependent economies.

THEORETICAL FRAMEWORK AND RESILIENCE THINKING

The theoretical framework used in this chapter for evaluating the future of waters relies heavily on a complex adaptive systems framework (CAS), and its framing towards self-organizing systems. A CAS is a complex, self-similar collectivity of interacting, adaptive agents. Complex Adaptive Systems show evidence of moving towards resilience when perturbed (Wikipedia, 2018) and are highly adaptive to new conditions that exist even at the boundaries of previous norms. Other traditions of multi-agent systems (MAS) focus on multiple interacting agents, but what distinguishes a CAS is the focus on top-level properties and features like self-similarity, complexity, emergence, and self-organization. CAS are dynamic systems that are closely linked with all other related systems making up an ecosystem; they can adapt in and evolve with a changing environment because there is no separation between the system and its environment. When change occurs as a co-evolution with all other related systems rather than as adaptation to a separate and distinct environment, certain individual systems' behaviors are triggered at tipping points to restore equilibrium to the larger ecosystem, but these may incur the cost of radical changes to the system. As catastrophic events continue to stretch and exceed current ecosystem and social system boundaries, the conditions of humanity's common future require serious considerations about what adaptation really means, and whether adaptive responses designed to return a system to its former state are truly resilient responses.

Auerbach (2016) tells the story of how Murray Gell-Mann, Gian-Carlo Rota, John H. Holland, Heinz Pagels, and Melanie Mitchell came together at the Santa Fe Institute to envision an interdisciplinary framework encompassing theories and knowledge of everything from biology to physics to economics to sociology in a wholistic and interwoven approach. As the 2004-2005 Pardee Visiting Professor of Future Studies and Distinguished Fellow of the Santa Fe Institute, Gell-Mann described his lecture as "a crude look at the whole" (Brockman, 1995) and with it he advanced complexity theory as the antidote to reductionism. Gell-Mann influenced such notable authors as Edward O. Wilson, to entertain notions of emergence in which even the simplest parts can interact in complex ways to generate positive emergent behavior out of simple base biological, and by extension social, materials (Wilson, 1998).

As climate change escalates and the effects of global weirding become increasingly apparent at every scale, the ways in which science informs resource decisions means looking beyond last year's flood, ice storm, or record-breaking drought. From personal to societal, from the community to global level, active and purposeful elasticity is necessary to attend to the needs of smallest, the largest, and back to the smallest scales. Applications of this aspect of resilience are evident in the guiding philosophies of organizations that depend upon the most recent and relevant science while embracing a multiple futures/ multiple scales approach (Intergovernmental Panel on Climate Change, 2014). For NGO SeaTrust Institute, *Local to Global: and Back Again* *(SM)* evolved from field experiences into an organizational moniker, research paradigm, and educational development directive derived from academic and practical experience with the changing global social/political/environmental system, particularly as it relates to climate change (SeaTrust Institute, 2018; Frumkin, 2016).

Making decisions for the future under the emerging conditions requires more than transboundary considerations and being purposefully multi-scalar; the new process requires an ability to address the many cultural and social frameworks within a variety of political contexts. It means using science reverently as knowledge within a broader, moral standpoint as philosopher and theologian Paul Woodruff (2001) suggests in claiming that "reverence is the well-developed capacity to have the feelings of awe, respect, and shame when these are the right feelings to have" (p. 8). Being informed by science does not mean being constrained by reductionist thinking when that is not the right response for decisions about transformative resilience. Marston and Marston define transformative resilience as the ability to "move away from a tendency to catastrophize" (Marston, 2018) and pursue success despite facing upheaval, crisis, and change. At an individual or a social policy level, this approach requires constructivist-inspired social responses that explicitly recognize the known implications of decisions made at local, national and global levels, and account for unintended consequences of actions, the known unknowns and the "unknown unknowns."

Resilience thinking must enable, rather than restrict, alternative future scenarios by moving beyond catastrophic narratives to creative discussion that develops and reinforces positive incentives for multiple pathways of resilience action. As awareness of vulnerabilities and the effects of catastrophic events has multiplied, one global sustainability response has been towards defining and working towards a preferable future through foresight, defined as "a systematic, participatory, future-intelligence-gathering and medium-to-long-term vision-building process aimed at enabling present-day decisions and mobilizing joint action" (Anderson and Anderson, 2012, p. 13). SDG implementation framework approaches like the UNDP Global Center for Public Service Excellence's *Empowered Futures Initiative* (UNDP, 2018) use foresight as a process through which new insights and capabilities are developed. Foresight does not predict or forecast the future, but rather works with alternative futures. Other efforts towards modeling and multiple futures approaches attempt to capture the inherent complexity and uncertainties while encouraging futures thinking by building options and flexibility into each stage. Gidley, (2017) offers the Swinburne approach in conjunction with the longer-horizon strategic foresight works of Richard Slaughter as an example of a workable multiple futures approach that captures those elements. However, a disconnect exists between these flexible and innovative processes and current sustainability success measurements.

MEASURING SUSTAINABILITY PROGRESS

Measurements guide decisions made on the pathway towards the future. Those decisions lead either towards or away from resilience. Creating measurements that are meaningful for specific present purposes, yet flexible enough to encompass current and future complexities, is no simple task. Pragmatic approaches that move beyond the continued paradigm war which polarizes the field between two alternatives (quantitative or qualitative methods; objectivist or constructivist norms; summative or formative evaluation) challenge assumptions held by mainstream measurement and evaluation experts (Bengston & Fan, 1999; Mohr, 1999; Pratt McGuigan, & Katzev, 2000) and continues to be contentious. For example, current measurement methods embedded within the global SDG indicators as adopted by the 2017 resolution by the General Assembly on Work of the Statistical Commission pertaining to the 2030 Agenda for Sustainable Development (General Assembly, 2017) do not necessarily reflect alternative futures at local levels such as those for Small Island Developing States in that the global mechanism fails to appropriately integrate measures of appropriate regional, topical, or cultural significance (United Nations Environment Programme, 2014). In these high-stakes contexts, metrics and indicators must be carefully considered for their maladaptive potentials as well as for their value in benchmarking and reporting. Indicators that are locally unresponsive or inappropriate may have unintended effects in rewarding actions that are harmful over longer time periods. To avoid such pitfalls, short term measures must be augmented with longer term processes that take a complex adaptive systems approach to resilience.

The Flexible Balance

Indicators are the main approach used to track and evaluate progress towards sustainability goals. At the global level, sustainability indicators present as mostly quantitative measures (General Assembly, 2017). They were developed by the Inter-Agency and Expert Group on SDG Indicators (IAEG-SDGs) at a scale and with content that often loses relevance for countries attempting to use them in measuring their progress towards their registered Nationally Determined Contributions (NDCs) and actions towards the Paris Agreement. UN Member States (countries) continue to request assistance in developing indicators capable of assessing the States' responsiveness to laws and policies at the level of overall global commitments that are at the same time compatible with what States have entered into as part of the agreement. Specific content areas related to sustainability such as health, energy, and water have developed thematic indicators meant to augment the global set. This is sometimes useful, but just as frequently it mirrors the global indicators but with a sectoral focus. Perhaps the most fully developed are the health indicators, based on significant quantitative research on causal mechanisms linking climate with health impacts and local scales (Hales, Kovats, Lloyd, & Campbell-Lendurm, 2014). The health indicators have roots in health and environmental frameworks dating back over 20 years on climate change (Patz, Epstein, Burke & Balbus,1996; Colwell 1996), social epidemiology (Subramanian 2004), and environmental health (Black, 2000; Parkes, Panelli & Weinstein, 2003). Still, countries and communities, are finding it difficult to meaningfully link their actions and intentions to these indicators that are largely quantitative and frequently fail to acknowledge significant cultural values, norms, and alternative ways of thinking. While gigatons of CO2 or millimeters of water per hectare may be described by a number, evaluating women's empowerment is challenging to quantify and the contribution of clean water availability to that empowerment proves even more difficult to numerically represent. The fallback response to date is to suggest that these numbers are proxies for action which injects significant subjectivity into the process, further threatening the integrity of the quantitatively biased process.

Particularly when everything is falling apart, a critical piece of making good decisions under stress is understanding when measurements do or do not lead to meaningful local action. Broader scope and new types of thinking are required, especially when threats require immediate action, poor information or few data are available, or underlying social or political issues have a disproportionate effect on outcomes. What is needed when facing a postnormal future is to go beyond the practice of disciplinary experts working together but retaining their disciplinary paradigms to an interweaving of both disciplinary content and process to create an innovative new way to measure efforts to address complex issues. Specifically, recent scholarship on a more complex approach to sustainability measurement (SA) outlines the need for research to help move the indicator development process from multidisciplinary, via interdisciplinarity, towards transdisciplinarity (Sala, Ciuffo, & Nijkamp, 2015).

In a world in which the past is no longer a good guide to the future, what motivators and indicators can better engage public policy makers and administrators in enabling positive ocean sustainability and water security outcomes under the conditions produced by our changing climate and associated social instability characterized by a post-normal future?

It is instructive to apply the new SA thinking when considering the water indicators that are being developed for 5 cities representing the range of water challenges facing cities around the world (Amman, Cape Town, Mexico City, Greater Miami and the Beaches, and Hull). The indicator developers are consultants from Arup, with support from The Rockefeller Foundation for The City Water Resilience Framework (CWRF) (Arup, 2018). Arup proposes to use infrastructure and ecosystems, economic benefits and social integrity, health and wellbeing, and governance and strategy in developing the indicators. This approach follows from their 2016 Cities Resilience Index, also created with Rockefeller support, which used both qualitative and quantitative measures. The Rockefeller Foundation and Arup (2016, p.9) list the development of the Index as being based on these research questions keyed to specific bases of measurement.

The Rockefeller/Arup approach converts qualitative (or subjective responses) to 156 questions asked of participating cities into numerical scores that can then be aggregated in order to communicate key strengths and weakness based on the 12 goals. It is during this phase of the process that the critical new thinking is threatened and often lost in aggregation, commonly used to support the confirmation bias of probability theory (Taleb, 2007). It appears likely that the water indicators development

Figure 1. Rockefeller and Arup key to research questions and measurement

Research Questions	Basis of Measurement
1. What matters What contributes to city resilience?	• Goals
2. How can this be observed? What to measure?	• Indicators/variables
3. How to measure?	• Metrics/scenarios

process may follow a similar pattern in converting qualitative data to numbers, which loses the richness of outliers with extreme impact that may represent Taleb's "Black Swans," those inevitable outcomes of complex adaptive systems. CWRF is now in the early planning phase so it remains to be seen if a more alternative futures-based complex adaptive systems framework might be entertained.

Identifying system components and agents at smaller scales that change system states might be another alternative framework through which to approach resilience and sustainability indicator development. The most effective agents in new and uncertain contexts tend to be new or emerging agents whose actions can correct a system on the edge of collapse. The first step in this alternative approach is to define goals for indicators in a new way that reinforces resilient behavior and seeks to change maladaptive behavior, especially in these emerging agents. Owen, et al (2014) advocates blending horizon scanning with foresight to explore potential opportunities and threats in the construction of ocean health/human health scenarios that go beyond forecasts to include a range of possible alternative futures. Some of those futures, if properly measured, may catalyze those emerging agents. Measuring change in a complex adaptive system is part of a larger iterative process that requires recognition followed by encouragement for the underlying values and agents that promote resilience.

For those working at the policy level, it is not uncommon to think that if the policy is right, correct actions will follow. But under uncertain conditions, it takes more than the policies themselves; it takes courageous simultaneous actions and policies, tried under the new uncertainties of climate change and fragmented societies, to move together towards a new, resilient future. How might decisions be made at different scales so that they act in resilient and transformational ways that transcend traditional adaptation? Small actions give cause for hope. Specific identified individual components may be scalable into policies that help provide real resilience. Positive bottom-up actions encourage policy makers to ask, "why aren't we doing more of this in more places," and to create policy that follows actions and encourages more positive outcomes. But scalable policies must be chosen carefully and with a long-term perspective.

How are those working in resilience development adapting strategies and processes to make them truly useful in addressing the urgencies of people under siege from climate and other environmental change? First, guiding principles for changes in approaching adaptation must move the needle towards the integrity required by true resilience. The dynamic and changing conditions require a deeper understanding of social learning to transition from awareness to action. Using human health as a gauge for positive action allows direct query of social learning in a measurable, transdisciplinary way to test these principles. Health as a gauge works in part because it levels the conversation to include all social, generational, gender and cultural groups, and because, if it is defined to do so, can take into account cultural and other values that influence resilience behavior and actions. The move towards considering health and well-being together is a movement in that direction (WHO, 2018).

As part of this thought experiment, it may be instructive to analyze a field research project that was conducted just as the SDGs were being developed in order to consider characteristics that align with resilience within a postnormal future as defined in this chapter. Scholars and practitioners in adaptation strategy and capacity building speak of working with the intention to create resilient communities. In climate change adaptation, resilience is often associated with adaptation in the same way sustainability is associated with mitigation – as an outcome rather than a process. Instead, in the 2013 project conducted along a stretch of the Lower Zambezi River by a SeaTrust Institute student intern and a senior researcher, the team consciously defined resilience as the process of maintaining core principles and integrity – of all kinds - under dramatically changed circumstances. Being intentional about all aspects of that process, not only in an internal sense, but also as communicated through word choices and definitions, provided clarity about the exchange between researchers and participants who were consciously aware of the connections between climate pattern changes, extreme events and the health of their communities and citizens. At the center of the conversation was water: water scarcity; too much water at the wrong times; and changes in precipitation patterns and seasonality. While research participants had not read the Intergovernmental Panel on Climate Change (IPCC) report or other climate science and policy reports, they offered empirical evidence from observing the effects of their own fragilities under increasing stress about water: increasing encounters with elephants raiding their winter gardens for food due to drought; ears of maize half the size they had been before the persistent droughts; water releases from the nearby dam for hydroelectric power that flooded their villages; and dry bore holes causing people to walk up to a day from their villages to reliable water sources. Introducing information and data through a socially and culturally sensitive way that respected and included these realities and blended relevant science with local data was critical. Health as a gauge emerged as a vehicle for this complementarity that combined the seemingly disparate elements in such a way as to enhance or emphasize the qualities of the other. But it was about more than just bringing information about fragilities and options and walking away with research;

the research team was committed to searching for keys to unlock at least a small part of the participants' system to incorporate that awareness into knowledge that could lead to sustained and iterative action that kept the core of that system in-tact.

This example is but one that stresses the need for more and better local data, and for regionally appropriate analyses that show the dynamic interconnected changes in which water figures heavily in adaptation decisions. Single point indicators often fail to account for the dynamics, particularly where data is scarce or poor quality. For example, chemical and microbial water pollution that overtly affects health, but for which poor or nonexistent baseline information exists, makes drawing the connections between changing climates and socioeconomic / political contexts and decisions difficult to document. In the Zambian example, bore holes access deep groundwater sources but water quality is marginal. Further, testing is irregular and unreliable so the problem of microbial pollution in small scale systems like ground wells compounds other alternatives like using the heavily polluted Zambezi river water. The emergence of new strains of waterborne diseases in addition to dryland vivax malaria needed to be documented and accounted for in creating dynamic adaptation strategies.

Intensified and appropriate data surveillance and global communications that use all available resources and technologies, such as crowdsourcing used during the earthquakes in Haiti (Zolli, 2012) or telehealth and reporting in rural settings for fast life-saving decisions during disasters show how locals are in the best position to create parameters and populate data that can be used together with expert knowledge to construct the baseline and navigate the patterns of climate variability. The challenge then becomes understanding how this data and knowledge becomes part of the fabric of communities at the threshold of a crisis. Here, the third principle of resilience, dynamic reorganization, is useful. How the elements reorganize can be difficult to measure, but again human health may be able to serve as a measurable proxy for the impact of those reorganized relationships.

That resilience strategies imposed from above most often fail makes sense from a systems point of view, particularly as it relates to learning and resilience development in complex social-ecological systems (Krasny, Lundholm, & Plummer, 2010). Resilience attitudes and actions must be embedded authentically into the relationships that mediate people's everyday lives. Systems that self-organize during crises involve new and unusual relationships including those that cross scales (Walker, Gunderson, Kinzig, Folke, Carpenter, & Schultz, 2006). When integrity driven self-organization operates using the guiding principles of reorganization and complementarity in action, the results can astonish. The emergence of what biologists call sleeping functional groups, something that exhibits a behavior that only emerges when needed to keep a system from annihilation, is an example of self-organization that moves from attempted adaptation to resilience. Such was the case with the batfish on Australia's Great Barrier Reef that in 2006 switched its diet from invertebrates to algae to save the reef system on which it and many other organisms depend. If this behavior is found in the obscure batfish, where else in nature does it occur and what characteristics are key to this type of resilience action that we may encourage in human society? Intentionality, complementarity and other special characteristics of sleeping functional units such as diversity, countercyclical action, and trust are required for encouraging and catalyzing the intervening behaviors by those emerging agents that only exhibit as thresholds are being reached. Because resilience is not a final state, but a delicate, constantly renegotiated balanced reaction to changing conditions, the unusual behavior lasts long enough to disrupt calamity to keep the system from flipping into disaster as thresholds are approached, after which those agents return to their former disconnected or even opposing positions.

Perhaps lessons from these waters could inform communities creating resilience to climate and other environmental change about how to locate and nourish the "batfish" for rehabilitating good waters that have gone bad. In human society, certain groups of youth exhibit batfish behavior to "flip" a system at a threshold moment in both policy and action.

When empowered with appropriate knowledge, these youths can exhibit emergent innovation that can flip the system. They are vested in long term outcomes, so enabling youth to co-develop the message and deliver it helps sensitize decision makers to opportunities for reframing related governance initiatives in finance, technology transition, and inclusion of local knowledge in ways that position governmental officials to act in their own best interest and in the interest of current and future generations. By providing a meaningful context for all actors in the iterative processes of policy creation, learning, and research, education as both local capacity building and policy support is truly "Local to Global - and Back Again(SM)" (SeaTrust Institute, n.d.). Armed with these principles, we can better help knit the sometimes apparently opposing constituencies, encouraging corrective action by individual system components for the benefit of the complex adaptive system with potentially drastically changed emergent system properties, true to the definition of resilience.

CONCLUSION AND AREAS FOR FUTURE STUDY

Waters, both fresh and saline, are complex adaptive systems. Humans are an integral part of that system. Educating and engaging youth about waters and nature-based solutions to developing resilience may contribute to avoiding the thresholds of freshwater scarcity and ocean changes that irrevocably alter the environment for human well-being, but are likely insufficient to ensure resilience. Awareness is only a first step. Action is required across generations and that necessarily engages a multigenerational effort in both education and action.

An alarm sounds as attention to SDG #14 (Life Below Water) that aims to "conserve and sustainably use the oceans, seas and marine resources for sustainable development," has received a small amount of philanthropic funding and is one of the lowest priority SDGs as indicated in country self-reports. Future research is needed to more closely identify and link the co-benefits of the oceans with the other 16 SDGs. From analyzing the results from a series of expert workshops, Singh et al. (2018) tested a framework for evaluating co-benefits and trade-offs among directional relationships, showing positive relationships between Goal #14 (oceans) and the achievement of the other goals and concluded that "…ending overfishing was positively related to the largest number of other SDG targets at approximately twice the number of obligate relationships as increasing economic benefits to SIDS (small island developing states) - the ocean target with the second-most obligate relationships" (p. 225). Environment, society and economy are tightly coupled as evidenced by these authors' evaluation of the relationships among the SDGs. At a time when concern is high that the SDGs may not be reached, new frameworks that can rapidly characterize relationships among not only SDGs but other policy goals such as Aichi targets are important to integrating water more fully into policy decisions. One criticism is that such frameworks are qualitative and therefore cannot show the degree of impact one SDG target has on another, nor do they represent uncertainty metrics in the selection of relationship categories.

Freshwater, SDG #6, also does not stand alone. With global groundwater diminishing (WWAP, 2015), water studies could benefit by using a more integrative framework among SDGs and other policy goals and targets such as the relationship of clean, available water to health, cities, peace, and more. Water demand is predicted to rise by 50% between 2000 and 2050, with most of it being consumed by agriculture (Smedley, 2017). Youth are particularly affected in ways not often considered in water policy discussions. For example, even mild levels of dehydration impair cognitive functions in young people with life-long effects (Popkin, D'Anci, & Rosenberg, 2010). SDG#6 is also about sanitation. Supplying adequate clean water and educating people about its use and conservation is key to resilience development. Research in the key catalysts for youth leadership in resilience development is needed.

Quantifying future conditions is always problematic; better data would aid in more reliable predictions. However, an overreliance on quantitatively-biased indicators leads to concerns such as those voiced by both scholars and policy practitioners that the current SDG global indicator sets fail to capture crucial factors for achieving the SDGs in taking numbers out of context, reflecting data constraints, and being deployed as measures in absence of the type of intense conceptual and methodological work that takes them beyond producing new environmental, social and economic statistics (Hak, Janouskova, & Moldan, 2016). When SDG indicators are siloed by individual SDGs and targeted towards near-term outputs, they do not take into account the longer view that reflects the world that today's youth will inherit. Human health and well-being offers a possible proxy in such cases, so long as the health measures themselves are defined in ways that are meaningful and operational at the scales and across geographies that policy makers can use. Alternative methods of measurement such as SA need to be more thoroughly developed and tested within complex adaptive systems research. Youth need to be cultivated to make those bold and meaningful moves, mentored into their roles as emerging adults, and explicitly considered beyond SDG #4 (education) as integral in working across all the SDGs and associated policy to integrate good water and oceans decisions for their future resilience and for ours.

POSTSCRIPT

In November 2017, while relating a version of the batfish story during a formal UN presentation in the Fiji Pavilion at COP23, I gestured to one of our panelists, Council of Youth (COY) Delegate Timothy Damon, whom I had known since he was a very young attendee when he was excluded from attending events inside the UNFCCC COP negotiation halls. I had followed Timothy through his development as a leader in the newly UNFCCC recognized formal COP Youth Constituency and his transformation from effervescent young man into a suited, eloquent spokesperson who held his own with the battalion of negotiators. I called him a batfish and told him why he was one. He took his place at the podium, laid aside his presentation notes, loosened his tie, and embraced his new identity, saying "I like being a batfish!" I hope he nibbles at the roots of the algae infested water, oceans, and sustainability indicators to induce multiple preferable futures for us all.

REFERENCES

Andersen, A. D., & Andersen, P. D. (2012). Innovation-system foresight: Explicating and systemizing the innovation-system foundations of foresight and exploring its implications. *Technical University of Denmark*. Retrieved from http://orbit.dtu.dk/files/10590515/Innovation_system_foresight.pdf

Auerbach, D. (2016). *The Theory of Everything and Then Some*. Slate.

Bengston, D. N., & Fan, D. P. (1999). An innovative method for evaluating strategic goals in a public agency: Conservation leadership. *Evaluation Review*, *23*(1), 77–10. doi:10.1177/0193841X9902300104

Black, H. (2000). Environmental and public health: Pulling together the pieces. *Environmental Health Perspectives*, *108*(11), A513–A515. doi:10.1289/ehp.108-a512

Bowen, R. E., Depledge, M. H., Carlarne, P., & Fleming, L. E. (2014). *Oceans and human health; implications for society and well-being*. Chichester, England: John Wiley & Sons.

Brockman, J. (1995). *The third culture: Beyond the scientific revolution*. New York, NY: Simon & Schuster.

Bruntland, G. H. (1987). *Our common future: Report of the 1987 World Commission on Environment and Development*. Oxford, UK: Oxford University Press.

Cape Town, on verge of running out of water, braces for "chaos" (2018, January 19). *CBS News*. Retrieved from https://www.cbsnews.com/news/cape-town-drought-water-shortage-south-africa/

Cassar, G. (2018). "The Davos 2018 environment agenda - what you need to know and how to follow online." World Economic Forum. Retrieved from https://www.weforum.org/agenda/2018/01/the-davos-2018-environment-agenda-online-what-you-need-to-know

Chege, N. (2016). The Implications of the implementation of UNCLOS: A review of the approach to resolving environmental issues in African nations. Manuscript submitted for publication.

Colwell, R. R. (1996). Global climate and infectious disease: The cholera paradigm. *Science*, *274*(5295), 2025–2031. doi:10.1126cience.274.5295.2025 PMID:8953025

Wikipedia. (n.d.). Complex adaptive systems. Retrieved from https://en.wikipedia.org/wiki/Complex_adaptive_system

Centre for Research on the Epidemiology of Disasters, & United Nations Office for Disaster Risk Reduction (2015). *The human cost of weather related disasters, 1995-2015*. Retrieved from www.unisdr.org/files/46796_cop21weatherdisastersreport2015.pdf

Csala, J. (2017). *Water and sustainable development* [Policy brief]. Port Townsend, WA: SeaTrust Institute.

Diez-Roux, A. V. (2000). Multilevel analysis in public health research. *Annual Review of Public Health, 21*(1), 171–192. doi:10.1146/annurev.publhealth.21.1.171 PMID:10884951

Ellis, B., & Herbert, S. (2011). Complex adaptive systems (CAS): An overview of key elements, characteristics and application to management theory. *Informatics in Primary Care, 19*(1), 33–37. PMID:22118334

Folke, C. (2007). Social-ecological systems and adaptive governance of the commons. *Ecological Research, 22*(1), 14–15. doi:10.100711284-006-0074-0

Frumkin, H., & Patz, J. (2016). Climate change and human health. In Frumkin (Ed.) Environmental health: Global to local. Hoboken, NJ: Wiley.

Gell-Mann, M. (1995). *Quark and the Jaguar*. New York, NY: W.H. Freeman.

United Nations. (2017, July 6). General Assembly resolution 71/313, Work of the Statistical Commission pertaining to the 2030 Agenda for sustainable development. Retrieved from undocs.org/A/RES/71/313/

Gidley, J. M. (2017). *The future: A very short introduction*. Oxford: Oxford University Press. doi:10.1093/actrade/9780198735281.001.0001

Hák, T., Janousková, S., & Moldan, B. (2016). Sustainable Development Goals: A need for relevant indicators. *Ecological Indicators, 60*, 565–573. doi:10.1016/j.ecolind.2015.08.003

Hogge, K. (2018, February 2). Not cool: Climate change turning 99% of these sea turtles female [blog post]. *Ocean Conservancy*. Retrieved from https://oceanconservancy.org/blog/2018/01/25/not-cool-climate-change-turning-99-sea-turtles-female/

International Standards Organization. (2018). *Water and sanitation. ISO-Focus, 126.*

Krasny, M. E., Lundholm, C., & Plummer, R. (2010) Resilience in social–ecological systems: the roles of learning and education. *Environmental Education Research, 16*(5-6), 463-474. doi:10.1080/13504622.2010.505416

Hales, S., Kovats, S., Lloyd, S., & Campbell-Lendurm, D. (2014). *Quantitative risk assessment of the effects of climate change on selected causes of death, 2030s and 2050s*. Geneva: World Health Organization.

Intergovernmental Panel on Climate Change. (2014). Summary for policymakers. In *Climate change 2014: Impacts, adaptation and vulnerability. Contribution of Working Group II to the Fifth Assessment Report of the Intergovernmental Panel on Climate Change*. Geneva, Switzerland: IPCC.

Lloyd, D., & Wallace, J. (2008). Imaging the Future of Science Education: The Case for Making Futures Studies Explicit in Student Learning. *Studies in Science Education, 40*(1), 139-177. doi:10.1080/03057260408560205

USAID Office of Population and Reproductive Health. (2014). Marine Ecosystems and Management. Interview with Heather D'Agnes. Seattle: Marine Affairs Research and Education.

Marston, A., & Marston, S. (2018). *Type R: Transformative resilience for thriving in a turbulent world.* New York: Public Affairs.

Mesfin, M., & Arjen, Y. (2016). Four billion people facing severe water scarcity. *Science Advances, 2*(2). doi:10.1126ciadv.1500323

Mohan, V., & Shellard, T. (2014). Providing family planning services to remote communities in areas of high biodiversity through a population-health-environment programme in Madagascar. *Reprod. Health Matters, 43,* 93-103. doi:10.1016/S0968-8080(14)43766-2

Mohr, L. B. (1999). The qualitative method of impact analysis. *The American Journal of Evaluation, 20*(1), 69–84. doi:10.1177/109821409902000106

Montanari, S. (2017). Plastic garbage patch bigger than Mexico found in Pacific. *National Geographic.* Retrieved from https://news.nationalgeographic.com/2017/07/ocean-plastic-patch-south-pacific-spd/

Munich, R. E. (2018). *Natural disasters: The year in figures.* Retrieved from https://www.munichre.com/topics-online/en/2017/topics-geo/overview-natural-catastrophe-2016

Owen, R., Depledge, M., Dr Lurio, J., Bardsley, S., & Palomino, M. (2014). Emerging issues in oceans and human health: Managing uncertainty and new knowledge. In R. E. Bowen, M. H. Depledge, P. Carlarne, & L. E. Fleming (Eds.), *Oceans and human health; implications for society and well-being* (pp. 267–287). Chichester, England: John Wiley & Sons.

Patz, J. A., Epstein, P. R., Burke, T. A., & Balbus, J. M. (1996). Global climate change and emerging infectious diseases. *Journal of the American Medical Association, 275*(3), 217–223. doi:10.1001/jama.1996.03530270057032 PMID:8604175

Paul, D. (2018, February 8). ISO journal highlights need to meet SDG6. *IISD SDG Knowledge Hub.* Retrieved from www.iso.org/files/live/sites/isoorg/files/news/magazine/ISOfocus%20(2013-NOW)/en/2018/ISOfocus_126/ISOfocus_126_en.pdf

Pilke, O., Pilkey-Jarvis, L., & Pilkey, K. (2017). *Retreat from a rising sea: Hard choices in an age of climate change*. New York, NY: Columbia University Press.

Popkin, B. M., D'Anci, K. E., & Rosenberg, I. H. (2010). Water, hydration and health. *Nutritional Review, 68*(8), 439-548. doi:10.1111/j.1753-4887.2010.00304.x

Pratt, W. M., McGuigan, C. C., & Katzev, A. R. (2000). Measuring program outcomes: Retrospective pretest methodology. *The American Journal of Evaluation, 21*(3), 341–349. doi:10.1177/109821400002100305

Serenella, S., Ciuffo, B. I., & Mijkamp, P. (2015). A systemic framework for sustainability assessment. *Ecological Economics, 119*, 314–325. doi:10.1016/j.ecolecon.2015.09.015

Scanlon, B. R., Faunt, C. C., Longuevergne, L., Reedy, R. C., Alley, W. M., McGuire, V. L., & McMahon, P. B. (2012, June 12). Groundwater depletion and sustainability of irrigation in the US High Plains and Central Valley. *Proceedings of the National Academy of Sciences of the United States of America, 24*(24), 9320–9325. doi:10.1073/pnas.1200311109 PMID:22645352

Singha, G.G., Cisneros-Montemayora, A.M., Swartz, W., Cheung, W., Guy, J.A., Kenny, T.A. (2018). A rapid assessment of co-benefits and trade-offs among Sustainable Development Goals. *Marine Policy, 93*, 223-231. doi:10.1016/j.marpol.2017.05.030

Smedley, T. (2017, April 12). Is the world running out of fresh water? *BBC*. Retrieved from www.bbc.com/future/story/20170412-is-the-world-running-out-of-fresh-water

Taleb, N. (2007). *The black swan: The impact of the highly improbable*. New York: Random House.

Taylor, A. H. (2011). *The dance of air and sea: How oceans, weather and life link together*. Oxford, UK: Oxford University Press.

Radford, T. (2018). Coral reefs face infection risks from plastics. *Climate Change Network*. Retrieved from https://climatenewsnetwork.net/coral-reefs-face-infection-risk-plastic/

Redman, C. L. (2014). Should sustainability and resilience be combined or remain distinct pursuits? *Ecology and Society, 19*(2), 37. doi:10.5751/ES-06390-190237

Redman, C. L. (2015). Adaptation, transformation and transition. In K. C. Seto, W. D. Solecki, & C. A. Griffith (Eds.), *The Routledge Handbook of Urbanization and Global Environmental Change*. London, UK: Routledge.

Resilience. (n.d.). Oxford English dictionary. Retrieved from https://en.oxforddictionaries.com/definition/resilient

The Rockefeller Foundation. (2016, March). *Cities resilience index research report (Vol. 4)*. Retrieved from www.cityresilienceindex.org/wp-content/uploads/2016/05/Vol4-MeasuringCityResilience.pdf

SeaTrust Institute. (n.d.). Current Projects. Retrieved from www.seatrustinstitute.org/projects

Schuster, P. F., Schaefer, K. M., Aiken, G. R., Antweiler, R. C., Dewild, J. F., Gryziec, J. D., ... Zhang, T. (2018). Permafrost stores a globally significant amount of mercury. *Geophysical Research Letters*, *45*(3), 1463–1471. doi:10.1002/2017GL075571

Subramanian, S. V. (2004). The relevance of multilevel statistical methods for identifying causal neighborhood effects. *Social Science & Medicine*, *58*(10), 1961–1967. doi:10.1016/S0277-9536(03)00415-5 PMID:15020011

Taleb, N. (2007). *The black swan*. New York: Random House.

UNESCO. (2018). *Ocean Acidification*. Retrieved from http://www.unesco.org/new/en/natural-sciences/ioc-oceans/focus-areas/rio-20-ocean/blueprint-for-the-future-we-want/ocean-acidification/

University of Leicester. (2015, December 1). Failing phytoplankton, failing oxygen: Global warming disaster could suffocate life on planet Earth. *ScienceDaily*. Retrieved from www.sciencedaily.com/releases/2015/12/151201094120.htm

United Nations Environment Programme. (2014). *Emerging issues for Small Island Developing States: Results of the UNEP Foresight Process*. Nairobi, Kenya: UNEP.

United Nations Development Programme. (2018). *Foresight manual: Empowered futures for the 2013 Agenda*. Singapore: UNDP Global Centre for Public Service Excellence.

United Nations Framework Convention on Climate Change. (2015, December 15). *Adoption of the Paris Agreement* (Report No. FCCC/CP/2015/L.9/Rev.1). Retrieved from http://unfccc.int/resource/docs/2015/cop21/eng/l09r01.pdf

U.S. Geological Survey. (2016). *The world's waters*. Retrieved from: http://water.usgs.gov/edu/earthwherewater.html

Wagner, L., & Pasini, O. (2018). 28th UN-Water Meeting. *UN Water Bulletin*, *82*(32). Retrieved from http://enb.iisd.org/water/un/28/html/enbplus82num32e.html

Walker, B. H., Gunderson, L. H., Kinzig, A. P., Folke, C., Carpenter, S. R., & Schultz, L. (2006). A handful of heuristics and some propositions for understanding resilience in social-ecological systems. *Ecology and Society*, *11*(1), 13. Retrieved from http://www.ecologyandsociety.org/vol11/iss1/art13/ doi:10.5751/ES-01530-110113

Wilson, E. O. (1998). *Consilience: A unity of knowledge*. New York: Vintage Books.

Wilson, L. (2014, October 22). Running dry in a changing world [Commentary]. *Westlaw Journal Thomson Reuters Blog*. Retrieved from http://thomsonreuters.com/index.php/running-dry-in-a-changing-world

Woodruff, P. (2001). *Reverence: Renewing a forgotten virtue*. Oxford, UK: Oxford University Press.

World Economic Forum. (2015). *WEF Outlook 2015: Outlook on the global agenda*. Retrieved from http://reports.weforum.org/outlook-global-agenda-2015/top-10-trends-of-2015/3-lack-of-leadership/

World Health Organization. (n.d.). *Synergies for beating NCDs and promoting mental health and well-being*. Retrieved from http://www.who.int/ncds/governance/high-level-commission/synergies-beating-ncds/en/

WWAP (United Nations World Water Assessment Programme). (2015). *Facing the challenges; Case studies and indicators*. Paris, France: UNESCO.

Zolli, A. (2012). *Resilience: Why things bounce back*. London: Headline Publishing Group.

Chapter 3
Nature-Based Solutions for Water Security and the Role of Education for Enhancing the Implementation of the 2030 Agenda

Janos Csala
SeaTrust Institute, USA & TU Delft University, The Netherlands

Jennifer Wanjiku Mwangi
SeaTrust Institute, USA

ABSTRACT

Water security is a central sustainable development challenge. Billions of people lack access to clean and reliable water, while global hydrological changes and increasingly common extreme weather events pose serious risks. However, current issues are mainly driven by unsustainable management and ensuing ecological degradation. Nature-based solutions restore, enhance and safeguard ecosystems that provide water for people and the rest of nature. They also buffer the impacts of natural hazards and provide other critical benefits. Global policy frameworks on sustainable development, disaster prevention, climate change, biodiversity, wetlands and desertification offer holistic objectives toward water security. Education and capacity development is one of their central connective tissues, and as a mean to enhance their implementation. In spite of this, major gaps remain that require novel approaches. This chapter explores these and discusses strategic considerations and innovative approaches that can leverage existing knowledge and foster context specific innovation for transformative solutions.

DOI: 10.4018/978-1-5225-7727-0.ch003

INTRODUCTION

Despite progress in the 21st century, 844 million people still do not have access to basic water services, while for another 2.1 billion, access is constrained by distance and quality issues. In 2015 2.3 and 4.5 billion people lacked access to basic sanitation, and facilities with safe treatment respectively. Water quality and water stress, i.e., inadequate water supply or quality (EEA, n.d.) remain serious issues across and within countries. Currently up to 3.6 billion people confront temporary water scarcities, i.e. "physical shortage due to" insufficient infrastructure and/or governance failures (UN-Water, n.d.). Pollution and land degradation significantly exacerbate these issues. Sustainable access to potable water is essential for human health and well-being. Fresh water is also the mainstay of economic development and ecological sustainability (UN, 2015).

Virtually all human activities rely on water including food and energy production and manufacturing. As a result, water availability and quality issues present adverse socio-economic implications. For instance, agriculture and energy generation depend on reliable water supply (UN, 2015). Both water quantity and quality are supplied and regulated by ecosystem services (ESS). Ecosystems are self-organizing communities of living organisms and "the nonliving environment" they interact with. Humans are part of ecosystems, and the benefits they and other species derive from nature are conceptualized as ESSs. Biodiversity i.e., the richness and diversity of life plays a central function in sustaining nature's life enabling benefits as illustrated in Figure 1 (MEA, 2005 pp. V). While biodiversity conservation is progressing globally, it is still insufficient to halt the alarming rates and magnitudes of extinctions (CBD, 2014; Ceballos, Elrich & Dirzo, 2017).

Water also connects and sustains all life (UN, 2015) which is reflected by UN Water's (2013) definition of water security. That is "the capacity to safeguard sustainable access" to water that fulfills all human and ecological needs in a manner that promotes participation and cooperation. Simultaneously it conserves underlying ecosystems and reduces the impact of water-related hazards like droughts and floods (pp. vi). Current water availability and quality problems are predominantly driven by unsustainable development and management practices that lead to biodiversity loss and other forms of ecological degradation. The estimated loss of 70% of global wetland ecosystems over the 20th century is one of the root causes of the current water crisis (UN-Water, 2018; UNWWAP, 2018). Natural wetlands are "a wide variety of inland habitats such as marshes, peatlands, floodplains, rivers and lakes, and coastal areas such as saltmarshes, mangroves, intertidal mudflats and seagrass beds, and also coral reefs and other marine areas no deeper than six metres at low tide" (Ramsar Convention Secretariat, 2016, p. 2).

These systems play critical functional roles in both water quantity and quality regulation, and water-related hazard mitigation (Ramsar, 2015). Climate change is observed not only to increase the incidence and intensity of such hazards like floods and drought, but also to undermine the supply and quality aspects of water security. In particular, climate change is altering the distribution and quality of fresh water over geographic locations and time which coupled with exponentially growing demand threatens access (UN, 2015). These phenomena are increasing disaster risk (DR) i.e., the scale of "potential loss of life, injury, or destroyed or damaged assets". Ecosystem degradation across local and global levels is one of the root causes of disaster risk. It exacerbates exposure to hazards e.g., loss of vegetation results in rainwater runoff damaging infrastructure because the vegetation that slows or controls the speed and volume of rainwater runoff in watersheds or populated areas is no longer present (UNWWAP, 2018).

Another consequence of ecosystem degradation is increased vulnerability, i.e. susceptibility to be harmed by hazards e.g. flood water pollutes drinking water sources and destroys food and income sources (UNGA/OEIWG, 2016, p. 14). Land degradation and drought-related disasters result in the highest loss of life and economic cost today impacting approximately 1.8 billion people (Ibid). Therefore, enhancing water security is essential for sustaining human life as highlighted in the United Nations 2030 Agenda for Sustainable Development. Sustainable Development Goal (SDG) 6, which aims to "Ensure availability and sustainable management of water and sanitation for all by 2030" has a mutually reinforcing relationship with virtually all other 16 SDGs and interlinks different sectors.

Advancing SDG 6 targets is necessary for reducing poverty and inequalities including gender, increasing food security and agricultural sustainability (UN-Water, 2016). It is also critical for improving health and well-being, education, access to sustainable and modern energy. Good governance is the "Effective, accountable and inclusive" institutional foundation of water security. Economic, infrastructural and ecological sustainability, and resilience i.e., the ability to withstand and recover from hazards (UNGA/OIEWG, 2016) are the other institutional and biophysical bases of SDG 6. As illustrated earlier all of these have a reciprocal relationship with water security, and are intertwined with climate change resilience, and the conservation, sustainable use of ecosystems (UN-Water, 2016). Water security is integral to the advance of other global policy frameworks related to sustainable development. Foremost to the other 2015 landmark agreements of the Sendai Framework which inter alia seeks to minimize disaster losses and reduce risks, and the Paris Agreement which aims to limit climate change and adapt to inevitable impacts.

On the other hand, water security and these frameworks hinge on ecological sustainability which is the primary objective of three other global frameworks. These are the Convention on Biological Diversity which seeks to halt species extinctions amongst other sustainability targets, the Ramsar Convention on Wetlands which aims at sustainable management, and the Convention to Combat Desertification which aims to halt and reverse land degradation. These frameworks and their linkages will be discussed later on. Nature or ecosystem-based approaches (NBS) to water security, climate change adaptation, disaster risk reduction, and development have been increasingly applied across the world over the past two decades (Renaud et al., 2016; UNWWAP, 2018). Overall, these approaches aim to restore, preserve and enhance ecosystems and their services and underlying biodiversity. Thus, they deliver holistic socio-ecological benefits. Interventions that integrate climate action with DRR and ecosystem management have the potential to address a multitude of pressing development challenges (MEA, 2005; Estrella et al., 2016; UNWWAP, 2018; Keesstra et al., 2018).

Despite the many lessons learned and good practices NBS is still 'chronically' underutilized, most investment in the water sector in line with SDG 6 favor conventional grey infrastructures (UNWWAP, 2018). Awareness of these critical ecosystem and biodiversity values, and of steps individuals and communities can take to regenerate them are essential for sustainable development. Education and capacity development therefore plays central role in advancing them. Education and capacity development are also core linkages across the corresponding global frameworks and the 2030 Agenda. This chapter focuses on the Sustainable Development Goals of:

- 6 - Clean water and sanitation
- 13 - Climate action
- 14 - "Life below water"
- 15 - "Life on land"

The SDG 4, which focuses on "quality education" will be examined as a mechanism for mainstreaming and enhancing implementation. Mainstreaming entails that through changes in "institutional and personal behaviours and practices" sustainable development (in this case) "becomes regarded as standard policy and institutional practice" (Estrella et al., 2016, p. 572). The chapter first discusses central concepts and issues around the climate change adaptation and disaster risk reduction aspects of sustainable development, and how nature-based solutions can enhance water security. Subsequently it examines linkages between global policy frameworks relevant to water security. It proceeds with an inquiry into how education is being framed in these policy frameworks. The overall objective of this chapter is to shed light on ways education can facilitate the mainstreaming i.e., and implementation of nature-based solutions to water security, and to promote further engagement with this question.

NATURE-BASED SOLUTIONS FOR WATER SECURITY

Global Water Security Challenges, Climate Adaptation, and Disaster Prevention Interventions

Climate change is a threat to water security because it is projected to exponentially increase water scarcity. Furthermore, it is observed to already strengthen both the incidence and the potency of extreme climate and weather events e.g., storms, droughts and floods (ECOSOC, 2016, Schipper et al., 2016; Howard et al., 2016). While all countries and populations confront growing challenges low- and middle-income countries of the Global South are at the most immediate risk due to different biophysical and economic factors and financial and human resource constraints (ECOSOC, 2016). Research indicates that these impacts will continue to increase and intensify in case international efforts succeed with the 2015 Paris Agreement's goals on limiting global temperature increase between 1,5 and 2 Celsius degrees i.e., climate change mitigation (Howard et al., 2016; UNFCCC, 2015). Therefore, it is essential to simultaneously pursue the global goals on climate change adaptation (CCA) and Disaster Risk Reduction (DRR) crystallized in the aforementioned policy frameworks (UNISDR, 2015; UNFCCC, 2015).

CCA can be an incremental or a transformational process which either intends to preserve, or fundamentally change existing systems in order to reduce the detrimental impacts of climate change on them. Exploiting potential benefits in the process is another basic goal of these interventions (IPCC, 2014). DRR seeks to address a multitude of different hazards including climate-related and natural ones. It aims to reduce existing and prevent new disaster risks through various measures e.g., dykes, early warning systems, risk insurance. However, certain disaster risks cannot be managed even with effective DRR; therefore, minimizing disruption and facilitating quick recovery as well can be integral objectives. All of these three objectives contribute to sustainable development by enhancing social, ecological, health and economic resilience i.e., the ability to withstand and recover from different shocks and stresses (UNGA/OIEWG, 2016).

Both CCA and DRR interventions can range from low- or no-regret interventions, such as early warning systems, livelihood diversification and enhancement, to changing physical infrastructures, and e.g., decision-making procedures in natural resource management. Facilitating development is another shared objective of these two types of interventions (ECOSOC, 2016). Such measures improve water security by mitigating hazards, for instance early warning systems, and infrastructural changes can prevent loss of life and assets during floods. While livelihood diversification can enhance the capacities of communities to cope with hazards e.g., varied income sources can mitigate the economic and food security impacts of droughts. CCA,

DRR and sustainable development are therefore essential components of each other. Integration within broader sustainable development frameworks are called for both in academia and practice (Aitsi-Selmi et al., 2016; ECOSOC 2016; Uitto & Shaw, 2016; Schipper et al. 2016; Andrade, 2018; Morchain, 2018).

The relationship between development and disaster risk is a circular loop. Poor development pathways, such as increasing greenhouse gas emissions that drive climate change, and environmental degradation increase vulnerability and consequently disaster risk (Thomalla et al., 2015). Climate- and water-related hazards can obstruct and/or entirely eliminate development progress (UN-Water, 2016). On the other hand (economic) development can reduce risk and vulnerability for example, through enhancing the resilience of physical infrastructures which then reduce exposure and vulnerability (Thomalla et al., 2015).

Nature-Based Solutions for Global Water Security Challenges

These cross-cutting challenges require holistic approaches and solutions. As discussed earlier water insecurity is primarily driven by unsustainable management, and ensuing ecological degradation. Therefore, the remediation of ecosystems ought to be an integral facet of any sustainable development intervention. Ecosystem services (ESS) are benefits that humans derive from nature as introduced earlier. Four different kinds of ESSs are distinguished: provisioning services, e.g. water, fiber, food, regulating services that shape climate, treat wastes, purify water, control flooding and drought. Cultural services furnish humans with aesthetic, educational, spiritual, and recreational values. All of these are dependent on the fourth category of supporting services such as, nutrient cycling, photosynthesis, and soil formation (MEA, 2005). Nature-based Solutions (NBS) for water security enhance water management by either utilizing or mimicking ESSs or applying both processes. They can range from micro like household, to macro like landscape level (UNWWAP, 2018).

Such approaches have been deployed for CCA and DRR as well. Ecosystem-based CCA (EbA) and DRR (Eco-DRR) entails the use of ecosystems and biodiversity for adaptation and risk reduction. Eco-DRR acknowledges that ecosystems facilitate communities to "...prepare for, cope with, and recover from disaster situations" (Reid, 2016; Sudmeier-Rieux and Ash, 2009 as cited by Uy et al., 2016, p. 126). This framework integrates the present and future needs of ecosystems and human livelihoods into decision-making considerations. Ecosystem health and underlying biodiversity facilitate resilience to natural hazards both directly and indirectly. Coastal vegetation buffering storm surges, or native plants and well-draining soils retaining excess rainfall are examples of direct hazard mitigation. An indirect benefit of healthy soils is that they not only provide higher agricultural yields, but also capture and store carbon dioxide which contributes to food security and climate mitigation, (Uy et al., 2016; Keesstra et al., 2018).

Regulating and provisioning ESSs like these facilitate resilience in human communities and the broader ecosystem as well. The decline of these services driven by unsustainable management, and natural hazards can therefore undermine socio-ecological adaptive capacity and resilience and exacerbate vulnerabilities and exposure to climate-related hazards. Ecosystem-based approaches to DRR and adaptation are 'low-regret' options due to their cost-effectiveness which leads to less reliance on external assistance, facilitation of local ownership of the DRR intervention and the potential to reverse current and prevent further ecological degradation (Uy et al., 2016). On the contrary, and to the complementary of conventional grey infrastructure e.g., pipes and dams, NBS for water security provide a multitude of benefits regardless of targets as illustrated by Figure 1. Appropriateness, and opportunities are context and purpose specific, for instance potable water provision in households require grey infrastructure, while for tackling desertification and land degradation ecosystem restoration is central or only practical option (UNWWAP, 2018).

Figure 1 illustrates water security-related ecosystem services of freshwater, terrestrial and marine and coastal ecosystems, and lists the three management

Figure 1.

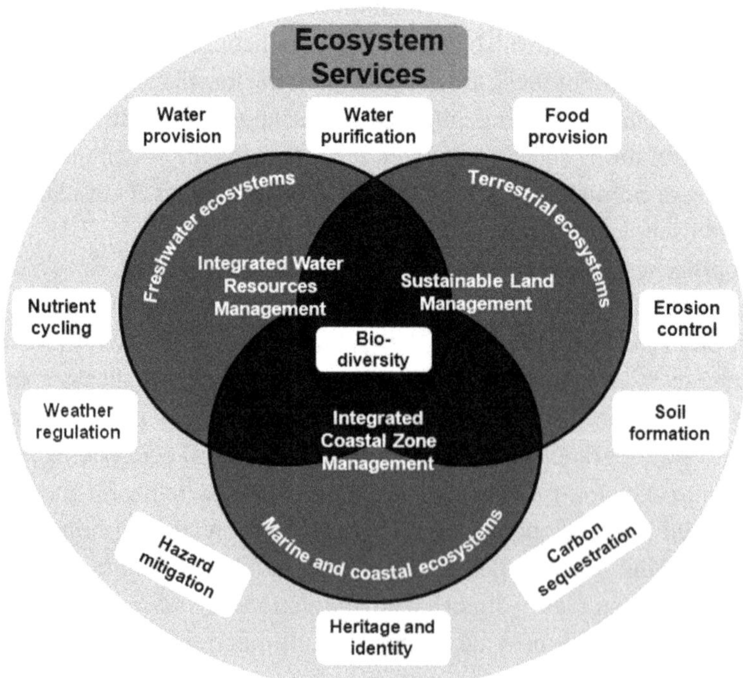

approaches of Integrated Water Resources Management (IWRM), Integrated Coastal Zone Management (ICZM) and Sustainable Land Management (SLM) that provide frameworks for Nature-based Solutions including Eco-DRR and EbA to sustainable development. Note, the list of ecosystem services and management approaches are not exhaustive, the figure merely provides examples for illustrative purposes (Keesstra et al., 2018; UNWWAP, 2018; Liquete et al., 2013).

However, the realization of sustainable water provision in households depend on healthy ecosystems that provide and purify water. Optimally green and grey infrastructures can be combined to enhance each other. For instance, constructed wetlands or other such eco-engineered solutions can be deployed for decentralized on-site wastewater treatment. Engineered coastal defense systems can also be enhanced by interventions that for instance restore upstream wetlands, or coastal vegetation to buffer hazards. As highlighted earlier water security is multifaceted, foremost it consists of the interrelated aspects of sustainable and adequate water supply, quality and resilience to water-related hazards (UN-Water, 2013). NBS that focus on quantitative aspects and corresponding spatial and temporal dimensions of water security deploy measures that enhance water infiltration, flow, storage, and revolve around precipitation and humidity management. Such measures include enhancing groundwater recharge via, for example, improving infiltration by adding organic matter to the soil and thereby utilizing wetlands for excess water storage (UNWWAP, 2018).

The above NBS measures could enhance biodiversity and various ESSs by providing habitat for various species for instance. Furthermore, they are more cost effective because appropriately managed or not degraded ecosystems enhance their benefits over time, while traditional grey infrastructures such as dams maintenance cost and utility decrease over time. The soil improvement and habitat provision examples are also strongly relevant to increasing the productivity, sustainability, resource efficiency and resilience of agricultural systems as well. Interrelated management practices include conservation agriculture that enhances ESSs through improved soil management. NBS are also being pioneered in cities to enhance water availability through green infrastructure such as parks and green roofs. The second, qualitative dimension of water security can be addressed by measures aimed at enhancing the quality of water sources, as well as at managing pollution that originates from different sources. They include sustainable wetland, forest and grassland management, and the aforementioned agricultural practices that enhance the water purification ESSs provided by such landscapes. For instance, recycling nutrients and filtrating pollutants via natural systems can improve agricultural productivity and reduce fertilizer input as well as corresponding nitrogen pollution (UNWWAP, 2018; Keesstra et al., 2018).

Consequently, these approaches provide a multitude of benefits through restoring, preserving and improving ESSs. In cities the aforementioned green infrastructures e.g. (constructed) wetlands, rain gardens, green roofs etc. can play an important role in both the quantitative and qualitative aspects of stormwater management. Both natural and constructed wetlands are also deployed for treating industrial wastewater, however, their appropriateness and effectiveness compared to grey infrastructures depend on the characteristics of such pollutants. Water quality and availability are intrinsically linked. Enhancing water quality through efficient water filtration techniques for instance could guarantee that the available quantity can be reused. These measures and their potential combination with grey infrastructures provide many opportunities for taking advantage of such synergies. The third aspect of water security is resilience to water-related hazards, which is also intrinsically linked with both quantity and quality. Floods and droughts can undermine both water quality and quantity, which can lead to increases in the incidence of water-borne diseases that aggravate disasters (UNWWAP, 2018; Keesstra et al., 2018).

As highlighted earlier, climate change is increasing the frequency and intensity of water-related hazards, most notably droughts and floods. The above outlined NBS approaches aimed at enhancing water availability provide flood and drought hazard mitigation co-benefits. For instance, enhanced natural storage capacity was underpinned by landscape connections can manage excess rainfall runoff, and store it for periods of drought. Context specific combination of the different nature-based and conventional CCA and DRR approaches is necessary to reduce and prevent disaster risks (UNWWAP, 2018). Furthermore, these should be complemented by more traditional DRR approaches like early warning systems, contingency planning, structural and grey infrastructure measures such as dykes, where and how appropriate (Uy et al., 2016). EbA's long-term climate horizon and biodiversity protection aspects are also integral for solutions (Renaud et al., 2016).

NBSs should also converge (corresponding) more tangible short-term benefits e.g., livelihood diversification and income generation opportunities through for instance sustainable agricultural practices (Uy et al., 2016). To support resilience and overall sustainable development objectives, another crucial aspect is the need for both grey and green infrastructures to be: flexible/upgradeable, socio-culturally relevant, scalable, robust, but with minimum ecological impact and financially sustainable (Hostettler, 2016; McNamara and Buggy, 2016; Sovacool et al., 2012). The 2030 Agenda for sustainable development seeks such holistic outcomes. To identify the ways forward first it is necessary to grasp which global policy frameworks are relevant for water security and how nature-based solutions can be catalyzed through them.

Global Policy Frameworks and Nature-Based Solutions for Water Security

As introduced earlier the following global policy frameworks are directly related to water security

- The Sustainable Development Goals of 2030 (SDG)
- The Paris Agreement on climate change
- The Sendai Framework for Disaster Risk Reduction of 2020 and 2030 (SFDRR)
- The Convention on Biological Diversity (CBD) and its 2020 Aichi Targets (ABT)
- The Ramsar Convention on Wetlands and its strategic plan 2016-2024
- The UN Convention to Combat Desertification (CCD) and its Strategic objectives 2018-2030

The SDGs, the SFDRR and the Ramsar are voluntary, non-binding frameworks, and the Paris Agreement, the CBD and the CCD are binding. Countries are to develop their own strategies, policies and action plans form implementation (UNFCCC, 2015, *FCCC/CP/2015/L.9/Rev.1*; CBD, 2010, COP 10 Decision X/2; UNISDR, 2015). The SDGs are 17 goals with 169 targets for 2030 and 2020 (UNGA, 2018). As highlighted previously, this chapter focuses on SDG 6, SDG 13, SDG 14, SDG 15. SDG 4 is analyzed as a mechanism for mainstreaming and enhancing the implementation of the SDGs themselves and the aforementioned other frameworks. Nevertheless, all other SDGs are directly related to water security as underlined in the introduction (UN-Water, 2016). The NBS (UNWWAP, 2018) are also directly related to water security but their discussion are beyond the scope of this paper.

The second global framework analyzed in this paper is the Paris Agreement, which sets the aforementioned global goals on mitigating and adapting to climate change (UNFCCC, 2015). The third, the SFDRR has seven targets, and four "Priorities for Action" which are geared towards mainstreaming DRR in all sectors and corresponding policies, plans etc. to alleviate existing, and avoid new disaster risks. Its targets aim to reduce:

- "...global disaster mortality by 2030…"
- "...the number of affected people globally by 2030"..., (c) "...direct disaster economic loss… by 2030"
- "...disaster damage to critical infrastructure and disruption of basic services… by 2030"

Furthermore the framework aims to increase:

- "...the number of countries with national and local DRR strategies by 2020"
- "... international cooperation to developing countries... by 2030"
- "...the availability of and access to multi-hazard early warning systems... by 2030" (UNISDR, 2015).

The third is the CBD which 20 ABTs for 2020 are embedded in its five strategic goals. These aim to tackle the root causes of "biodiversity loss by mainstreaming biodiversity values across government and society", enhance "sustainable use" and mitigate "direct pressures", protect ecosystems to enhance biodiversity, improve "benefits" derived "to all", and "enhance implementation through participatory planning, knowledge management and capacity building" (CBD, n.d.). The fourth framework is the Ramsar Convention on Wetlands, which aims to preserve and judiciously manage "all wetlands" "as a contribution towards achieving sustainable development". The convention defines wetlands broadly including virtually all natural and artificial ecosystems from "flood plains", "coral reefs", to "wastewater treatment ponds", "rice paddies". In 2015 there were 2,208 designated Ramsar Sites around the world "covering 210.73 million hectares" (Ramsar Convention Secretariat, 2016, pp. 2, 13).

The fifth framework is the CCD which set five strategic objectives toward "a land degradation-neutral world consistent with the 2030 Agenda" by halting and reversing land degradation, and managing drought in 'drylands'. These are to enhance the status of "affected ecosystems", tackle desertification and degradation, and "promote sustainable land management". Furthermore, the Convention aims "to improve the living conditions of affected populations", and to "enhance the resilience of vulnerable populations and ecosystems" by drought management. In addition, the Convention aims to engender "global environmental benefits", and to realize these via "effective partnerships" that can pool both the needed "financial and non-financial resources" (UNCCD, 2017, pp. 3, 4).

Both the Paris Agreement and the SFDRR recognize the role of ecosystems and biodiversity for resilience, climate change adaptation and mitigation, and call for integration. As discussed earlier, the SDGs crystallize goals and targets around ecosystems and biodiversity (14 and 15), resilience and DRR (including 13 and 15). They also provide springboard for NBS like Eco-DRR and EbA (including 6, 13,14, 15) (Estrella et al., 2016). While the latter three conventions revolving around biodiversity, wetlands and land degradation provide robust foundations for catalyzing NBS for water security through their holistic visions, and comprehensive and actionable objectives and targets. These are complementary to the SDGs and in many cases critical enabling factors for progress. Such links are explicitly highlighted by the respective Secretariats (CBD Secretariat, 2016; Ramsar Convention Secretariat, 2016; UNCCD, 2016) and call for Eco-DRR and EbA (Ramsar, 2015; CBD, 2016; UNISDR, 2017; Estrella et al., 2016).

Furthermore, all of these frameworks are people centered (UNFCCC, 2017), i.e., explicitly aim to enhance livelihoods and participatory governance as shown by the above introduced objectives. Thus, their instruments and implementation can deliver the aforementioned holistic benefits around water security addressing ecological, societal and economic challenges, needs, priorities and opportunities. Incorporating the global policy frameworks discussed in education programs is instrumental in equipping populations with the know-how to incorporate NBS in their communities. The following section will address the role of education in realizing the aforementioned benefits as well as how education and capacity development are framed in the global policy frameworks.

EDUCATION AND CAPACITY DEVELOPMENT

The Framing of Education Across the 2030 Agenda

The overarching goal of SDG 4 is to "Ensure inclusive and equitable quality education and promote lifelong learning opportunities for all." UN flagship publications highlight the following direct links between SDG 4, and SDG 6, SDG 13, and SDG 15. Both the availability and sustainable management of water are achieved through education. Awareness on water conservation and on links between pollution and health are instrumental in improving water availability and quality. Thus, awareness is a critical enabling factor of more sustainable water management at all levels. On the other hand, reliable and accessible safe water is a pivotal aspect of reducing "time poverty" which can hinder especially women's ability to participate in formal education. At the same time educational facilities equipped with safe WASH are necessary for enhancing education as well. Education's role is also well recognized in enabling and catalyzing climate change mitigation, adaptation and risk reduction actions at all levels. It is also underlined as a central tool for awareness raising, behavioral and value changes, as well as for capacity development around halting biodiversity loss and ecosystem degradation (Vladimirova & Le Blanc, 2016, p. 259).

Consequently, education plays a central role in enhancing environmental governance. At the same time loss of ecosystem services stemming from environmental degradation can increase risk to educational facilities, as well as hinder vulnerable children from attending education (Ibid). These are direct synergies with all of the quantitative, qualitative and resilience aspects of water security discussed previously. However, they are not referenced across SDG 4, 6 and 15. The focus of SDG 4.a is on on educational facilities including building upgrading and water and sanitation access. However, SDG 6 itself does not reference education. On the other hand, SDG 13 explicitly reflects these synergies. SDG 13.3 calls for "improved education, awareness-raising and human and institutional capacity on climate change mitigation, adaptation, impact reduction and early warning". Corresponding indicators of SDG 13.3.1 and 13.3.2 measure the integration of mitigation, adaptation and DRR into "secondary and tertiary curricula", as well as the communication of initiatives aimed at enhancing capacity building at different levels to implement these actions. The previous goals of SDG 13.1 and 13.2 revolve around enhancing adaptive capacity and resilience, and the integration of mitigation and adaptation into "national policies, strategies and planning, respectively (UNGA, 2018, p. 14).

Corresponding SDG 13 indicators examine reported disaster casualties, adoption and implementation of national DRR strategies, as well as strategies on the local level, in line with Sendai Framework's target (e). While progress on SDG 13.2 is measured by the "establishment or operationalization of an integrated policy/strategy/plan" (p. 14). Education and awareness were also underscored as an essential element of local and national DRR strategies (UNISDR, 2017). Despite the lack of explicit references to education, SDG 6 and 15 provide significant entry points through SDG 4.7 "By 2030, ensure that all learners acquire the knowledge and skills needed to promote sustainable development, including, among others, through education for sustainable development and sustainable lifestyles…" (Ibid, pp. 5). As highlighted above education is critical for

- "universal and equitable access" (SDG 6.1)
- Water, sanitation and hygiene (SDG 6.2)
- Water quality improvement (SDG 6.3)
- Water use efficiency (SDG 6.4)
- Integrated Water Resources Management (SDG 6.5)
- conservation and restoration of water-related ecosystems (SDG 6.6) (UNGA, 2018 A/RES/71/313, pp. 16)

Education also plays a central role in the sustainable management, conservation and restoration of:

- "terrestrial and inland freshwater ecosystems and their services"
- Forests (sdg 15.2)
- Tackling land degradation (sdg 15.3)
- Mountain ecosystems (sdg 15.4)
- Halting habitat and biodiversity loss (SDG 15.5)

The integration "ecosystem and biodiversity values into national and local planning, development processes, poverty reduction strategies and accounts" that are most directly linked to water security as discussed in the previous sections also require capacity development (p. 17). Awareness and education are also the backbone of the 2020 Aichi Biodiversity Targets (ABT) of the CBD. ABT 1 calls for strengthening awareness of people on "the values of biodiversity and the steps they can take to conserve and use it sustainably" (CBD, n.d.). This is a central aspect of "sustainable development and sustainable lifestyles" (SDG 4.7) (CBD, 2016), and there is a need to diffuse it into national educational curricula (CBD, 2014).

However, it is critical to enhance awareness of governmental, non-governmental, private-sector decision-makers and communities as well to elicit behavioral changes required for halting biodiversity loss. Comprehension and appreciation of the multitude of benefits and different biodiversity values, including intrinsic, can compel people to act (Schultz et al., 2016). Yet as ABT 1 emphasizes stakeholders need to be aware of what they can do in their capacity to halt biodiversity loss and ensure the regeneration of critical values (CBD, n.d.). Formal and informal education, public awareness and capacity development are some of the integral means through such learning and awareness can be facilitated (Schultz et al., 2016). The Ramsar Convention's "Programme on communication, capacity building, education, participation and awareness (CEPA) 2016-2024" is also framed as one of the key implementation mechanisms (Ramsar, 2015, p. 109).

This programme addresses what education on wetlands should deliver, to whom and how, as well as calls on governments to converge it with their "broader environment, biodiversity, wetland and water management, education, health, and poverty reduction policy instruments and mainstreamed in relevant programmes…" (Ibid, 110). Therefore, education itself is one of the most explicit links between the aforementioned frameworks, as well as the highlighted specific targets. Mainstreaming, i.e. institutionalization is consistently called for throughout objectives and targets in the discussed frameworks. This entails that e.g., disaster prevention, NBS etc. "becomes regarded as standard policy and institutional practice" underpinned by changes in "institutional and personal behaviours and practices" (Estrella et al., 2016, p. 572). These frameworks provide mandate and enormous, transformative action space for education and capacity development. However, to address these opportunities and corresponding challenges it is necessary to inquire how education and capacity development can facilitate the mainstreaming and implementation of nature-based solutions to water security.

Education as Means of Mainstreaming and Enhancing Implementation: Ways Forward

As emphasized in the previous sections the 2030 Agendas are moving into implementation phase. At the same time windows of opportunities for halting the loss of ecosystems and biodiversity, and regenerating them are shrinking due to global degradation and increasing uncertainty around climatic and ecological dynamics (Andrade, 2018; Steffen et al., 2015). Yet as illustrated NBS solutions embedded in locally relevant and context specific holistic approaches provide transformative opportunities toward sustainable and resilient development pathways (Keesstra et al., 2018; UNWWAP, 2018). Such challenges and opportunities require strategic considerations and re-newed approaches to education and capacity development (Estrella et al., 2016). As mentioned earlier NBS are rooted in systems thinking which entails "thinking in wholes rather than thinking only in the properties of individual elements and how they interact" (Keesstra et al., 2018, p. 999). This

Figure 2. Enhancing water security
Source: authors

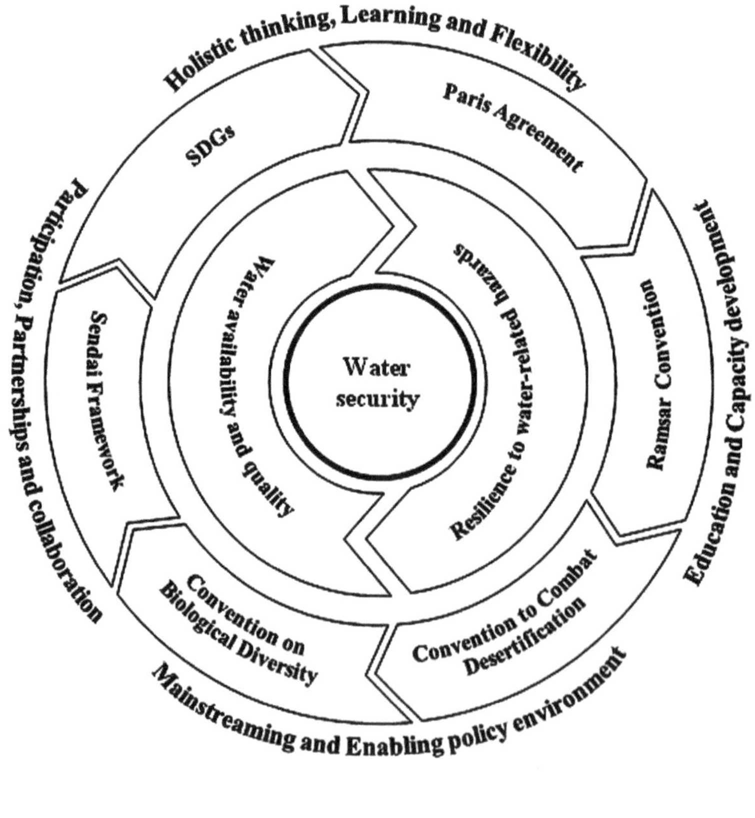

requires skills such as examining, evaluating, and inventing, which include more than just a recall of facts (Lee et al., 2017). Furthermore, capacity development should be a deliberative long-term process that needs to be guided by stakeholders' needs and feedback and rely on knowledge management and transfer, leadership and incentives (UN-Water, 2013).

With respect to this, as well as overall education and capacity development are equally important and needed both across the Global South and North around NBS and underlying challenges. Yet interdisciplinary education and research remain limited in both of these contexts, as well as across different institutional levels. These need to be addressed through context specific work that reflects and attains to local capacities, needs and priorities. However, the characteristics of conventional "in-classroom trainings or courses" can constrain mainstreaming and upscaling potential. For instance, capacity development of stakeholders in public and private sectors, assembling people with various backgrounds from different places is limited by cost and time-related factors. Another issue is that while a significant amount of good practices and lessons learned exist, these are predominantly documented in "international organizations and agencies rather than in national institutions" which can constrain access (Estrella et al., 2016, p. 584).

It is critical to disseminate such existing knowledge to different stakeholders in accessible ways including individuals, communities and professionals to foster 'champions' who can promote, implement and innovate context specific holistic solutions (Estrella et al., 2016). One way of fostering environmental governance locally and nationally could be through the creation of groups that consist of interdisciplinary and multicultural stakeholders from local communities, higher education institutions, the private sector and governmental and non-governmental institutions who help to inform as well as update the environmental content in school textbooks or community handouts using guidelines from the global policy frameworks discussed. By informing multiple generations on environmental governance, nations can build more resilient populations (to impacts of climate change) in the long-term.

Collaboration and partnerships with organizations that have consistent financial, legal etc. capacities and authority to educate is another important facet of capacity development for different stakeholders. Actions from national to local levels can also be enhanced by support of applied and field-based research. This could also address potential gaps in various sectors e.g., agriculture that require further attention. The internet provides significant and novel opportunities for applying these, it can provide access to relevant knowledge for various stakeholders in different contexts. It can also be combined with conventional teaching methods, and "solution oriented-approaches" underpinned by inquiry and "discovery-based learning" (Estrella et al., 2016, p. 585). Despite these novel opportunities offered by the internet, online approaches also have inherent limitations.

According to the Internet Society's (2017) Global Internet Report, 53 percent of the world's population still lacks internet access. The UNESCO General Conference proclaimed February 13 as World Radio Day and stated that radio reaches the widest audience in the world; specifically, remote communities and vulnerable people and plays a critical role in emergency communication and disaster relief (United Nations Educational, Scientific and Cultural Organization, 2018). In developing countries, at least 75% of households have access to a radio (EFA Global Monitoring Report, 2012). In order to reach the most vulnerable communities, it is important for educators to utilize the communication media that is most accessible to them. On a global scale, radio programs could reach a larger population where internet access is not available. One way communities could utilize radio could be through production of recurring daily or weekly education programs that cover content from new research findings on topics such as CCA, DRR, mitigation and water management. The information could be presented in various formats such as storytelling or plays.

Another central challenge of education and capacity development, as well as of implementation is integration. It is called upon consistently across the above reviewed frameworks, the academic literature and practitioner discussions as mentioned before. However, what does it mean, in which context? Could it be inviting expert guest speakers for in-classroom education; developing partnerships between different institutions of research and practice; establishing cross-sectoral bodies? There are practical limitations to scientific and non-scientific knowledge and (policy) integration such as, cost, time and knowledge gaps. Not all stakeholders need to understand for instance complex science like ecological interaction models, however, they should be able to 'connect the dots' and access relevant knowledge. (Scott, 2017; Schipper et al., 2016). Nevertheless, as highlighted throughout this chapter today's challenges require holistic solutions which demand integration. Part of "the answer lies in long-term commitment and support for knowledge and capacity development" (UN-Water, 2018, p. 9). Knowledge should build on systems thinking and be transferred through strong partnerships connecting people that have the necessary knowledge be it ecological, engineering or indigenous for example.

Education and capacity building programs that engage stakeholders in areas that have a personal meaning for them (for example, culturally or socially), have a higher likelihood of engendering long-term commitment within the communities the stakeholders belong. Despite a lack of adequate financial resources, when community members individually value the crucial functions of their local natural resources, they may mobilize other community members to partner with one another and perhaps other communities and or institutions to conserve and manage and therefore protect their shared natural resources. An example of such a program is SeaTrust Institute's AWARE^{SM} (Action Within a Resilient Environment^{SM}) suit of educational and capacity development programs. AWARE^{SM} incorporates climate, environmental and social science with local and traditional knowledge through inquiry-based and experiential learning that engenders intergenerational interaction.

This promotes learning from the stakeholders' personal capacity (ability) about the issues most relevant to them, while linking them with global policy frameworks. For instance, students get to learn more about their communities and the ecosystems they inhabit through applying science via activities such as, community mapping, and developing projects for increasing resilience. Such activities foster students' abilities to recognize or discover elements of systems, as well as how these operate. They acquire further systems thinking skills e.g., to recognize and map how the respective elements of the systems under their investigation interact. This feeds into knowledge about how such dynamics could be managed for desired outcomes (Lee et al., 2017). In this process they get to learn about biodiversity and other ecosystem values and engage more in stewarding these benefits across community and policy levels as required by ABT 1 for instance.

In 2017, students at Kobujoi Institute, an associate program college in Kenya incorporated the AWARE^{SM} Kenya program into their studies. The program was facilitated by Donmas Felix, an instructor at the Institute. The students applied the information within their college community and through the inquiry-based activities in the course, they experienced the biodiversity and ecosystems in their surroundings in a way that led them to value the latter for themselves. At present, some of the students are proceeding to engage others into making informed decisions regarding the conservation of natural resources within their communities. Donmas stated, that before AWARE^{SM} Kenya, he knew that the environment was important but after completing the program, he now values the environment. One of the challenges faced by the students included a lack of internet access that would have enabled them to acquire more knowledge on the topic and the ability to share with others their discoveries that could count as anecdotal data. This is a challenge experienced by many students in developing countries, which slows down the pace for resilience building in vulnerable areas.

In light of the above discussed educational and capacity development issues and opportunities it is also necessary to highlight that there are a multitude of research gaps related to NBS (Estrella et al., 2016). Furthermore, as it was emphasized throughout this chapter NBS should be part of a context specific holistic approach that attains to short-term stakeholder needs and priorities as well as longer-term sustainability and resilience (Uy et al., 2015). This requires participatory approaches as stressed by the discussed global frameworks. Major gaps remain in finance, technology and capacities around not only NBS, but also adaptation, mitigation, DRR, water management and other aspects of sustainable development, which can constrain or even obscure implementation and sustainable outcomes. Furthermore water-related infrastructures are aging across the North, and in the South existing systems are often insufficient to meet basic needs. These challenges provide entry points for complementary NBS, as well as innovation in grey infrastructures. However, water governance is frequently fragmented (UNWWAP, 2018), as well as is plagued by the lack of participation, transparency and accountability around the world which can hinder sustainability (Das et al., 2016). As explored previously SDG 6 is critical for advancing all other SDG targets and vice versa. For effective implementation "targets must be localized and adapted to the country context" (UN-Water, 2018, p. 13). These issues ought to be addressed as well to enhance water security.

CONCLUSION

This chapter outlined the lack of access of billions of people to clean and reliable sources of water, underlying unsustainable management and corresponding ecological degradation, changing global hydrological cycle, and increasingly common extreme weather events as pressing global water security challenges. Nature-based Solutions restore, enhance and safeguard ecosystems that purify, store and make water accessible for humans and the rest of nature, as well as buffer the impacts of natural hazards and provide multitude of other critical benefits. Irrespective of their objectives these interventions provide holistic benefits including climate change mitigation, food security, heritage etc. These are underscored by firm and expanding evidence and knowledge bases for example, in ecosystem-based climate change adaptation and disaster risk reduction. However, lack of awareness and capacity gaps hinder further deployment, replication and the upscaling of these solutions.

Nevertheless, the global policy frameworks of the Sustainable Development Goals, the Paris Agreement, the Sendai Framework on Disaster Risk Reduction, the Convention on Biological Diversity, the Ramsar Convention on Wetlands, and the Convention to Combat Desertification offer holistic visions, and comprehensive and actionable objectives and corresponding targets. These are underpinned by people centered visions and objectives and promote participatory governance and partnerships. Synergies across the SDGs and between the aforementioned frameworks have been identified and communicated by respective Secretariats, and communities of practice and technical expertise. SDG 6 on water is critical for advancing all other SDG targets, as well as the other frameworks, and vice versa. Yet effective implementation requires context specific translation of targets.

Education emerges as a mean for mainstreaming i.e., making sustainable development, disaster prevention and nature-based solutions, etc. as standards in decision-making, operation, behavior etc. Consequently, it is a mean to enhance the implementation of the discussed global frameworks. At the same time, like water, education is also one of the key connective tissues across these frameworks.

Awareness on the values of nature, and on the merits of corresponding management are an integral aspect of sustainable development and sustainable lifestyles called upon by SDG 4. Capacity development is the other backbone of fostering stakeholders' agency by shedding light on steps they can take to address complex challenges, as emphasized by the aforementioned frameworks. Yet considerable funding, capacity etc. gaps remain across all of the discussed areas of sustainable development. Research gaps in NBS, as well as governance challenges in the water sector require further engagement.

Challenges and opportunities around water security and NBS, as well as overall sustainable development require strategic considerations and renewed approaches to education and capacity development. NBS are based on systems thinking that can enable stakeholders to 'connect the dots'. Partnerships and collaboration can provide access to necessary knowledge. Yet despite a significant existing knowledge base access is lacking and is hindered by insufficient interdisciplinary education and research. Online technologies offer novel ways to advance solution oriented-approaches underpinned by inquiry and discovery-based learning. Such are integral to address knowledge and capacity gaps, and reach a broader set of stakeholders. However, internet-based approaches also reproduce the limitations of traditional in-class education and capacity development, as large segments of the population lack access. It is critical to reach individuals, communities and professionals in culturally sensitive and contextually appropriate ways and through media that is most accessible to them such as radio, to foster 'champions' who can promote, implement and innovate holistic solutions. The ways forward are an open question for the reader to engage on.

REFERENCES

Aitsi-Selmi, A., Murray, V., Wannous, C., Dickinson, C., Johnston, D., Kawasaki, A., ... Yeung, T. (2016). Reflections on a Science and Technology Agenda for 21st Century Disaster Risk Reduction Based on the Scientific Content of the 2016 UNISDR Science and Technology Conference on the Implementation of the Sendai Framework for Disaster Risk Reduction 2015–2030. *International Journal of Disaster Risk Science*, *7*(1), 1–29. doi:10.100713753-016-0081-x

Andrade, A. (2018). *Technical Expert Meeting on Adaptation 2018: Session #1 Adaptation planning for vulnerable ecosystems* [Webcast]. Retrieved from http://tep-a.org/technical-expert-meetings-on-adaptation/agenda/

Ceballos, G., Elrich, P. R., & Dirzo, R. (2017). Biological annihilation via the ongoing sixth mass extinction signaled by vertebrate population losses and declines. In *Proceedings of the National Academy of Sciences*. doi:10.1073/pnas.1704949114

Pachauri, R.K. & Meyer L.A. (Eds.). (2014). *Climate Change 2014: Synthesis Report, Contribution of Working Groups I, II and III to the Fifth Assessment Report of the Intergovernmental Panel on Climate Change*. Geneva, Switzerland: IPCC.

Convention on Biological Diversity. (2010). *COP 10 Decision X/2: X/2.Strategic Plan for Biodiversity 2011-2020.* Retrieved from https://www.cbd.int/decision/cop/?id=12268

Convention on Biological Diversity. (2016, December 10). *Decisions adopted by the Conference of the Parties to the Convention on Biological Diversity XIII/4. Biodiversity and climate change* CBD/COP/DEC/XIII/4. Retrieved from www.cbd.int/doc/decisions/cop-13/cop-13-dec-04-en.pdf

Convention on Biological Diversity. (n.d.). *The Aichi Biodiversity Targets.* Retrieved from: https://www.cbd.int/sp/targets/

Convention on Biological Diversity Secretariat. (2014). *Global Biodiversity Outlook 4. A mid-term assessment of progress towards the implementation of the Strategic Plan for Biodiversity 2011-2020.* Montreal, Quebec, Canada: CBD Convention on Biological Diversity.

Das, B., Fernández Fernández, C., van der Gaag, N., McIntyre, P., & Rychlewski, M. (Eds.). (2016). *Water Integrity Global Outlook 2016*. Berlin, Germany: Water Integrity Network.

ECOSOC. (2016). Climate change resilience for sustainable development. In *World Economic and Social Survey 2016: Climate change resilience – an opportunity for reducing inequalities* (E/2016/50). UN Economic and Social Council. Retrieved from: http://www.un.org/en/development/desa/policy/wess/

Estrella, M., Renaud, F. G., Sudmeier-Rieux, K., & Nehren, U. (2016). Defining New Pathways for Ecosystem-Based Disaster Risk Reduction and Adaptation in the Post-2015 Sustainable Development Agenda. In F. G. Renaud, K. Sudmeier-Rieux, M. Estrella, & U. Nehren (Eds.), *Ecosystem-Based Disaster Risk Reduction and Adaptation in Practice. Advances in Natural and Technological Hazards Research* (Vol. 42, pp. 553–593). Switzerland: Springer. doi:10.1007/978-3-319-43633-3

European Environmental Agency. (n.d.). *Water stress*. Retrieved from: https://www.eea.europa.eu/themes/water/wise-help-centre/glossary-definitions/water-stress

Hostettler, S. (2015). Technologies for Development. What Really Matters? In S. Hostettler, S., Hazboun, E. & Bolay, J.-C. (Eds.) Technologies for Development: What is Essential? Paris, Heidelberg, New York, Dordrecht, London: Springer. doi:10.1007/978-3-319-16247-8

Howard, G., Calow, R., Macdonald, A., & Bartram, J. (2016). Climate Change and Water and Sanitation: Likely Impacts and Emerging Trends for Action. *Annual Review of Environment and Resources, 41*(1), 253–276. doi:10.1146/annurev-environ-110615-085856

Internet Society. (2017). Paths to Our Digital Future. Retrieved from https://future.internetsociety.org/wp-content/uploads/2017/09/2017-Internet-Society-Global-Internet-Report-Paths-to-Our-Digital-Future.pdf

Keesstra, S., Nunes, J., Novara, N., Finger, D., Avelar, D., Kalantari, Z., & Cerdà, A. (2018). The superior effect of nature based solutions in land management for enhancing ecosystem services. *The Science of the Total Environment, 610-611*, 997–1009. doi:10.1016/j.scitotenv.2017.08.077 PMID:28838037

Lee, T. D., Jones, M. G., & Chesnutt, K. (2017). Teaching Systems Thinking in the Context of Water Cycle. *Research in Science Education*. doi:10.100711165-017-9613-7

Liquete, C., Piroddi, C., Drakou, E. G., Gurney, L., Katsanevakis, S., Charef, A., & Egoh, B. (2013). Current Status and Future Prospects for the Assessment of Marine and Coastal Ecosystem Services: A Systematic Review. *PLoS One, 8*(7), e67737. doi:10.1371/journal.pone.0067737 PMID:23844080

McNamara, K. E., & Buggy, L. (2016). Community-based climate change adaptation: A review of academic literature. *Local Environment, 21*, 1–18. doi:10.1080/13549839.2016.1216954

MEA. (2005). *Ecosystems and Human Well-being: Synthesis*. Washington DC: Island Press. Retrieved from http://millenniumassessment.org/en/Synthesis.html

Morchain, D. (2018). *Technical Expert Meeting on Adaptation 2018: Session #2 Adaptation planning for vulnerable communities* [Webcast]. Retrieved from: http://tep-a.org/technical-expert-meetings-on-adaptation/agenda/

Ramsar. (2015). *12th Meeting of the Conference of the Parties to the Convention on Wetlands, Compilation of COP12 Resolutions*. Retrieved from https://www.ramsar.org/sites/default/files/documents/library/cop12_resolutions_pdf_e.pdf

Ramsar Convention Secretariat. (2016). *The Fourth Ramsar Strategic Plan 2016–2024. Ramsar handbooks for the wise use of wetlands* (5th ed., Vol. 2). Gland, Switzerland: Ramsar Convention Secretariat.

Schipper, E. L. F., Thomalla, F., Vulturius, G., Davis, M., & Johnson, K. (2016). Linking disaster risk reduction, climate change and development. *International Journal of Disaster Resilience in the Built Environment, 7*(2), 216–228. doi:10.1108/IJDRBE-03-2015-0014

Schultz, M., Tyrrell, T. D., & Ebenhard, T. (2016). The 2030 Agenda and Ecosystems - A discussion paper on the links between the Aichi Biodiversity Targets and the Sustainable Development Goals. *SwedBio at Stockholm Resilience Centre*, Stockholm, Sweden. Retrieved from http://swed.bio/wp-content/uploads/2016/11/The-2030-Agenda-and-Ecosystems_spread.pdf

Scott, A. (2017). Making governance work for water–energy–food nexus approaches (working paper). *Overseas Development Institute and Climate and Development Knowledge Network.* Retrieved from https://cdkn.org/resource/working-paper-making-governance-work-water-energy-food-nexus-approaches/?loclang=en_gb

Secretariat. (2016). *Biodiversity and the 2030 Agenda for Sustainable Development [Technical Note].* Montreal, Quebec, Canada: CBD, FAO, The World Bank, UNEP, UNDP

Sovacool, B. K., D'Agostino, A. L., Meenawat, H., & Rawlani, A. (2012). Expert views of climate change adaptation in least developed Asia. *Journal of Environmental Management, Vol., 97,* 78–88. doi:10.1016/j.jenvman.2011.11.005 PMID:22325585

Steffen, W., Richardson, K., Rockström, J., Cornell, S. E., Fetzer, I., Bennett, E. M., ... Sorlin, S. (2015). Planetary boundaries: Guiding human development on a changing planet. *Science, 347*(6223), 1259855. doi:10.1126cience.1259855 PMID:25592418

Thomalla, F., Johnson, K., Bharwani, S., Johannessen, A., & Butterfield, R. (2015). *Transforming disaster risk reduction for more inclusive, equitable and sustainable development.* Global Sustainable Development Report. *Briefs.* Retrieved from https://sustainabledevelopment.un.org/content/documents/616462-thomalla%20et%20al%20-%20transforming%20disaster%20risk%20reduction%20for%20more%20inclusive%20equitable%20sd.pdf

Uitto, J. I., & Shaw, R. (2016). Sustainable Development and Disaster Risk Reduction: Introduction. In J. I. Uitto & R. Shaw (Eds.), *Sustainable Development and Disaster Risk Reduction* (pp. 1–12). Japan: Springer. doi:10.1007/978-4-431-55078-5_1

UN. (2015). *Water for a sustainable world, The United Nations World Water Development Report 2015*, United Nations. Paris, France: UNESCO on behalf of UN-Water. Retrieved from http://unesdoc.unesco.org/images/0023/002318/231823E.pdf

UN General Assembly (2016, December 1). *Report of the open-ended intergovernmental expert working group on indicators and terminology relating to disaster risk reduction* A/71/644. Retrieved from undocs.org/A/71/644

UN General Assembly. (2018). *Global indicator framework for the Sustainable Development Goals and targets of the 2030 Agenda for Sustainable Developmen*t A/RES/71/313 E/CN.3/2018/2. Retrieved from https://unstats.un.org/sdgs/indicators/indicators-list/

UN-Water. (2013). *Water Security & the Global Water Agenda A UN-Water Analytical Brief.* Canada: United Nations University Institute for Water, Environment & Health. Retrieved from https://collections.unu.edu/view/UNU:2651

UN-Water. (2016). *Water and Sanitation Interlinkages across the 2030 Agenda for Sustainable Development.* Geneva, Switzerland: UN-Water. Retrieved from http://www.unwater.org/publications/water-sanitation-interlinkages-across-2030-agenda-sustainable-development/

UN-Water. (2018). *Executive Summary – SDG 6 Synthesis Report 2018 on Water and Sanitation.* UN-Water. Retrieved from http://www.unwater.org/publications/executive-summary-sdg-6-synthesis-report-2018-on-water-and-sanitation/

UN-Water. (n.d.). *Water scarcity.* Retrieved from http://www.unwater.org/water-facts/scarcity/

UNCCD. (2016). A Natural Fix: A Joined-up Approach to Delivering the Global Goals for Sustainable Development. Retrieved from https://www.unccd.int/publications/natural-fix-joined-approach-delivering-global-goals-sustainable-development

UNCCD. (2017). *The future strategic framework of the Convention, The UNCCD 2018–2030 Strategic Framework ICCD/COP(13)/L.18 (14 September 2017).* Retrieved from https://www.unccd.int/official-documents/cop13-l-documents/iccdcop13l18

UNFCCC. (2015). *Adoption of the Paris Agreement (FCCC/CP/2015/L.9/Rev.1)* United Nations Framework Convention on Climate Change, Retrieved from: https://unfccc.int/resource/docs/2015/cop21/eng/l09r01.pdf

UNFCCC. (2017). *Opportunities and options for integrating climate change adaptation with the Sustainable Development Goals and the Sendai Framework for Disaster Risk Reduction 2015–2030* (Report No. FCCC/TP/2017/3). Retrieved from http://unfccc.int/resource/docs/2017/tp/03.pdf

UNISDR. (2015). *Sendai Framework for Disaster Risk Reduction 2015 – 2030*, United Nations International Strategy for Disaster Reduction. Retrieved from http://www.unisdr.org/we/coordinate/sendai-framework

UNISDR. (2017). *2017 Global Platform for Disaster Risk Reduction Proceedings.* United Nations International Strategy for Disaster Reduction. Retrieved from https://www.unisdr.org/we/inform/publications/55465

United Nations Educational, Scientific and Cultural Organization. (2012). Youth and skills: Putting education to work (EFA Global Monitoring Report). France. Retrieved from http://unesdoc.unesco.org/images/0021/002180/218003e.pdf

United Nations Educational, Scientific and Cultural Organization. (2018). World Radio Day 2018. Retrieved from http://www.diamundialradio.org/home

UNWWAP. (2018). *2018 UN World Water Development Report, Nature-based Solutions for Water.* World Water Assessment Programme. Retrieved from: http://www.unesco.org/new/en/natural-sciences/environment/water/wwap/wwdr/2018-nature-based-solutions/

Uy, N., Delfino, R. J. P., & Shaw, R. (2016) Ecosystem-Based Disaster Risk Reduction: Experiences, Challenges, and Opportunities in the Post-2015 Development Agenda. In J. I. Uitto & R. Shaw (Eds.), *Sustainable Development and Disaster Risk Reduction* (pp. 119–142). Japan: Springer. doi:10.1007/978-4-431-55078-5_8

Vladimirova, K., & Le Blanc, D. (2016). Exploring Links Between Education and Sustainable Development Goals Through the Lens of UN Flagship Reports. *Sustainable Development*, *24*(4), 254–271. doi:10.1002d.1626

Wood, S.L.R., Jones, SK., Johnson J.A., Brauman, K.A., Chaplin-Kamer, R., Fremier, A., … DeClerk, F.A. (2018) Distilling the role of ecosystem services in the Sustainable Development Goals, *Ecosystem Services*, *29*(Part A), 70-82. doi:10.1016/j.ecoser.2017.10.010

KEY TERMS AND DEFINITIONS

Biodiversity: The richness and variety of life in a given area. It plays a critical role in sustaining life and development enabling benefits from nature to people.

Climate Change Adaptation: An intervention that aim to minimize climate-related risks e.g., extreme weather event impacts on water systems, and simultaneously promote development benefits. It can either seek to maintain existing systems by smaller changes, e.g. land use planning, or fundamentally change their central features, e.g. decision-making procedures.

Convention on Biological Diversity: A binding international treaty in effect since 1993. It has the following three objectives: sustain biodiversity, manage and use it sustainably, and distribute the positive gains stemming from the exploitation of genetic resources in equitable ways.

Convention to Combat Desertification: A binding international framework which aims to halt and reverse the degradation of dry lands. Its central objective is to prevent further loss of land productivity and underlying ecological functions by 2030.

Disaster Risk Reduction: An intervention that aims to reduce existing and prevent new disaster risks, i.e. potential loss of life and assets. Various measures can be deployed including dykes, early warning systems, risk insurance, and emergency response. Such interventions promote development co-benefits.

Ecosystem Services: Benefits that humans derive from nature. Four categories are distinguished, provisioning services, e.g. water, food, regulating services that shape climate, purify water, control flooding etc. Cultural services furnish humans with educational, spiritual, etc. values. All of these are dependent on the fourth category of supporting services such as, photosynthesis, nutrient cycling, etc.

Ramsar Convention on Wetlands: A non-binding voluntary framework which aims to preserve and judiciously manage "all wetlands" as a mean for enhancing sustainable development. The convention defines wetlands broadly including virtually all natural and artificial ecosystems.

Sendai Framework for Disaster Risk Reduction: A 15-year-long (2015-2030) voluntary agreement which aims to reduce existing and prevent new disaster risks.

Chapter 4
Building Sustainability Through Environmental Education:
Education for Sustainable Development

Ediola Pashollari
World Assembly of Youth, Malaysia

ABSTRACT

Education is an important informative tool used to maintain the prevailing values of a society. It is the best thing anyone can acquire; it is an asset, an act of attaining knowledge, developing sense of analyzing and perception in preparing oneself. Quality education is one of the 17 Global Goals that make up the 2030 Agenda for Sustainable Development. There are three types of education, namely formal, non-formal, and informal education. Vulnerable young people are often excluded from educational systems. Inclusive polices are needed to ensure access to education for poorest youth in cities and remote areas, youth affected with HIV, refugee youth, and migrant youth. This chapter explores education and sustainable development.

INTRODUCTION

Climate change has become a global and curious issue to scientists and incidentally continues to be part and parcel of rules, regulations, and laws drafted by governments and other organisations. Climate change has been identified as "the major" environmental issue that has overridden all other environmental issues of our time, even to the extent of becoming the single greatest challenge facing environmental

DOI: 10.4018/978-1-5225-7727-0.ch004

regulators .The scope of its critical impact extends to economy, health and safety, food production, security, etc.

The loss of ice and ice sheets from glaciers continues to have significant effects across the global. These losses has significantly contributed to the varied patterns of weather in recent times resulting in increased unpredictable precipitation which adversely poses a threat to the production of food and the increased catastrophic risk of flooding by rising sea levels caused by contaminated coastal freshwater reserves .The news keep getting worse by the day. This melting is a major contribution to the rates and a huge extent to the rise in sea-level that could outstrip by far the most recent global scientific assessment anticipations. There abounds very alarming evidences that point to important factors directed towards irreversible changes in major ecosystems and the planetary climate system .This most probably might already have been attained or passed. This goes to point the fact that as diverse as ecosystems may be, it might be getting to its threshold of dramatic change triggered by drying and warming.

Mountain glaciers are in alarming retreat and the downstream effects of reduced water supply in the driest months will have repercussions that transcend generations. Climate feedback systems and environmental cumulative effects are building across earth systems demonstrating behaviours' we cannot anticipate. The potential for runaway greenhouse warming has never been more present and real. The most dangerous climate changes may still be avoided if we transform our hydrocarbon based energy systems and if we initiate rational and adequately financed adaptation programs to forestall disasters and migrations at unprecedented scales. The tools are available, but they must be applied immediately and aggressively.

EFFECTS OF CLIMATE CHANGE

Listed according to the seven (7) UNPFII's indigenous regions is a brief overview of the climatic climate.

Africa

One of the major areas to be affected by climate change in Africa is the Kalahari Desert. There are 2.5 million kilometres of dunes in southern Africa which are covered in vegetation and used for grazing. However, the rise in temperatures and the expected dune expansion along with increased wind speeds will result in the region loosing most of its vegetation cover and hence, becoming less feasible for indigenous peoples living in the region. As their traditional resource base diminishes, the traditional practices of cattle and goat farming will no longer survive. There are already areas where indigenous peoples are forced to live around government drilled bores for water and depend on government support for their survival. Food security is a major issue for indigenous peoples residing in the deserts and they are on the frontline of global climate change.

Asia

In the tropical rainforests of Asia, temperatures are expected to rise 2-8 degree Celsius and further climatic variation will include decrease in rainfall, crop failures and forest fires. Tropical rainforests are the haven for biodiversity, as well as indigenous peoples' cultural diversity and forest fires will threaten this heritage of biodiversity. People in low lying areas of Bangladesh could be displaced by a one-meter rise in sea levels. Such a rise could also threaten the coastal zones of Japan and China. The impact will mean that salt water could intrude on inland rivers, threatening some supplies of fresh water.

In the high altitude regions of the Himalayans, there are glacial melts which affect hundreds of millions of rural dwellers who depend on the seasonal flow of water; there might be more water in the short term, but less in the long run as glaciers and snow cover shrink. The warming of the high altitude regions are likely to mean that population growth, settlement expansion and encroachment are likely to become a major management challenge and these external influences are likely to have an impact on indigenous peoples and their lands.

The poor, many of whom are indigenous peoples are highly vulnerable to climate change in urban areas because of their limited access to profitable livelihood opportunities and limited access to areas that are fit for safe and healthy habitation. Consequently, the poor sector will be exposed to more risks from floods and other climate related hazards in areas where they are forced to live.

Central and South America and the Caribbean

This region is very diverse from the Chilean deserts to the tropical rainforests of Brazil and Ecuador to high altitudes of the Peruvian Andes.

Like elsewhere in the world, peoples' use of biodiversity is central to environmental management and livelihoods. In the Andes, alpine warming and deforestation will threaten peoples' access to plants and tuba crops for food, medicine, grazing animals and hunting. Once these food crops are replaced by trees that will grow in the region, peoples will be deprived of important traditional resources which are central to their livelihoods.

The warming of the earth's surface is forcing everyone in this region to farm at higher altitudes to grow their staple crops which adds to further deforestation.

Not only does this affect the water sources and leads to soil erosion, it also has a cultural impact. The displacement of Andean cultures to higher lands means the loss of the places where their culture is rooted, putting its survival at risk. Indigenous communities in the Imbakucha Basin in Otavalo in Ecuador, the unexpected frosts and long drought periods affect all farming activities. The older generations say they no longer know when to sow because the rains do not come as expected. Migration offers one way out but represents a cultural nemesis and the human and social price to pay is high.

In the Amazon, the effects of climate change will include deforestation and forest fragmentation and as a result there will be more carbon released into the atmosphere exacerbating and creating further changes. The droughts of 2005 resulted in fires in the western Amazon region and this is likely to occur again as rainforest is replaced by savannah thus, having a huge effect of the livelihoods of the peoples in the region.

Many communities in the Caribbean are in coastal locations which are often the centre of government activities, ports and international airports hence there is dependence on coastal resources for subsistence living. As a result there is rapid and unplanned movement of rural and outer island residents to the major centres. This puts enormous pressures on urban resources to meet the most basic needs and hence creates social and economic stresses and vulnerability to hazardous weather conditions such as cyclones and diseases. Also in the Caribbean the relationship between climate change and water security will be a major issue as access to safe water already eludes the populations of several Caribbean countries which are dependent on rainfall and groundwater. At the same time, pollution of ground water is a major problem, especially for low-lying islands. Poor water quality affects human health and carries water-borne diseases.

Arctic

The Polar Regions are now experiencing some of the most rapid and severe climate change on earth, which contribute to environmental and socio-economic changes. Indigenous peoples, their culture and the whole ecosystem that they interact with is very much dependent on the cold and the extreme physical conditions of the Arctic region. Indigenous peoples depend on hunting for polar bears, walrus, seals and caribou, herding reindeer, fishing and gathering not only for food to support the local economy, but also as the basis for their cultural and social identity. Some of the concerns facing indigenous peoples include the change in species and availability of traditional food sources, perceived reduction in weather predictions and the safety of travelling in changing ice and weather conditions. All these provide serious challenges to human health and food security.

According to indigenous peoples, the Arctic is becoming an environment at risk because the sea ice is less stable, unusual weather patterns are occurring, vegetation cover is changing, and particular animals are no longer found in traditional hunting areas during specific seasons. Local landscapes, seascapes and icescapes are becoming unfamiliar, making peoples feel like strangers in their own land.

People across the Arctic region report changes in the timing, length and character of the seasons including more rain in autumn and winter and more extreme heat in summer. In several indigenous villages in Alaska, entire communities may have to relocate because of erosion due to the thawing of permafrost and large waves slamming against the west and northern shores. Coastal indigenous communities are severely threatened by storm related erosion because of melting sea ice. Hence, up to 80% of Alaskan communities, comprised mainly of indigenous peoples, are vulnerable to either coastal or river erosion.

In Nunavut the elders can longer predict the weather using their traditional knowledge because the weather has become so unpredictable and extreme. Due to drop in water levels, indigenous hunters are no longer able to travel by boats to caribou hunting grounds because of shallow waters. Hence many important summer hunting grounds cannot be reached. Storage of traditional foods for the winter months is also a major issue for indigenous peoples in the region especially in the Northwest Territories due to warmer weather. For example, drying and smoking foods is more difficult because the food is pre-cooked in the summer heat.

Central and Eastern Europe, Russian Federation, Central Asia and Transcaucasia

Like the Polar Regions, Siberia and the far north-east are now experiencing some of the most rapid and severe climate change on earth which contribute to environmental and socio-economic changes. The survival of indigenous peoples who depend on fishing, hunting and agriculture also depends on the success of their fragile environment and its resources. As bears and other wild game disappear the local villages and the people that live in them will suffer particular hardships. Inferior, unique indigenous cultures, traditions and languages will face major challenges in maintaining their diversity.

Indigenous peoples have noticed the arrival of new species of plants which were never seen in the region previously. There is a view the hotter summers have provided the conditions for the new plants to thrive in rivers and lakes where the small flowered duckweed. This had made it difficult for the fish hence; people's fishing opportunities have declined due to closure of lakes because of the new plant growth.

Changes in migration and foraging patterns of reindeer herds, sparked by fluctuating weather patterns will also cause problems for many northern communities.

Those who depend on hunting walrus will bear the brunt of melting ice caps and glaciers. One of the main observations in the region has been the changing seasonal weather patterns, the increased unpredictability and instability of the weather as well as shorter winters and the fall–winter transition is occurring later and spring weather arriving earlier.

North America

Climate change is likely to have a major impact on indigenous peoples and their communities that are dependent on natural resources. About 1.2 million tribal members live on or near reservations and many pursue lifestyles with a mix of traditional subsistence activities and wage labour. Many reservation economies and budgets of indigenous governments depend heavily on agriculture, forest products and tourism.

Due to global warming there will be less snowfall and more droughts in many parts of North America which will have a significant impact on indigenous peoples. For example water resources and water quality may decrease with less precipitation. Further extended heat waves will increase evaporation and deplete the underground water resources. There may be impacts on health, plant cover, wildlife populations, tribal water rights and individual agricultural operations and a reduction of tribal services due to decrease in income from land leases.

Natural disasters such as blizzards, ice storms, and floods, electric power outages, lack of transportation, fuel depletion and food supply shortages will isolate indigenous communities. Poor housing conditions and high energy costs as well as limited access to off-reservation emergency assistance will contribute to the risks faced by indigenous peoples. Also livestock loss due to severe blizzards can force tribal ranchers out of business due to lack of financial resources. Tribal governments are dependent on lease income for their operations. With extreme weather events the risk of land transfer to non-tribal use and ownership will increase.

Higher temperatures over extended periods will result in the loss of native grass and medicinal plants, and erosion that allows the invasion of non-native plants. The zones of semi-arid and desert shrubs, cactus and sagebrush will move northward. Finally fire frequency could also increase with more fuel and lightning strikes degrading the land and reducing regional bio-diversity.

Pacific

Most of the Pacific region comprises small island states and are affected by rising sea levels due to climate change. Environmental changes are prominent on islands where volcanoes build and erode; coral atolls submerge and reappear and the islands' biodiversity is in flux. The region has suffered extensively from humankind disasters such as atomic testing, pollution including shipping-related pollution, hazardous chemicals and hazardous wastes (Persistent Organic Pollutants or POPs) and solid waste management and disposals. These issues as well as the threats of climate change have severely affected the ability of island ecosystems to maintain a healthy and pristine environment for the enjoyment of indigenous peoples.

The effects of climate change have included high tides which flood causeways linking villages, forcing cars, buses and trucks to drive through seawater. This has been particularly noticeable in Kiribati and a number of other small Pacific island nations that could drown during this century. High tides and stormy seas have also caused problems recently in the Marshall Islands, Cook Island, Tuvalu and low-lying islands of Papua New Guinea (PNG).

Migration will become a major issue as flooding (due to rise in sea level) forces families to move from their homes. For example the people living in PNG's Bougainville's atoll island of Cartaret have asked to be moved to higher ground on the mainland. Also the people of Sikaiana Atoll in the Solomon Islands have been migrating away from their atoll primarily to Honiara the capital.

Similarly there has been internal migration from the outer islands of Tuvalu to the capital Funafuti. In the case of Tuvalu this migration has brought almost half of the national population to Funafuti atoll with negative environmental consequences including a demand on local resources.

In addition warmer temperatures have led to the bleaching of the Pacific Island's main source of survival the coral reefs. Bleaching occurs when reef-building corals, reacting to stress such as warm water, loosen the algae that help feed them because the algae give them colour, the starved corals look pale, thus the term "bleaching, continued bleaching ultimately kills corals and reef-building corals provide most of the primary productivity of coral reefs and an important shelter for the coral reef organisms. Reduction of abundance and diversity of reef-building corals is thus very likely to have a major influence on the surrounding biodiversity. Tropical fishery yields are on the decline worldwide and it is now clear that the conditions may become critical for the local fish population.

Agriculture in the Pacific region especially in small island states is becoming increasingly vulnerable due to heat stress on plants and salt water incursions. Therefore food security is of great concern to the region.

ROLES OF NGO

Climate change is a broad idea of which global warming is just one aspect, NGOs ought to take the main lead in updating the world about the complex reality of climate change.

History shows that NGOs have acted as powerful pressure groups and made a difference in societies in the past. They can make a valuable contribution in tackling the threat of climate change. Localized approaches are needed in order to find a serious solution to this global challenge. With a broad base of public support for working on behalf of environmental causes, NGOs can be tremendously helpful in the process. Owing to their close interactions with people they can inform policy-makers at both national and international levels about the core realities of climate change at different strata.

Education and Counselling

Environmental education and counselling are needed in the school because these systems provide children the opportunity to participate equally with adult stakeholders in addressing this ugly menace. Environmental education according to Stapp (1997) is aimed at producing a citizenry that is knowledgeable concerning the biophysical environment and to associated problems, aware of how to help solve these problems and motivated to work towards their solution. In addition Oghone (2012) defined environmental education as a process aimed at developing a world population that is aware of and concerned about the total environment and its associated problems and has the attitudes, motivations, knowledge, commitment and skills to work individually and collectively towards solutions of current problems and the prevention of new ones.

On the other hand counselling is a process of helping individuals through their own efforts to discover and develop their potentialities both for personal happiness and social usefulness. Cobia and Hendson (2003) suggested that counselling is an enabling process designed to help an individual come to terms with his life as it is and ultimately to grow to greater maturity through learning to take responsibility and to make decisions for himself. Bassey, Ejue and Bassey (2005) defined counselling as a range of counselling services provided with the aim of enhancing positive self-concept, attitude, perceptions, and overall social development of the human person. Counselling is to help an individual to make his own decisions and choices in the light of his feelings and needs. Counselling would help the students become more aware of the realities of climate change to be able to make informed decisions and take responsibilities for mitigating climate change within their environment. Environmental education and counselling should be integral parts of the education process aimed at practical problems of an interdisciplinary character to build a sense of values and contribute to public wellbeing. The focus should reside mainly in the initiative of the learners and their involvement in action and guided by both the immediate and future subjects of concern. From the above definitions the main objectives of environmental education and counselling are:

1. Awareness – to help individuals and social groups become aware of and sensitivity to the total environment and its allied problems.
2. **Knowledge:** To help individuals and social groups gain variety of experiences with the total environment and to acquire a basic understanding of the environment, its associated problems and humanity's critical responsible presence and role in it.
3. **Attitudes:** To help individuals and social group acquire social values, strong feelings of concern for the environment and the motivation for actively participating in its protection and improvement.
4. **Skills:** To help individuals and social groups acquire the skills for working towards the solution of environmental problems such as digging holes, non-littering of school compound, using sand bag method to wage flood, planting trees (palm trees and domestic shrubs) compost keeping, indiscriminate throwing of paper among others (Women action committee,2010).
5. **Participation:** To help individuals and social groups develop a sense of responsibility and urgency regarding environmental problems through personal and community involvement to ensure appropriate action to help solve these problems and avoid future problems.

General Education and Awareness

Many actions can be employed by NGOs to educate and raise environmental awareness in the community. These actions will empower people to participate effectively in democratic change towards a better environment for all. Recognition on the complexity of environmental issues will be realized and the need to develop solutions collectively accepted. In environmental education everyone has something to learn and something to contribute.

Listed below are some of the actions that cut across between various themes and community groups that will be supported by this strategy:

1. Regular interaction with community groups and other relevant stakeholders to promote local environmental projects and local initiatives. Mobilizing and build on learner's knowledge and competencies.
2. Offer financial support via grants schemes for projects / initiatives that raise community awareness and that promote a positive benefit on their local environment.
3. Support the Green Flag Schools Programs
4. Develop and disseminate educational materials and facts sheets for water protection, waste prevention, litter, climate change and biodiversity.
5. Support environmental award programs and competitions.

6. Support environmental talks and information seminars that target the general public and businesses.
7. Sponsor such events and programs that will increase waste minimization, water and energy conservation.
8. Encourage private companies to have environmental education programs as part of company policy.

Objectives of Climate Change

Climate Change and Biodiversity the issue of climate change is a very complex one, covering all areas of life, involving all sectors and all aspects of the environment. Climate change presents an immediate and significant threat to our natural and built environment and to the ways of life which co-exist with this environment. For most people climate change issues and habitat damage are new and not fully understood. A general lack of knowledge about the effects of certain actions and decisions on the natural environment means that people continue unabated with harmful activities.

Table 1. Reducing the CO2 Emissions

Issue	Reduce the rate of increase of CO2 emissions generated in the society.
	Reduce the accelerating loss of habitats and species decline.
Target Groups	assist in the provision of public awareness programs on climate
Strategy Objectives	To change & habitat protection.
	To assist with the implementation of the national climate change strategy at local level.
	To assist with the implementation of public awareness of Biodiversity
	Disseminate information regarding best practice e.g. in relation to watercourse management; the control of invasive species etc.). Support demonstration projects undertaken by the community / NGO's groups / stakeholders.
	Provide energy awareness information assistant to the public.
	Through planning & licensing policies and planning & licensing authorization, promote a sustainable society. Promote the use of low carbon emission technologies, high levels of insulation and the use of renewable energy fuels in all new developments and renovations.
	Reduce CO2 emissions from Council owned public buildings and promote the use of renewable energy sources.
	Reduce CO2 emissions arising from the provisions of public services i.e. water and waste water treatment facilities, through green procurement.
	Promote the use of renewable fuels in Council service vehicles.

Water Conservation and Water Pollution

- **Issue:** Water is a finite resource and vast quantities of treated purified water are needlessly wasted. A conservation awareness program is needed across all sectors of the society. A greater appreciation for this finite resource as a necessity for life is required. The introduction of a fixed rate payment for water for domestic users may have a negative consequence on water conservation and increase wastage.
- **Target Groups:** Domestic users, Non domestic users (schools, general public).
- **Strategy Objectives:**
    - Education, information and public awareness initiatives are essential in the support of water management and water conservation.
    - Promote the reporting of leaks to the local authority.
    - Promote through education programs the sustainable use of water in all sectors and in all buildings.
    - Support the water conservation module of the Green Flag program in Schools.
    - Support the Green Hospitality Program for Hotels and Catering sectors.
    - Promote sustainable recreational waterborne sporting activities.
    - Support Global Action initiatives on sustainable practices in the home.
    - Promote rain harvesting.
    - Encourage sustainable garden practices.

Water Pollution

- **Issue:** The lack of awareness across all sectors of society of the frailty of our water ecosystems, the vulnerability to pollution, the requirement to protect aqua wildlife habitats and abstraction sources for drinking water supplies. Nutrient enrichment causing eutrophication is a significant threat to rivers and some coastal areas.
- **Target Groups:** General public, Licensee and permit holders, Schools, Business, Farmers, Anglers, Yacht Clubs.
- **Strategy Objectives:**
    - Support the public participation in awareness-raising programs and initiatives, to be implemented through the Library Council and delivered across the society.
    - Promotion of water protection measures that can be undertaken on farms, in businesses and commercial sectors.

- Promote the reporting of pollution incidents to the Local Authority and Fisheries Board.
- Promote the correct disposal by householders of hazardous household and gardening chemicals.
- Promote pollution awareness initiatives and voluntary accreditation schemes, for clubs / individuals and commercial entities, where water is used by them for recreational purposes.

Waste Prevention Awareness

- **Issue:** To assist with the implementation of policies on the reduction of waste to landfill, the Council will continue to promote and support Reduce, Reuse, & Recycle programs in all sectors. In addition, develop waste prevention initiatives and incorporate these initiatives into the above programs.
- **Target Groups:** Domestic householders, Businesses, Schools, Community groups.
- **Strategy Objectives:**
    - Promote the waste hierarchy of waste avoidance and elimination.
    - Educate the public, Council staff, commercial entities in waste avoidance, diversion of waste streams and the correct separation of waste.
    - Promote waste prevention and minimize the production of harmful waste.
    - Encourage and support the recovery & reuse of waste.
    - Work in partnerships with organizations and address the polluters pay principle in relation to waste disposal.

Waste Prevention

- **Issue:** To assist with the implementation of policies on the reduction of waste to landfill, the Council will continue to promote and support Reduce, Reuse, & Recycle programs in all sectors. In addition, develop waste prevention initiatives and incorporate these initiatives into the society.
- **Target Groups:** Domestic householders, Businesses, Schools, Community groups.
- **Strategy Objectives:**
    - Promote the waste hierarchy of waste avoidance and elimination.
    - Educate the public, commercial entities in waste avoidance, diversion of waste streams and the correct separation of waste.
    - Promote waste prevention and minimize the production of harmful waste.

- Encourage and support the recovery & reuse of waste.
- Work in partnerships with organizations and address the polluters pay principle in relation to waste disposal.

Litter Awareness

- **Issue:** Litter is present along rural and urban roads, foreshores, amenity parks and most areas utilized our residents and visitors. Along with the unsightly visual affect, litter is polluting our environment and poses health and safety issues.
- **Target Groups:** Residents, Businesses, Schools, Community groups
- **Strategy Objectives:**
  - Increase awareness of the problems associated with litter.
  - Support the education of our children on antilitter initiatives via the Green Schools program.
  - Work in partnerships with local community organizations to encourage neighbourhood clean ups, reporting of illegal dumping, participation in Anti-litter Challenge initiative for towns and villages.
  - Encourage day trippers to beaches, parks & public places to take their rubbish home and their participation in non- government organization programs.

2018 UPDATE

Education is an important informative tool used to maintain the prevailing values of a society. It is the best thing anyone can acquire; it is an asset, an act of attaining knowledge, developing sense of analysing and perception in preparing oneself. Quality education is one of the 17 Global Goals that make up the 2030 Agenda for Sustainable Development.

A good primary and secondary education is exceptionally important, both for academic and social development, economic development. Over the years of education children are taught reading and writing skills as well as sports and creativity. There has been enormous progress in achieving the target of universal primary education. The total enrolment rate in developing regions reached 91 percent in 2015, and the worldwide number of children out of school has dropped by almost half. The new Sustainable Development Goal to achieve universal primary and secondary education by 2030 marks a victory for activists fighting for the right to education.

Attaining a quality education is the basis to improving people's lives and sustainable development. Major progress has been made towards increasing access to education at all levels and increasing enrolment rates in schools mostly for women and girls. Basic literacy skills have improved enormously, yet bolder efforts are needed to make even greater steps for achieving universal education goals.

SDG 4 states that ensuring inclusive and quality education for all and promotes lifelong learning. This goal certifies the need for both boys and girls in inclusive of free of charge for both primary and secondary education by 2030.It also assures all youth to be significant proportion in order to attain literacy and numeracy and seeks to offer the same right of entry to inexpensive vocational training, and to eradicate gender and wealth differences with an aim of accomplishing collective access to a valuable learning.

There are three types of education namely formal, non-formal and informal education. Formal education is the hierarchically structured, sequential graded education system, running from primary school through the university. Non-formal education is classified as any structured educational activity outside the established formal system, whether operating separately or as an important feature of some broader activity that is intended to serve identifiable learning clienteles and learning objectives. Informal education is a constant process whereby every individual acquires attitudes, values, skills and knowledge from daily experience and the educative influences and resources in his or her surroundings.

Through education, literacy levels go high, and human resources are improved because of the needs of the education sector, most governments need to engage with NGOs in meeting their needs. This will makes the process sustainable, by bringing together all the stakeholders, including then civil society.

Vulnerable young people are often excluded from educational systems. Inclusive polices are needed to ensure access to education for poorest youth in cities and remote areas, youth affected with HIV, refugee youth, and migrant youth. Data from 2005-2008 indicates that in developing countries, the percentage of nonliterate youth is 13%, with Sub-Saharan Africa's percentage standing at 29% and 11% of the world's youth (15-24 years old) are non-literate.

Education is vastly an important tool for everyone to thrive in life. It helps in diminution the challenges of a difficult life. Knowledge achieved through education period facilitates each and every individual's confident about life and the future. Edification opens a variety of doors to the opportunities of achieving better prospects in life to uphold career growth.

The World Assembly of Youth, believe that education is an empowering process that allows and guides youth to develop their fervour, unlocks human potential development, critical thinking, empathy, and orientation towards knowledge for opportune action. In line with that, our 14th Melaka International Youth Dialogue (MIYD) we tackled a theme "Youth and Education: Taking Action, Getting Results"'. The aim was to fulfil the principal opinions and roles of young people towards the

Development Agenda on education. In fact, one of the main topics discussed was on how to improve literacy rate which also includes in the six goals of Education for All (EFA).

WAY recognises that there is no better advocacy for education than young people. We believe in young people making their voices heard and playing a key role in shaping and influencing education policies. WAY will always take the necessary actions to ensure all youth acquire knowledge and skills needed to promote sustainable development and other youth issues. Increase by 30% the number of youth who have relevant skills, including technical and vocational skills, for employment, decent jobs and entrepreneurship.

SOLUTIONS AND RECOMMENDATIONS

- Create more programs aimed at raising awareness among the youth regarding climate
- Take action in combating climate change such as tree planting, riverbanks and seaside cleaning, species protection,etc
- Obtain a commitment from public and private education systems to include climate change and environmental issues in the curriculum, particularly in primary school
- Engage with government on local, national, regional and international level to ensure it to include the sustainable use of economic development, particularly in the developing countries with threatened natural resources and species, such as eco-tourism
- Promote the idea that the protection of biodiversity and natural resources on a local level is often supportive of economic development, particularly in the developing countries with threatened natural resources and species, such as eco-tourism
- Develop sustainable and lasting networks, such as the IYFCC, among national youth organisations to promote the diffusion of best practices and exchanges of knowledge on climate change.

CONCLUSION

The climate change has emerged as a new challenge and its becoming an integral priority concern of various state holders. Innovation, rapid development policies have great impact on environment and climate. Climate change is happening, that is due to human activities, and that it will have significant social, economic and environmental consequesnces.

To a greater extent, NGOs play a big role in ensuring that both international and local level action is crucial to combat climate change. However, the concern of climate change is a very complex one, covering all areas of life, involving all sectors and all aspects of the environment. Climate change presents an immediate and significant threat to our natural and built environment and to the ways of life which co-exist with this environment. This notion proposes roles played by the NGOs to promote sustainable living by improving school educational standards and the climate change.

Therefore addressing climate change is the challenge for young generation, and requires the action and cooperation of people from all cultures and countries. Young people have a responsibility to take care of the earth and commit themselves to working towards a just and sustainable future.

REFERENCES

Anagbogu, M. A., Ada, C., & Nwokolo, A. (2014). *School children's perceptions of climate change and possible remediation*. Retrieved from www.academicresearch

Bartosh, O. (2003). *Environmental Education*. Retrieved from http://www.seer.org

Bassey, A., Ejue, J., & Bassey, I. (2005). Attitudinal patterns of counsellors and teachers towards counselling curriculum for schools. *Journal of World Council for Curriculum and Instruction Nigeria Chapter Forum*, *5*(2), 48–44.

Biodiversity and Climate Change. (2007). *International Day for Biological Diversity*. Convention on Biological Diversity.

Buddemeier, Kleypas, & Aronson. (2004). *Coral Reefs and Global Climate Change: Potential Contributions of Climate Change to Stresses on Coral Reef Ecosystems*. Pew Centre.

Cobia, D., & Henderson, D. (2003). Handbook of school counselling. Academic Press.

Impacts of a Warming Arctic: Arctic Climate Impact Assessment. (2004). Cambridge University Press.

Ogbone, E. N. (2012). *Water, Sanitation and hygiene Education: Children and adolescents leading the way in Niger Delta*. Academic Press.

Report of Intergovernmental Panel on Climate Change. (2007). *Working Group 2: Climate Change Impacts, Adaptation and Vulnerability*. Author.

Salick & Byg. (2007). *Indigenous Peoples and Climate Change*. Tyndall Centre for Climate Change Research.

Stapp, W. B. (1997). *The concept of environmental education*. Academic Press.

Tagicakibau. (2007). *"Pollution in Paradise": The Impact of Nuclear Testing and Radio-Active Pollution on Indigenous Peoples in the Pacific and Strategies for Resolution*. Pacific Concerns Resource Centre.

Up in Smoke? Latin America and Caribbean: The Threat from Climate Change to the Environment Human Development. (2006). The Third Report for the Working Group on Climate Change and Development.

Women Action Committee (WAC). (2010). *Using indigenous method to fight environmental disaster in schools*. Paper presented to secondary and primary schools in Aghamelum local Government area.

Chapter 5
Conferences as Learning Spaces for Advancing Knowledge and Action for the SDGs:
Insights From Youth Experiences

Mona Betour El Zoghbi
Independent Researcher, Lebanon

ABSTRACT

The international community is increasingly recognizing the importance of youth as key stakeholders in achieving the Sustainable Development Goals and Agenda 2030. Young people need to be continuously empowered and provided with opportunities to enhance their competences and networks. There are multiple capacity-building and networking events that seek to engage youth, yet the learning and impact generated in such spaces remains less understood. This chapter explores the value of youth-targeted and youth-led conferences and events centered on sustainability themes in advancing the learning experiences of youth participants. Testimonials from young people provide key insights into their engagement. The findings highlight the need for such platforms to be more empowering through focusing more directly on fostering collaborative actions amongst youth and organizations rather than merely on capacity-building on the spot; and through garnering financial or technical support for advancing youth action on the SDGs, especially at the local community level.

DOI: 10.4018/978-1-5225-7727-0.ch005

INTRODUCTION

In today's era of heightened global interest in crafting development pathways that are sustainable, equitable, and seek to 'leave no one behind,' the Sustainable Development Goals (SDGs) and Agenda 2030 are driving dialogue and action for a more promising future. In this regard, the international community is increasingly recognizing the importance of youth's role as key stakeholders in this development process. Today's youth are living and learning in a vastly globalized world and green-oriented economy whilst facing rising global challenges to their daily lives, well-being and career prospects. Questions are increasingly being raised regarding the extent to which current education and learning experiences are equipping young people, who constitute key stakeholders, decision-makers and leaders of future society, with the skills to efficiently manage such contemporary and long-term complexities. Current literature is often focused on expert reviews and analysis of the potentials of formal learning spaces such as schools and universities for enhancing students' knowledge and skills in leading the transition towards a more sustainable future, or for integrating education for sustainability into academic curricula (Sterling, Glasser, Rieckmann, & Warwick, 2017). A growing body of research is also exploring informal and non-formal learning spaces and the importance of sharing knowledge and action on environmental issues through social ventures, media projects and national and international conferences and forums (Van Poeck., Læssøe, & Block, 2017; Wals, Mochizuki, & Leicht, 2017). Less insight is provided into the experiences, perspectives and priorities of youth themselves as collaborators, participants or attendees at conferences and events around the SDGs and sustainability pathways. Scholars are increasingly emphasizing the importance of informal educational spaces and social learning and networking platforms for enhancing understanding of sustainable development and promoting collective and collaborative action on sustainability (Dlouha, Barton, Janouskova, & Dlouhy, 2013; Wals; 2010).

This chapter seeks to highlight the educational value for youth of informal learning and networking spaces on sustainability issues, such as conferences, youth forums and regional and international networking and policy consultation events. The chapter reports on the insights of diverse young people who have participated in national and international sustainability events and who have shared their perspectives and experiences for this study on the advantages and drawbacks of their participation at these events. The chapter therefore discusses the strengths and weaknesses of such platforms as learning spaces on sustainability issues through providing testimonials from youth participants in various national and international conferences and forums that the author has attended.

Aims and Content of the Chapter

The aim of the chapter is to explore the benefits of, and ways to enhance youth-initiated and led and youth-targeted conferences, workshops and forums around themes of sustainability and climate change, from the perspectives and experiences of youth themselves. The chapter discusses the strengths and weaknesses of such events for empowering youth to contribute to sustainability and to achieving the SDGs. It delineates the findings generated through a qualitative study which aimed to investigate the diverse levels and forms of youth engagement with climate change and sustainability and to explore the influence of the environmental, socio-economic, political and cultural context on such youth engagement.

The study fieldwork was undertaken in the Netherlands and South Africa, chosen to meet the study's aims for a diverse data set across different cultural and contextual settings. The Netherlands is geographically vulnerable to climate change, with 70% of its land below sea level. Yet, as a rich country, it holds the human, financial, and technical resources to manage the risks posed by climate change (Stive, Fresco, Kabat, Parmet & Veerman, 2011). South Africa is facing multiple developmental challenges including social and economic inequities and livelihood dependencies in several regions on scarce natural resources. Climate change further burdens the economically developing country with a long-term challenge which requires extensive investment in technical and financial resources and political commitment. (Madzwamuse, 2010). Therefore, the two countries presented diverse settings for exploring in-depth the contextual engagement of young people with climate change and sustainability. The study participants constituted of higher education youth from various geographic regions and academic disciplines in each country.

The findings presented in this Chapter are part of a wider data set that was generated through this study. This Chapter focuses on presenting insights on the strengths and weaknesses of sustainability-themes and youth-focused conferences, forums and similar platforms through the researcher's attendance as participant-observer at a total of six events, three events in each of the Netherlands and South Africa.

The events in the Netherlands encompassed a two-day youth sustainability workshop by a young nature group, an on-campus climate change awareness event organized by a student sustainability committee at a Dutch university, and an energy awareness public event organized by a national youth environmental association.

The events in South Africa encompassed an inaugural event for a regional youth-initiated and led sustainability organization, a national conference on youth empowerment through leadership training, and youth leadership workshops in a regional media conference.

All the attended events comprised both undergraduate and graduate students, between ages 18 and 30, from diverse socio-demographic and academic backgrounds from different universities and regions across the two countries. At only one particular event, all the youth participants shared the same academic major. This event was the media conference in South Africa, where all the youth participants were undergraduate students in journalism from different universities within the country,

The study approach used two main data collection methods. The first method constituted in-depth interviews with various young people participating in these events in order to elicit their personal views and experiences. The second method was participant-observation and the author's documentation of the young participants' accounts and interactions at discussion sessions and workshops throughout these events. In addition, the author occasionally shares her personal observations captured through keeping a research journal of the fieldwork phase. The generated data were organized and initially analyzed using N-Vivo software for qualitative data analysis, followed by inductive content analysis in order to identify more deeply key emerging themes and interconnections across themes (Denzin & Lincoln, 2011). The author maintained participant anonymity and data confidentially throughout the study.

The generated insights highlight the importance of inter-personal interactions and the power of envisioning and story-telling in enhancing youth efficacy and agency to contribute to change; the need for adequate mentoring and guided leadership for youth participants in local and international environmental events; the need for more reflective, engaging and innovative programs and strategies that promote meaningful and sustained youth engagement with sustainability, and the limitations posed by inadequate management of youth-targeted forums that do not engage youth from the start, and by events that tend to be selective and thereby may exclude a wider group of interested youth from different academic disciplines or socio-demographic backgrounds.

The knowledge generated from this study is useful for informing policy-making on strategies to facilitate and support youth's meaningful engagement with climate change and sustainability through the provision of adequate discussion forums and consultation platforms. It can also inform national and international youth and sustainability organizations, conference organizers and other practitioners on young people's priorities and preferences for climate change and sustainability communication and participation strategies. Advancing knowledge on such issues can also guide future planning and evaluation of environmental and youth conferences for generating meaningful and collaborative learning spaces for youth and other participating stakeholders.

BACKGROUND

The current global deliberations and national strategies around the operationalization of the 17 SDGs and moving forward Agenda 2030 increasingly emphasize the importance of engaging all relevant stakeholders in such development process. In particular, there is a need for creating diverse spaces for youth to actively engage in national and global consultations and decisions regarding the action plans and strategies for achieving the SDGs. In this regard, it is essential to ensure that young people are empowered to participate in this process, through adequate opportunities for quality education and learning spaces within and beyond the academic curriculum and campus. For instance, the SDG4 is focused particularly on improving quality education and improving life-long learning; recognizing the importance of quality education as a main driver for equitable and sustainable development.

Education is therefore essential for empowerment. Young people perceived self-efficacy and agency to influence positive societal change can be enhanced through participatory learning approaches that encourage shared knowledge production and building of their skills and capabilities. Numerous scholars are stressing the need for higher education institutions (HEIs) to focus on experiential and inter-disciplinary learning that can advance young people's competences through promoting skills in critical and reflective thinking and collaborative decision-making, as well as through applying 'leaning by doing' pedagogies and hands-on activities for sustainability literacy (Corcoran, Weakland, & Wals, 2017; Hayden et al., 2011). Young learners should be engaged in critical reflection and analysis of the causes and consequences of global issues such as climate change and encouraged to undertake creative and interactive ways of addressing these issues within their local setting. Developing these skills amongst young people promotes ownership of their visions for sustainability and enhances their capacities to contribute towards it (Tilbury, Stevenson, Fien, & Schreuder, 2002). It also promotes social justice and strengthens power relations between young people and their community and government (Edwards, 2012). For example, UNESCO has developed the Climate Change Education for Sustainable Development programme, which aims to make climate change education a more central and visible part of the international response to this global crisis. The programme targets an enhanced understanding of the impacts of climate change, particularly amongst young people, through strengthening the capacity of its Member States to provide quality climate change education, encouraging innovative teaching approaches to integrate climate change education in schools and universities, and enhancing non-formal education programmes through media and partnerships.

Scholars are increasingly calling for greater efforts by HEIs for linking the theory and practice of sustainability on-campus and modeling sustainability initiatives (Jones, Selby, & Sterling, 2010; Frisk & Larson, 2011). A critical aspect within such modeling practices is the engagement of the student body in direct and creative ways in the decision-making and operationalization of these initiatives. This can greatly enhance the students' learning and understanding on sustainability by providing a practical landscape through which students and staff could learn together and experience sustainability on-campus (McMillin & Dyball, 2009). The involvement of college and university students in campus sustainability initiatives can be an empowering and capacity-building scheme. It develops their skills to identify and manage sustainability challenges in critical, creative and collective ways, and promotes their environmental citizenship by developing meaningful relationships with their physical and social environment (Kezar & Lester, 2011). Gaining a sense of ownership and connection to sustainability initiatives on-campus helps students recognize the significance and contribution of their individual and collective actions and activities and their role as key agents of change.

Furthermore, the provision of non-formal and informal learning experiences outside the direct academic arena also fosters youth participation and leadership in national, regional and global developmental decision-making and action (Hopkinson, Hughes, & Layer, 2008; Lipscombe, Burek, Potter, Ribchester, & Degg, 2008). Conferences, workshops and policy debate forums present valuable platforms for young people to enhance their knowledge and leadership skills on global governance and decision-making processes. The opportunities that such events present for youth to interact, connect with, and learn from participants from diverse disciplines and expertise enables youth to develop important skills in critical thinking, dialogue and group dynamics and decision-making, which become increasingly important as they take on key roles in society as current stakeholders and future parents, citizens and professionals. For example, the Sustainable Development Solutions Network (SDSN) Youth network seeks to promote youth awareness and action around the SDGs. The SDSN Youth 'educates young people about the SDGs and provides opportunities for them to pioneer innovative solutions to address the world's biggest challenges.' Also, the Youth Caucus of the United Nations Commission on Sustainable Development (CSD Youth Caucus) forms an important avenue for youth participation in the inter-governmental processes within the UN system. The Youth Caucus is an international network constituting of almost 2,000 young leaders from over 1,000 youth organizations around the world. It aims to join together diverse youth who are interested in sustainability issues through providing a platform for voicing young people's perspectives and coordinating their communication and participation in high-level meetings and conferences on sustainability.

Academic conferences and events that engage young people can also enhance their intellectual development and their sense of belonging and connection to a wider community of scholars. Gumbhir (2014) highlights the experiences of undergraduate students who participated in regional academic conferences through presenting their own research and attending roundtable discussions, workshops and poster sessions along with other youth and with experts and professionals in various fields. He described the youth participants' experiences as 'transformational' in the sense that their interactions with the wider scholarly community empowered them to better understand their academic abilities and to recognize more clearly their contribution to the academic and social world.

Through authentic participation and leadership, youth develop skills in decision-making, communication and negotiation which can increase their ability to influence public affairs and contribute to societal growth (Checkoway & Gutiérrez, 2006). In addition, the enhanced knowledge, understanding, and skills that meaningful participation generates can boost young people's morale, improve their mental and emotional health, and expand their belief in their efficacy and power to drive social change (Stokols, Misra, Runnerstrom, & Hipp, 2009). Ultimately, youth empowerment encompasses multiple dimensions that not only relate to young people's cognitive and emotional features or involvement, but also to the role played by wider contextual factors in driving youth empowerment. Well-supported political learning experiences can enhance young people's sense of political efficacy, agency, and empowerment. Young people's access to political platforms for voicing their concerns and contributing to policy consultations enables them to contribute strong and creative strategies, as well as enhance their trust in public institutions (Weiss, Little, & Bouffard, 2005). Support from the community, private sector and government is essential for helping youth participate in civic life. Supportive institutional and community systems, which provide spaces for youth to collaborate with other stakeholders, can help develop young people's personal and social identity and awareness, and strengthen their confidence and sense of contribution (Jennings, Parra-Medina, Hilfinger-Messias, & McLoughlin, 2006).

Young people's participation and collaboration with other stakeholders also increases their sense of well-being over their value and contribution in academic and social communities, which ultimately influences more positive personal and societal development and enhances their productivity and performance. Schusler and Krasny (2008) emphasize the importance of youth participation in environmental events and initiatives as an act of empowerment in itself, focusing less on the extent of success of the outcome; "...regardless of whether or not their efforts are successful, engaging in collective action enables youth to think critically about the kind of world they want to live in. It also can enhance their understanding of social, economic, and political systems as they identify opportunities for and obstacles to realizing their vision..." (p. 272). Meaningful and authentic youth participation thus moves beyond mere tokenism or presence towards a process of shared power in decision-making and ownership of initiatives.

It is therefore essential that young people attain the necessary learning, training, and guidance in order to develop the knowledge and skills that can strengthen their sense of efficacy and agency; hence empowering them to meaningful engagement and participation for the SDGs. Although several mechanisms have been established that promote young people's contributions to local and global development programs, yet these mechanisms do not always promote meaningful youth participation and engagement in ways that truly influence the decision-making process. For instance, Simovska and Jensen (2009) argue that "consequential participation, which implies young people engaging in meaningful dialogue with adults and institutions and influencing decision-making processes in matters that concern them, is still in its infancy" (p. 5). Furthermore, the forms and depth of new learning that takes place at these national, regional and global conferences and forums is less fully understood. In particular, the extent to which young people, who are key stakeholders of future society, are empowered to contribute to the debates and decisions at these events remains underexplored. Several scholars even point out that current youth-targeted conferences and forums are unable to authentically engage youth in the design and development of these events, and that global policy-oriented conferences often lack adequate platforms for meaningful youth consultation on decision-making processes.

Delving deeper into these contentious issues highlights another key element that may be limiting young people's efficacy and agency to participate in these events, namely the lack of opportunities for adequate mentoring, guidance and capacity building for youth participants. Youth engagement that is targeted towards positive societal change requires that youth are equipped with adequate resources and support to build their competences and skills to contribute to long-term and efficient solutions. Yet despite the emerging focus and rising efforts for youth empowerment on global environmental issues, the literature points out to the lack of meaningful pathways for youth active engagement in ways that authentically influence decisions and policies (Bourn & Brown, 2010; UNDESA, 2010).

For example, the author has attended a youth masterclass and facilitated discussion sessions at the 2014 Global Landscapes Forum (GLF), held on the sidelines of the United Nations Framework Convention on Climate Change (UNFCCC) Conference of the Parties (COP20) in December 2014. The GLF sought to engage young people in key policy debates and decisions by holding roundtable youth discussions at the event and then pitching the youth's voices and ideas into a panel of science and policy experts. A key aspect of the GLF was the provision of expert monitoring and skills-building workshops and webinars that preceded the event in order to empower the young participants to meaningfully engage in the discussions and consultations. On their website, the organizers indicated that;

Despite increasing commitments for youth representation in committees and at conferences, there's not been a space for sufficient mentoring and skill building opportunities so young people can effectively contribute to discussions, network and become better professionals... We want to build the skills of youth to become leaders and give them a space to effectively contribute to innovative cross-cutting solutions to land use, climate change and sustainable development issues.

Building on these insights, this Chapter seeks to enhance the knowledge of the benefits that young people gain from participating in national, regional and global conferences, youth forums, workshops, and similar academic, civic and policy events on sustainability and development issues. It also seeks to explore the negative aspects or drawbacks of these events from the perspectives and experiences of youth themselves, and identify core areas for improvement so that young participants can more meaningfully contribute to debates, decision-making processes and actions towards achieving the 17 SDGs.

MAIN FOCUS OF THE CHAPTER

In the below sections, the Chapter presents the main findings of a qualitative study that investigated the diverse forms, opportunities for, and challenges of youth engagement with climate change, sustainability, and overall development issues, with particular focus on higher education youth in the Netherlands and South Africa. The findings presented hereafter pertain particularly to the conferences, forums and similar events that the author attended as part of the fieldwork for this study.

Issues, Controversies, Problems

The study gained insight into the extent and forms of learning that young people experience in their participation in sustainability events. The diverse findings are presented based on two key themes regarding the influence of these events on the participants' personal and collective development (knowledge, skills, competences on sustainability), and on the participants' personal and collective contribution to sustainability. The findings within each theme are categorized and titled based on various study participants' quotes which capture the core argument within each major study finding.

Personal and Collective Development

At the Personal Level

The study captured important data which demonstrate the influence of the attended conferences and forums on youth participants' knowledge and competences on climate change and sustainability issues. On the one hand, the findings particularly highlight the participants' increased self-confidence and motivation for development work and their enhanced sense of connection and bonding to like-minded youth around the world. On the other hand, the complexity of topics addressed at these large-scale events left some young participants feeling overwhelmed and powerless. These findings are further elaborated below.

Increased Confidence and Motivation: Being Part of this Forum Changed My Mindset - I've Been Elevated to Another Level

Numerous youth participants in the attended events discussed the ways in which they benefitted from interacting with other youth and with experts in different fields and from various nations, backgrounds and disciplines. They found that the spaces created in these events for 'like-minded people' to share local experiences and contextual challenges and discuss possibilities on global environmental issues enhanced their sense of motivation and their belief in the significance of their actions.

This finding was most evident in the conferences and events that were centered on creating spaces for a wide diversity of participants to critically reflect on their personal opinions and beliefs, to share stories on local environmental problems and solutions, and to have in-depth discussions on the visions and pathways for a sustainable future. For example, the youth leadership conference in South Africa provided a unique platform for young people from diverse socio-demographic backgrounds and academic disciplines to share experiences on witnessing first-hand the impacts of climate change and the difficulties of influencing their local communities towards sustainable practices. One participant described his enhanced self-esteem and confidence after realizing that he was 'not alone in this fight':

I felt alone, no support, but then we come together and I see there are a lot of people working on this (climate change) issue. We have more confidence now because we are all here together.

Another participant in this conference reflected on the value of personalizing global environmental problems such as climate change by sharing personal stories, hopes and visions for a unified project:

We are not same people as we were when we first came here. We made it real, how climate change has affected us in our spheres of life, we made it (climate change) personal and alive, not just a statistic. We listened to each other's past influences and future visions on the environment we want to live in. It makes you feel like we are working on a unified project, it gives you hope.

This participant emphasized the sense of bonding that participants feel when they are given the opportunity to profoundly interact with others over shared concerns and interests. She expressed her enthusiasm over 'watching people open their hearts and people feel safe to connect.'

Similarly, participants in the two-day environmental workshop in the Netherlands highlighted the value of events that are centered around profound discussions and nature activities for generating inspiration to doing things differently and trying new ways of inspiring others on environmental conservation. One Dutch participant indicated that:

This event is very inspirational because you meet so many like-minded people who are interested in the same things as you so you feel there's a shared value for the environment and you can also pick up on something that can be useful to you when we share our stories. This group gives us a way of living that I was never taught before I came to these events but now it's part of my life.

The author's personal observations further manifest the strong influence that story-telling appeared to have on encouraging participants to engage with the 'personal side of global environmental issues. The following excerpt from the researcher's journal captures these insights:

This man is a banker, and now rather than start with his pre-prepared speech he has directly started with his story after listening to the young participants' stories. This shows the impact of story-telling, of sharing your story: you encourage others to do so. You open up the space for sharing, for trust and for emotion to lead the talk.

Furthermore, the sharing of personal stories and incidents appeared to have strong influence on the participants' sense of self as well as sense of belonging to a shared community of activists and leaders towards sustainable development. The in-depth and personalized discussions often brought back memories of personal incidents and childhood stories, which encouraged critical reflection that then sparked an idea, an inspiration, or a vision for the future. Indeed, the literature highlights the value of sharing positive stories on how various people and communities experience the impacts of climate change and take adaptive actions for fostering a more positive outlook and inspiring community-based environmental projects (Wibek, 2014). The importance of story-telling is being increasingly recognized as an effective tool for learning and exchange amongst people and communities. In addition to face-to-face story-telling, new media is also providing opportunities for engaging audiences online in such dialogue and cooperation. For example, the 'SDG Communicator' portal provides a platform for 'sharing stories and advice to engage citizens for sustainable development. The sharing of personal stories, good practices, innovations, and ideas on the SDGs communication is proving to be highly important as an inspirational and empowering tool amongst stakeholders, especially young women and men.

Overwhelming Magnitude: The Conference Was Too Broad - I Felt Overwhelmed

Several study participants indicated that the complexity and large magnitude in which global environmental issues such as climate change are addressed in conferences and large-scale events instigated feelings of confusion and powerlessness. Two young participants in the regional media conference in South Africa considered that the conference could have generated discussions on the environmental problems and solutions that are more pertinent to the region such as food insecurity, water shortages and development challenges rather than debating complex global arguments:

I'm a bit disappointed with this conference. They talk to us about ice caps melting, it's not relevant to us, we can't do anything about the ice caps. But if they discuss food shortage or water insecurity, these themes we can relate to in our lives and communities.

We could have used this conference to generate ideas. These are the people who have power to act on these ideas. I felt intimidated, not empowered, by the conference. I feel severely underused, I sat around for 2 days. We need to have more constructive discussions, not just colleagues talking together. It makes you feel like you don't even want to contribute to the discussion.

A youth participant in the energy awareness campaign in her university in the Netherlands shared a similar outlook on the broad-based environmental events. She suggested that large-scale and global conferences and forums that seek to address multiple environmental challenges can approach these problems a few at a time to maintain focus on the central theme and to generate greater impact:

We could focus on a few particular themes that extend for the whole year and have few discussions in each university or city and at the end convene one conference where all come together. Now it feels like a heavy burden and too many topics and problems to solve at once.

The insights captured through this study therefore demonstrate that empowering methods of youth engagement should be youth-led and informal and should resonate with young people's personal interests and daily experiences rather than with complex and larger-than-life policy discussions. Additionally, in the light of the SDGs, there is a growing need for promoting youth competences for designing and implementing local and global action plans that resonate with community needs and challenges, along with SDG-oriented evaluations to assess impact on the ground.

Language and cultural barriers also emerged as limitations to young people's meaningful participation in sustainability events. An interview with an undergraduate student during the inaugural event for the environmental network in South Africa criticized the use of technical terminology and scientific wording which confuses participants and creates a barrier to their understanding and inclusion in the discussions. The language barrier to youth environmental communication and engagement can be particularly challenging in a country with diverse cultures and communities such as South Africa, as explained by this participant:

The information at this event was too much, needs dictionary to understand. It wasn't useless but it was part of misunderstanding of us youth. It should also be in our own language. English is too big sometimes for us, and some bombard us with big English words. How will I understand the jargon. I will become excluded from the talk.

Other studies have similarly found that youth may feel overwhelmed and powerless in relation to global health and environmental problems (Jensen, Simovska, Larsen, & Holm, 2005). Moser (2007) and Randall (2009) argued that the negativity and complexity in which climate change is often conveyed may prevent people from seeking more information about the problem and lower perceived ability to influence changes given the magnitude of climate threats. Strazdins and Skeat (2011) indicated that conveying positive and pertinent images of the future, such as on potential solutions to local climate change risks, enabled people to identify with, and work towards collective goals and priorities beyond their individual concerns or fears.

At the Collective Level

The study findings highlighted the importance of interactive discussions and innovative facilitation techniques for encouraging event participants towards critical reflection and dialogue.

Interactive, Innovative Discussions: We Should Try to Format Meetings in Different Ways, More Organic, Reflective, Open to Sharing, Be Creative and Flexible

Many young participants pointed out that sustainability-themed conferences, forums and similar events should provide spaces for deeper reflection and critical thinking on complex global issues in order to promote more meaningful learning between different participants. Several youth participants even indicated a link between the lack of such reflective spaces and the ineffectiveness of discussions that take place at these events. For instance, a Dutch university student who was on the organizing committee for an environmental awareness event on campus shared her personal experience and lessons learnt:

People left right after the event, they did not stay for in-depth discussions. There was a lot of side-talk but no real profound conversation and reflection on the themes we were exposing. We learned for next time to try to integrate more reflection into the discussions and to integrate these discussions into the events rather than leave till the end, to get people talking and reflecting about sustainability. This is how they will begin to understand it.

Similarly, in the youth leadership forum in South Africa, a central theme was the creation of spaces for reflection, personal exploration and in-depth dialogue on sustainability themes. Several youth participants shed light on the importance of generating spaces for reflection and critical thinking in order to foster significant learning experiences between diverse participants. The following quotes highlight two youth participants' experiences:

We are all coming from our hectic lives and here we can reflect, be given space to reflect. In these events we need to always give the delegates this space to reflect, to make better decisions. To get them to think: why are we here?

We need more facilitated dialogue instead of presentations, us versus them. We need to have these events more engaged and interactive.

One participant even proposed that international events and negotiations such COP events incorporate critical reflection and more informal discussion sessions that can generate clarity and lessen the tension of such high-level meetings:

In events like COP, they are always sitting and listening and in meetings. We could do more reflective and engaging activities, even have cocktail party setting so that people can be at ease and relaxed.

Another young participant also shared her experience of organizing an awareness workshop in her community for high-school students. She described the ways in which she organized more innovative events and workshops in order to keep youth interested and motivated to participate:

I tried to make it fun for the students, called them rangers, and held contests and competitions among each other. These were students who were bullied at school. At first, I engaged them, then they became interested and started taking initiative and even went out to the community. I don't think they would be so interested if all I did was just lecture them on the environment.

In an example from the Netherlands, the participants in the two-day environmental awareness workshop had multiple chances to reflect on their values and experiences and to share ideas for future sustainable development projects in their communities. The workshop was held for two days in a nature reserve in which participants learned about the environment whilst going on nature walks and playing educational games and cooking vegetarian meals. The dynamic format and nature-based setting of this event greatly encouraged critical reflection and creative activities. One participant reflected on the benefits of taking part in this event for her third consecutive year:

For me I think this environmental workshop made a big difference for me and gave me chance to talk about lots of things related to my life and also how to live my life, otherwise I don't think I would have been so into sustainability and wouldn't know many things about being environmentally-friendly. But also it's about the social aspect, a lot of people are here because of the social relations and the group and being together. So you have to make the educational part fun and enjoyable.

A young participant who helped organize this event indicated that the workshop provided space for the organizing members to step back and reflect on their roles, an opportunity they do not often get in their hectic daily schedules:

It is good to have these discussions because in our own group we have fixed tasks, each focusing on something specific and sometimes it can get really stressed. No chance to sit and reflect on the whole picture and discuss together.

These insights highlight the value of encouraging the participants at sustainability events to reflect on their beliefs and experiences and to listen and share each other's personal stories, ideas and lessons learned. For instance, Wilson (2018) highlights multiple successful initiatives led by ordinary citizens, youth, and local NGOs to solve important sustainability challenges in their communities, describing such initiatives as catalysts for hope. The inspiration created through such open sharing of experiences and insights is of great value for learning, and dynamic spaces for such dialogue should become staple at each sustainability-themed gathering. An innovative discussion facilitation technique that was applied in one of the attended events was the Earth Forum, further elaborated below.

Earth Forum: An Example of an Innovative and Interactive Facilitation Platform

The Earth Forum is a clear example of a creative strategy for building people's capacity to reflect on their values and decisions and to imagine a more desirable future. It is a method of interaction that seeks to encourage different stakeholders to listen to and learn from each other's ideas and understandings of the environment and of development. The author critically observed this event in a regional youth-led conference in South Africa. In this activity, each of the participants picked out a specific piece of 'earth' such as a leaf, a flower, a rock or a handful of soil and held it in their hands for a few minutes, sitting quietly in a circle and reflecting on what that piece of earth personally means to them. Then each participant shared their experience of what they felt and thought as they held the piece of earth, the personal significance or reason for their choice of 'earth', and their hopes for a better future. The reflective spaces created at the beginning of this forum generated more in-depth and meaningful discussions about environmental sustainability, human responsibility, ethics and so on. The value of this reflection-listening centered activity can be demonstrated in the following excerpt from an 'Earth Forum' youth participant:

I was able to explain briefly and concisely what I felt holding 'earth', because we had the space to think and reflect.

For another participant, this activity enabled her to get in touch with her emotions over the environmental degradation that she sees taking place in her community, and provided a platform for her to share her feelings and commitments with official government delegates and climate change experts whom she hoped to inspire:

I want them (experts and policy-makers) to feel humbled by us. These people have the ability and power to make a difference in the climate fight. They have broader range of power and here we are grappling for enough power to make a difference in a small sphere, and we are so motivated and passionate.

The facilitator of the earth forum was a doctoral student in environmental education and the session served as part of his fieldwork process. He further explained that this activity was intended to create a new social space for people from all ages, cultural, educational and economic backgrounds and from both genders to reflect on their personal values and hopes for the natural environment and for themselves. He highlighted the opportunity that the forum provided for different participants ' *to listen to each other, share with each other, and learn from each other'*. Thus, the sense of overwhelm that some people may feel at the idea of 17 global goals and 169 targets of the SDGs could be relieved through highlighting and celebrating the personal as well as collaborative actions that diverse individuals and communities are undertaking for a more sustainable world.

Personal and Collective Contribution

In principle, the attended conferences provided important platforms for diverse youth to contribute their ideas and to collaborate with other stakeholders on solutions to various sustainability issues. Yet in reality, numerous participating youth expressed their lack of competence in effectively conveying their ideas, networking with stakeholders or moving from talk to action, as further discussed in the subsequent sections.

At the Personal Level

The Challenge of Moving From Talk to Action: I Feel Frustrated - How Do We Get the Message Across and Get People to Listen?

Numerous participants indicated their need for guided leadership so that their participation at these events can be made more meaningful and effective.

First, several participants discussed the need for such conferences to provide workshops and learning opportunities for developing their skills in public speaking, networking and leadership, as articulated by two participants in the youth leadership forum in South Africa:

We have great ideas. We feel there is something wrong and we see the need to do something. My question is: how do we go about to solve this? How do we get them to listen to what we have to say and act on that? I expect a set of skills from this conference.

How can I network ideas here and keep people contributing as they are doing now? We need continuity of these processes, but we don't know how.

Similar studies have further indicated that meaningful youth empowerment to contribute to environmental events and platforms should not merely take place within these events but also ahead of them so that youth are prepared to contribute to the discussions taking place. For instance, the need for adequate workshops and skills development for youth participants in conferences was highlighted in the assessment of the 7th UNESCO youth forum, in which 86% of the participating youth delegates found the capacity building sessions and preparatory on-line consultations and training very useful. The participants in the forum also suggested that the next youth forums provide greater time for debates and discussions and less time for speeches (UNESCO, 2012).

Second, the participants in this study also indicated the need to build their skills for taking leadership not only in the debates and consultations but also in the actual implementation of agreements and decisions. A young South African journalism student who was participating in the regional media event pointed out this challenge:

I feel we can't take it further than just talk. We don't know the process to take it further from talk to action. And even if there are structures to participate, I don't feel part of it.

This issue of youth action is increasingly important when linking it to the SDGs, given the importance of operationalizing the goals into concrete action plans that can be designed, implemented, monitored and assessed. Youth role in this process is pivotal, for advancing the action on the ground as well as for holding the accountability to responsible sides.

Building on this last idea, this study has also found that one of the main reasons for youth feeling dissociated or not involved in the implementation phases that result from such events and initiatives was the fact that they were not participating from the beginning. Several young participants indicated the importance of engaging youth peers from the start in organizing sustainability events and conferences and in identifying project aims and strategies. A Dutch university student who was participating in the environmental awareness campaign articulated this idea:

We want to ask youth: how do you want to organize these events? How would you like to be involved? In the past we organize and plan everything and then ask young people to attend and to sign, it's passive. Now we notice they want to do things for themselves. So we want to ask them what are things that you can and want to do, and how can we help you with that?

The author's reflections on the youth leadership forum, captured through the research journal, also shed light on the importance of helping young people take an active role in the design as well as implementation of the strategies and activities that materialize their ideas and visions:

At first these youth were researching solutions to environmental problems off the internet. What is needed is for them to come up with their own solutions and ideas, based on what is happening in their lives and what they believe they can do and want to do. They have the opportunity in these conferences to present their ideas to people who can help them. But they need platforms and guidance for developing their ideas right from the start and devising their messages to make them more concrete and relevant to solving their local situation.

In resonance with these insights, Head (2011) indicates that although youth consultation on policy-making is on the rise, a wide gap remains between written commitments and agreements and the actual application and operation of such commitments. Therefore, events and initiatives that promote positive youth learning and development require opportunities for skills-building, action-oriented workshops, and the integration of multi-stakeholder efforts to provide adequate guiding structures that enhance young people's competencies as leaders who can make change in their communities and who can meaningfully participate in decision-making processes nationally and globally. For example, the global survey 'My World 2030' garnered youth perspectives and priorities from around the world into informing the global discussions and policy deliberations regarding the articulation of the SDGs.

At the Collective Level

Lack of Understanding of, and Guidance for Youth Limits Their Contribution: I Participated in COP 15 – No Understanding of What Youth Want

The study captured important insights into the experiences of young people who have previously participated in international policy negotiation processes and events. Some of these youth, who were participants at events the author attended in the Netherlands and South Africa, shared their experiences and lessons learned from their involvement in these high-level policy consultation meetings. They emphasized young people's need for guidance and training for their participation, especially in global environmental events in order to empower youth to contribute meaningfully to the international policy debates and decisions. The author interviewed one of these youth participants, a young South African university student who participated in the Copenhagen Climate Change Conference 2009 (COP15). She highlighted the valuable platform that such high-level events provide for young people to interact with other youth and with professionals, experts and policy-makers from different countries and disciplines, stating that:

You feel all are equal. I could talk to the head of a company or to a minister and wouldn't even know it. Just a space for dialogue, no intimidation...Youth are so inspired by all the other youth who are present there and by their passion and commitment to the work being done. Youth passion is contagious.

She nevertheless pointed out to two limitations to the effectiveness of youth participation in these international events. First, she indicated that the main arena which was available for young people to engage with the official delegates was in the side events rather than in the official policy negotiations, which limited their ability to make their voices heard. Yet she highlighted the importance of benefiting from media presence to achieve greater impact:

They (official delegates) only interact with youth in the side events, and we can't enter all negotiations. There's also no guidance for us on how to approach the delegates or the best way to present our case...So our voice is not really heard. So we figured that the only way to get them to hear us is to work together. So we joined together 6 participants from southern Africa and we were making noise to attract the media. Then the media started asking questions.'

Second, she stressed that youth participants in these international events should be given adequate guidance, information and support so that they are able to contribute to the deliberations and decisions being made. She indicated that without proper guidance, young people who are inexperienced in participating in these events might feel overwhelmed and confused or might get caught up in the events and entertainment. She shared her own personal experience in COP15 and called for greater youth involvement and responsibility in the decision-making process:

I went to COP15, as part of the South African delegation. There were two delegates from Swaziland who were the youngest ever (15 & 16 years old). I started thinking, they didn't realize the power that they had. We need to move away from youth v/s adults. Give us a bigger role in making decisions. It's our future on the line and we need to work together. We need to make it fun and positive.

This young participant also had some valuable advice to give to other young participants in international environmental conferences:

COP provides you direct link with corporate leaders and with experts, scientists, funders. Try not to get caught up in the events and entertainment. Focus on the theme or area that you know best. Speak to people; network...You need to plan in advance. Come prepared. Know your topic, plan out your ideas...Get disposable income to hold side-events.

Another South African participant at the inaugural event for a youth-led environmental network shared similar sentiments regarding the benefits and disadvantages of participating in global environmental and developmental events:

I participated in the World Economic Forum. It was overwhelming, too much information and you need to sieve through lots of people, people who are not really listening to you, just exchanging business cards...But it's a good experience for getting subjected to a global event and for networking.

Nevertheless, not all the young people who participated in international sustainability conferences held positive views regarding the value and usefulness of such events. Some participants felt that these events tend to be too politicized which often stalls decisions or limits true implementation and meaningful actions on the ground. For instance, two Dutch participants in a national environmental awareness event expressed their doubt over the effectiveness of large-scale global events in solving environmental challenges, voicing greater support for small-scale and local or regional initiatives:

I personally don't believe in political solutions anymore, especially at that high international level. After Copenhagen nothing happened. It might be better if there is co-operation and steps for improvement within Europe. COP is too big and too many states and too many interests...There are already many decisions that have been made at the European level that are already more progressive. What did the U.N. ever decide?

Similar sentiments were echoed by a young South African participant in a regional media conference held in South Africa, who was vocal against the politicization of international environmental platforms. He further criticized what he considered to be the lack of adequate and non-political platforms for youth to participate in devising environmental solutions:

I feel frustrated as youth of South Africa. If you want to make changes, all is politicized. No other platform to make changes. Even media access is influenced by politics. We need to create a platform for youth to say even if we are not interested in politics we still want to participate. We want to be involved but not involved in politics.

Indeed, the scientific and policy world has yet to establish binding global agreements for green-house gas emissions reductions which global and UN-led negotiations have so far failed to achieve due to long-standing disagreements and power differentials between industrialized and developing countries (UNFCCC, 2006). It can be argued that such political shortcomings and failures have generated feelings of despair or disinterest amongst young people who see greater hope for localized and regional commitments and collaborations for solving environmental challenges. Ultimately, genuine forms of participation, especially on global sustainable development issues, need to be clear on the extent to which young people hold power and opportunity to engage in the consultations and decisions that take place at high-level meetings that are often dominated by policy-makers, experts and scientists. The Agenda 2030 and the national operationalization of the SDGs offer crucial spaces for young women and men to be directly involved in such decision-making processes; as long as genuine, meaningful youth participation is ensured through proper consultation processes and taking their voices into perspective in final decisions.

Exclusivity of Conferences Limits Participation and Representation: Some Conferences Are Very Exclusive, Only Privileged People Get to Be Part of It

The study captured data which indicate that the exclusivity of some sustainability or youth-targeted conferences and forums may pose representational issues. Several participants criticized the limitations posed by certain youth-targeted forums and events that tend to be selective to a particular age range, cultural background or educational level. These participants pointed out that such limitations or criteria may exclude a wider group of youth from different academic disciplines or socio-demographic backgrounds who would be interested in taking part in these events. For example, an interview with a youth participant in the inaugural event for the youth-led environmental network discussed the challenges to participation that he experienced with the environmental club at his university:

We need to create a non-discriminatory environment between different schools and students from different backgrounds to interact together... The problem is that it should not be a group thing because they isolate. I focus on my environment group and you focus on your group. There was an environmental campaign here at the university and I could not attend because I didn't know. I am isolated, it's a barrier to engagement. If I were in university grounds I would have known. So it's not well publicized.

Similar sentiments were expressed by a youth participant in the regional media conference in South Africa. She criticized the lack of platforms and forums that target participation from a wide range of academic disciplines:

We need more platforms for youth to engage on environmental issues. There are no platforms, only for engineers and medical students, because companies and hospitals support and want to train and they have money. Other students don't get these chances and even if it exists, we don't know about it.

The literature indicates that formal methods of youth participation that rely on policy consultation mechanisms might not always be inclusive or representative of a wide range of youth perspectives and experiences (Wibek, 2014). These events are often organized to include a limited number and profile of participants, thus posing potential risks of under-representation of the general youth population, of excluding young people who are not directly engaged in the field of sustainable development in their studies or activities, and of side-lining young people who are interested in addressing environmental issues from a more action-oriented approach.

SOLUTIONS AND RECOMMENDATIONS

This study has generated valuable insight into the experiences of young people who participate in national, regional and international sustainability events. The study findings emphasize that, in order to be meaningful, authentic, and impactful, youth engagement in such events should not be merely representational. Such events and initiatives should enable young people to develop a sense of shared responsibility over global and local complex challenges, and a collective identity and agency to analyze social and political power dynamics and to identify and mobilize community resources for driving safe and sustainable development in their communities. The Agenda 2030 provides an important platform for engaging youth more meaningfully in the development pathway; yet quality education remains key for ensuring true empowerment over the long term.

Prior to presenting the author's recommendations, it is of great value to share with the reader the suggestions and recommendations that youth mentors and trainers at the attended events provided to the participating young people. Two major themes that emerged from such recommendations were the importance of youth presenting a united and confident front when participating in large-scale sustainability conferences and of engaging delegates in emotive language to foster bonding and connection with real-life events rather than abstract descriptions of environmental problems.

The main advice to youth given by organizers, facilitators and youth mentors at some of these conferences are presented in Table 1.

Further to the breadth of insights gained in this study on youth participation in environmental conferences, the author presents the following solutions and recommendations to key audience and stakeholder groups:

- National and local governmental authorities should institutionalize mechanisms and platforms for youth participation in expert and wider stakeholder consultations and decisions on policies and programmes regarding the SDGs. This can be achieved though establishing national youth councils that grant such youth consultation processes more formal and legislative recognition and greater representational power over the larger youth spectrum in the community or country. For example, governments can create official platforms for youth to collaborate and to review processes for national legislations and strategies over the 17 SDGs and overall development agenda. In particular, it is recommended that formal platforms be established for fostering youth leadership in upcoming debates regarding processes for national legislations and strategies over the 17 SDGs and overall development agenda. Educational, socio-cultural, and political institutions are instrumental in increasing conscious power-sharing amongst youth and other stakeholders regarding the debates, decisions and actions towards a more secure and sustainable society. Academic institutions play a particularly important role in helping achieve this, and it is essential to integrate 'education for sustainable development' (ESD) into learning processes for more empowered learners who can contribute to the change process.

Table 1. Advice given to youth for more effective conference participation

Advice Given to Youth Participants	Quote from Conference Organizers and Youth Mentors
Connecting with delegates in a confident manner	'Don't look tired. Don't go there apologetically. Claim your space.' 'Make eye contact when questioning and talking with the official delegates, be confident and challenging.'
Encouraging critical reflection	'What happens inside you when you see this picture (of starving children/food insecurity)? Make sure people connect. Stop and give them time to reflect and to feel the reality. It's uncomfortable to facilitate silence. But give people the chance to slow down and connect to what you're speaking.'
Targeting emotions through sharing personal stories	'Draw people in. The real thing that speaks to all of us is emotion. They want something different. We have the capacity to deliver that through our personal stories (on climate change) and how they impacted our lives.'
Partnering with delegates from different disciplines	'The discussions should not come as mentoring but as partnership: this is a barrier now to be crossed: environmental activists as super greenie and the rest as bad.'

- Academic institutions should provide platforms for young people to meaningfully engage in the consultations and decisions that seek to incorporate environmental education and learning for sustainability into academic curricula, so that they take ownership of the knowledge creation process that can help them successfully manage the rising social and environmental challenges.
- Leaders, coordinators and practitioners in environmental organizations (UN-based and local and international NGOS) could involve youth participants from previous years, and more-experienced youth mentors, in supporting and guiding less experienced youth participants for more authentic engagement in environmental events. This can be achieved through integrating, into the preparatory phases and the actual events, adequate guidance and advice from experts, mentoring from more experienced youth leaders, and training workshops and master classes for developing young participants' skills in public speaking, dialogue facilitation, policy negotiation, and idea generation and activation so that they move beyond one-sided lectures and speeches to more interactive discussions and skills development.
- Youth, civic and local environmental organizations (local NGOs) should develop environmental programmes that promote social interactions and activities not only between youth but also across different stakeholders in society, and especially in natural settings which can foster critical reflection and participant-bonding over personal stories, and inspire creative and collaborative solutions amongst various stakeholders. Environmental and youth groups should also support the formation of online networks with the aim of fostering long-term collaborations and exchange of ideas and lessons learnt amongst the diverse participating youth and other participating stakeholders.
- Organizations and initiatives engaging young people should invest in participative evaluation and monitoring of the conferences and events that these youth attend in order to continuously assess the success of such events and improve their impact for effective youth leadership on environmental issues.

FUTURE RESEARCH DIRECTIONS

Future research should develop contextually-relevant indicators and monitoring and evaluation frameworks for assessing and advancing youth-led and youth-targeted events and leadership programs around global sustainability themes. The role of youth in monitoring and evaluation of SDGs implementation is particularly important;

hence the design of key development indicators and results-based management and monitoring. Research should also seek to identify and incorporate indicators that measure and investigate the role of affect, social interactions, and reflective spaces in enhancing the learning processes generated in these events. Future studies can also examine and improve processes of supporting higher education youth in balancing educational and leadership tasks and the positive and/or negative influences on their academic and professional performance and overall sense of well-being from their participation in sustainability conferences, workshops and similar events.

CONCLUSION

This study has sought to explore the advantages and drawbacks of sustainability conferences and networking events from the perspectives and experiences of young people who have participated in these events. The insights gained from this study emphasize three key findings regarding young people's potential and opportunity to learn and develop as well as to contribute at these events. First, the study highlighted the need for more reflective, interactive, and engaging approach to the design and conduction of sustainability and youth-focused events, workshops and projects to achieve better quality learning outcomes for youth. The approach should therefore foster critical reflection and dialogue amongst various participants, and should engage young people from the start in the objective-setting, design, and application of such initiatives in order to promote their continuous and enthusiastic involvement. Second, the study findings indicated the importance of emotional engagement, story-telling, and creative communication in encouraging youth to share visions on positive changes they would like to see for their future and reflect on ways to work towards it. Positive emotions with regards to the communication and confrontation of complex global challenges such as climate change can reduce feelings of powerlessness and help young people face their concerns in more optimistic perspectives and constructive actions. Youth who undertake sustainability actions in a group setting were also found to experience a more optimistic vision for the future and a belief in their individual and collective agency to influence and contribute to change. Third, and in line with the SDG17 on 'partnerships', the study indicated the importance of fostering partnerships with and for the youth, in order that governments, the private sector, academic institutions, civil society, media and other stakeholders altogether collaborate with young people to further support their brilliant initiatives that seek to advance sustainable development in their communities and globally.

Youth efficacy and agency to meaningfully participate in such conferences and events can therefore be enhanced through the provision of adequate spaces for peer-to-peer collaboration, and through mentorship on project management and networking. Many of these youths come to these conferences with innovative ideas and successful good practices that can not only be shared but also scaled-up or expanded for further collaborations amongst the youth and their institutions. Expert guidance should also be provided to these youth on the key SDGs that can be aligned with their local community needs and activities and how to design relevant projects. Proper institutional support is also needed for wide-ranging and organized youth representation at these events. Equally important, it is crucial that young participants in these conferences and events do not return back to their communities empty-handed, but rather with ideas, innovations, and investments that can further advance their sustainability initiatives. Accordingly, the organizers of such events should make sure to provide platforms that link youth with other stakeholders including private sector, businesses, and the media, and should provide support, whether financially though prizes or through linking to potential investors, as well as technically through expertise, resources, tools, and guides. This is of utmost importance for ensuring that youth-led sustainability initiatives are mobilized further and empowered for greater effectiveness and impact. Therefore, in order that conferences and events truly promote youth education and learning and foster their contribution towards the SDGs and Agenda 2030, it is essential that youth are provided with the adequate platform, guidance, mentorship and support to fully engage in the design, consultation, decision-making, implementation and evaluation of sustainability and developmental programs and events.

REFERENCES

Bourn, D., & Brown, K. (2011). Young People and International Development: Engagement and Learning. Development Education Research Centre (Research Paper 2). Institute of Education, University of London.

Checkoway, B., & Gutiérrez, L. (2006). Youth Participation and Community Change: An Introduction. *Journal of Community Practice*, *14*(1/2), 1–9. doi:10.1300/J125v14n01_01

Corcoran, P., Weakland, J., & Wals, A. E. J. (2017). *Envisioning futures for environmental and sustainability education*. Wageningen: Wageningen Academic Press. doi:10.3920/978-90-8686-846-9

Denzin, N., & Lincoln, Y. (2011). *The SAGE Handbook of Qualitative Research* (4th ed.). London: SAGE.

Dlouha, J., Barton, A., Janouskova, S., & Dlouhy, J. (2013). Social learning indicators in sustainability-oriented regional learning networks. *Journal of Cleaner Production, 49*, 64–73. doi:10.1016/j.jclepro.2012.07.023

Edwards, K. (2012). Moving beyond green: Sustainable development toward healthy environments, social justice, and strong economies. *New Directions for Student Services, 2012*(137), 19–28. doi:10.1002s.20011

Frisk, E. & Larson, K. (2011). Educating for Sustainability: Competencies & Practices for Transformative Action. *Journal of Sustainability Education, 2.*

Grant, M. (2009). Internationalising higher education for sustainability: Youth Encounter on Sustainability. In P. Corcoran & P. Osano (Eds.), *Young People, Education, and Sustainable Development: Exploring Principles, Perspectives, and Praxis*. Wageningen Academic Publishers.

Gumbhir, V. K. (2014, May). From Students to Scholars: Undergraduate Research and the Importance of Regional Conferences. *The American Sociologist, 45*(2-3), 298–300. doi:10.100712108-014-9212-2

Hayden, M., Houwer, R., Frankfort, M., Rueter, J., Black, T., & Mortfield, P. (2011). Pedagogies of Empowerment in the Face of Climate Change Uncertainty. *Journal for Activism in Science and Technology Education, 3*(1), 118–130.

Head, B. (2011). Why not ask them? Mapping and promoting youth participation. *Children and Youth Services Review, 33*(4), 541–547. doi:10.1016/j.childyouth.2010.05.015

Hopkinson, P., Hughes, P., & Layer, G. (2008). Sustainable Graduates: Linking Formal, Informal and Campus Curricula to Embed Education for Sustainable Development in the Student Learning Experience. *Environmental Education Research, 14*(4), 435–454. doi:10.1080/13504620802283100

Jennings, L., Parra-Medina, D., Hilfinger-Messias, D., & McLoughlin, K. (2006). Towards a critical social theory of youth empowerment. Journal of Community Practice, 14(1/2), 31-55.

Jensen, B. B., Simovska, V., Larsen, N., & Holm, L. G. (2005). *Young people want to be part of the answer*. European Network of Health Promoting Schools.

Jones, P., Selby, D., & Sterling, S. (2010). *Sustainability education: perspectives and practice across higher education*. EarthScan.

Kezar, A., & Lester, J. (2011). *Enhancing Campus Capacity for Leadership: An Examination of Grassroots Leaders in Higher Education*. Stanford University Press. doi:10.11126tanford/9780804776479.001.0001

Lipscombe, B. P., Burek, C. V., Potter, J. A., Ribchester, C., & Degg, M. R. (2008). An Overview of Extra-Curricular Education for Sustainable Development (ESD) Interventions in UK Universities. *International Journal of Sustainability in Higher Education*, *9*(3), 222–234. doi:10.1108/14676370810885853

Lozano, R., Lukman, R., Lozano, F., Huisingh, D., & Lambrechts, W. (2011). Declarations for sustainability in highereducation: Becoming better leaders, through addressing the university system. *Journal of Cleaner Production*, 1–10.

Madzwamuse, M. (2010). *Climate Change Vulnerability and Adaptation Preparedness in South Africa*. Heinrich Böll Stiftung Southern Africa.

McMillin, J., & Dyball, R. (2009). Developing a Whole-of-University Approach to Educating for Sustainability: Linking Curriculum, Research and Sustainable Campus Operations. *Journal of Education for Sustainable Development*, *3*(1), 55–64. doi:10.1177/097340820900300113

Moser, S. (2007). More bad news: The risk of neglecting emotional responses to climate change information. In S.C. Moser & L. Dilling (Eds.), Creating a climate for change: Communicating climate change and facilitating social change. (pp. 64-80). New York, NY: Cambridge University Press.

Randall, R. (2009). Loss and climate change: The cost of parallel narratives. *Ecopsychology*, *1*(3), 118–129. doi:10.1089/eco.2009.0034

Schusler, T., & Krasny, M. (2008). Youth participation in local environmental action: Developing political and scientific literacy. In A. Reid, B. B. Jensen, J. Nikel, & V. Simovska (Eds.), *Participation and Learning: Perspectives on Education and the Environment, Health and Sustainability* (pp. 268–284). Springer. doi:10.1007/978-1-4020-6416-6_17

Simovska, V., & Jensen, B. (2009). Conceptualizing participation – the health of children and young people. *Report Prepared For the European Commission Conference on Youth Health, Brussels, Belgium, July 9-10*. WHO Regional Office for Europe.

Stive, M., Fresco, L., Kabat, P., Parmet, B., & Veerman, C. (2011). How the Dutch plan to stay dry over the next century *Proceedings of the Institution of Civil Engineers (ICE) –. Civil Engineering (New York, N.Y.)*, *164*(CE3).

Stokols, D., Misra, S., Runnerstrom, M. G., & Hipp, J. A. (2009). Psychology in an age of ecological crisis: From personal angst to collective action. *The American Psychologist*, *64*(3), 181–193. doi:10.1037/a0014717 PMID:19348519

Strazdins, L., & Skeat, H. (2011). Weathering the future: Climate change, children and young people, and decision making: A Report to the Australian Research Alliance for Children and Youth. Canberra, Australia.

Tilbury, D., Stevenson, R., Fien, J. & Schreuder, D. (2002). Education and Sustainability. Responding to the Global Challenge. *The World Conservation Union, Commission on Education and Communication CEC.*

UNDESA. (2010). *World Youth Report: Youth and Climate Change*. United Nations Department of Economic and Social Affairs.

UNESCO. (2012). 7th UNESCO Youth Forum Assessment. United Nations Educational, Scientific and Cultural Organization.

UNFCCC. (2006). *United Nations Framework Convention on Climate Change: Handbook. Intergovernmental and Legal Affairs*. Bonn, Germany: Climate Change Secretariat.

Van Poeck, K., Læssøe, J., & Block, T. (2017). An exploration of sustainability change agents as facilitators of nonformal learning: Mapping a moving and intertwined landscape. *Ecology and Society*, *22*(2), 33. doi:10.5751/ES-09308-220233

Wals, A. (2010). Mirroring, Gestaltswitching and Transformative Social Learning: Stepping Stones for Developing Sustainability Competence. *International Journal of Sustainability in Higher Education*, *11*(4), 380–390. doi:10.1108/14676371011077595

Wals, A. E. J., Mochizuki, Y., & Leicht, A. (2017). Critical case-studies of non-formal and community learning for sustainable development. *International Review of Education*, *63*(6), 783–792. doi:10.100711159-017-9691-9

Weiss, H. B., Little, P. M., & Bouffard, S. M. (2005). More than just being there: Balancing the participation equation. *New Directions for Youth Development*, 105. PMID:15943135

Wibek, V. (2014). Enhancing learning, communication and public engagement about climate change – some lessons from recent literature. *Environmental Education Research*, *20*(3), 387–411. doi:10.1080/13504622.2013.812720

Wilson, D. (2018). *Designing the Purposeful World: The Sustainable Development Goals as a Blueprint for Humanity*. New York: Routledge.

ADDITIONAL READING

Earth Forum and University of the Trees. (n.d.). Retrieved from http://www.universityofthetrees.org/

Hatcher, T., Wiessner, C. A., Walker, J. S., & Chapman, D. (2006). *How a professional conference created new learning: A case study*. USA: North Carolina State University.

Percy-Smith, B., & Thomas, M. (2010). *A Handbook of Children and Young People's Participation: Perspectives from Theory and Practice*. New York: Routledge.

Rogers, T. (2013). *Conferences and Conventions: A Global Industry* (3rd ed.). New York: Routledge. doi:10.4324/9780203119402

Tilbury, D. (2012). Learning to Connect. Reflections along a Personal Journey of Education and Learning for a Sustainable Future in the Context of Rio + 20. *Journal of Education for Sustainable Development*, *6*(1), 59–62. doi:10.1177/097340821100600112

KEY TERMS AND DEFINITIONS

Agency: Agency is the ability and power that people have to undertake and impact positive changes in their personal lives and wider socio-political contexts.

Decade of Education for Sustainable Development: The Decade (2005-2014) seeks to mobilize educational resources and engage all stakeholders in critical, holistic and fore-sighted education and learning, and to encourage multi-stakeholder and inter-cultural dialogue and collaboration for promoting the pathway towards positive and sustainable societal change. UNESCO is the lead agency for the Decade.

Efficacy: Perceived self-efficacy involves people's belief in their ability to make a change or to succeed in a certain endeavour. It enhances their feelings as agents of change who are able to contribute to positive social and political change.

Empowerment: Youth empowerment mainly relates to young people's willingness as well as capacity to take action. It is the ability of youth to increase their participation in decisions regarding their own lives as well as wider community issues that are of importance or concern to them.

Engagement: Youth's meaningful engagement transcends the general notion of active participation or involvement to entail a deeper level of reflection, exploration, valuing, and understanding; and a sense of perceived power and control over issues that affect their future.

Envisioning: The envisioning technique, often applied within sustainability research, aims to engage people in a profound, critical and future-oriented thinking over their desired futures and their preferred pathways to sustainability. A critical aspect of envisioning is its ability to empower people by helping them to see themselves as owners of their visions and as vital contributors and responsible stakeholders to the process of change.

Chapter 6 Addressing the Sustainable Development Goals Through Environmental Education

Carolyn N. Stevenson
Purdue University Global, USA

ABSTRACT

The United Nations Sustainable Development Goal 4 Quality Education has increased awareness in pre-university environmental education efforts. Environmental education is important to not only creating awareness of world environmental issues, but taking action towards fostering positive change. Environmental education programs such as SeaTrust Institute's AWARE (Action Within a Resilient Environment) assist teens in learning about issues that directly impact their communities and their world. AWARE combines environmental education with hands-on experiential learning projects that help promote environmental awareness in their communities. Through education and experience with active scholars and professional practitioners, students gain an increased understanding of environmental challenges and ways to make a positive impact – both domestically and globally. This is especially critical to developing countries which lack the educational programs and resources to address the impact environmental changes have on their nations and communities.

INTRODUCTION

Research and a review of the literature conclude that there is a need for promoting environmental education at the pre-university level. This effort is also supported by the United Nations Sustainable Development Goals (SDGs), specifically SDG4 Quality Education (United Nations Division for Sustainable Development Goals,

DOI: 10.4018/978-1-5225-7727-0.ch006

2018). Youth have strong concerns about climate change issues and want to not only increase individual awareness but take steps toward improving their communities. The purpose of the case study presented in this chapter is to describe and explain pre-university perspectives on climate change issues and ways to take awareness into action. The findings of the study will inform pre-university administrators, teachers, environmental youth group leaders, and other environmental educators of ways youth can have an active role in creating positive change. The findings will also inform current and future environmental educators of ways to integrate environmentally-based service learning projects into the curriculum. The study will provide the basis for greater understanding of the ways a pre-university environmental programs can assist in promoting both individual and community awareness of climate change issues. Finally, the findings of the study can draw attention to taking action through the standards set forth by SDG4 in promoting quality education and access to all by increasing understanding of the need for creating sustainable communities.

The exploratory questions that guided the study are:

1. What elements constitute pre-university student perspectives on climate change issues and environmental education?
2. What variables influence this perspective on climate change issues and environmental education?
3. What beliefs do these pre-university students hold which support or negate this perspective?

This chapter discusses ways environmental education for pre-university students is essential for creating change and fostering awareness in local communities. A case study based on the perspectives of four teenage youth from a large urban city in the United States is presented. The teens' perspectives of not only creating awareness of climate change issues but taking action toward making local change is also provided. The chapter describes SeaTrust Institute's AWARE environmental education program and ways youth can directly impact their communities. While there is no panacea to resolving issues related to climate change issues, individuals can adapt to these changes and take action towards improving individual communities.

BACKGROUND

As the earth's temperature continues to rise, climate change is an issue of great concern for individuals of all countries. These changes can be seen in the large number of floods, drought, and extreme hot and cold temperatures. The results of climate change are great and impact human health, ecosystems, agriculture, transportation, forests, coastlines, adaptation, and migration of people. There is a need to become resilient or build "the capability to anticipate, prepare for, respond to, and recover from significant multihazard threats with minimum damage to social well-being, the economy, and the environment," (United States Environmental Protection Agency, 2014, p. 1). The world is faced with an increased vulnerability and is susceptible to, or unable to cope with, adverse effects of climate change, including climate variability and extremes (United States Environmental Protection Agency, 2014). As a result, environmental education for pre-university students is imperative to help plan and prepare for the environmental changes in the future. It is important to inform future generations on ways to meet the challenges of climate change issues through education and awareness.

Promoting action and awareness through environmental education also aligns with the United Nations SDGs. "The Sustainable Development Goals (SDGs), otherwise known as the Global Goals, are a universal call to action to end poverty, protect the planet and ensure that all people enjoy peace and prosperity, (United Nations Division of Sustainable Development, 2018).

Providing pre-university students with education, training, and field research can expose youth to the impact of climate change issues and learn first-hand how they can make a difference. As Priyanto, Fanani, and Sasmitojati (2013) state, "sustainable development can be achieved through education, because education is a means to change the perceptions, attitudes, and human behavior (p. 7). It is through education that youth can learn to shift behavioral patterns to help promote a more sustainable future. The U.S. offers many formal education programs for pre-university students. "More than 200 charter schools in the United States call themselves members of the Green Charter Schools Network, a consortium and clearinghouse that was founded in 2008," (Grayson, 2011, p. 26). The schools become members by adhering to the network's "green print," or list of core practices (Grayson, 2011, p. 26). "The network facilitates communication among educators, parents, and policymakers alike, connecting them to resources and other schools that might want to adopt their own eco-focused programs," (Grayson, 2011, p. 26). In addition to formal training in school, environmental clubs, community programs, and non-governmental agencies offer environmental education opportunities for youth. While there are many options for environmental education for youth, there is a strong need to promote participation and involvement in these programs.

CASE DESCRIPTION

The Organizations

The organizations selected for this case study were three public high schools in the city of Chicago, Illinois. The three high schools serve a diverse student body in terms of race, ethnicity, socio-economic status, religion, and geographic location.

For this case study, four high school students were selected. Each high school student received varying formalized environmental education in their current high school. The four high school students were of the same age, that being 16 years of age. All of the participating high school students agreed to discuss their perspectives on climate change issues and environmental education. Each participant agreed to use their first name in the study. The official names of the high schools were eliminated to ensure confidentiality of the institution. Due to space limitations, highlights of the interviews are represented in the vignettes that follow. Participants did go into greater detail in responses to the interview questions. The passages were selected based on the relevance to the topic.

Statement of the Problem

Literature and observation reveal there is a need to provide pre-university students with more environmental education and programming related t climate change issues. Additionally, while programs do exist, it is through first-have experience through activities such as field research that promote a stronger awareness and action within students' communities. This is especially important for high school faculty and administrators who can provide programming and environmental education opportunities for youth. The findings of the research illustrate there is a need to increase climate change awareness through the increased amount of environmental education.

It is reasonable to suggest there is a need for promotion of climate awareness through environmental education for pre-university students. This is a topic of interest for youth, teachers, administrators, and all members of a given community. The challenge for teachers, educators, and community members is how to offer environmental education programs and training with limited funds.

Teachers, educators, and community members must employ strong leadership skills as well as knowledge of climate change issues. Communicating concerns related to climate change issues is important and pre-university teachers and administrators must be able to motivate others to see the value of providing environmental programs within the school setting. Community members also need to play an active role in promoting awareness of climate issues among youth through neighborhood-based activities.

It is reasonable to suggest there is a need for qualitative researchers to capture the lived experiences of pre-university student opinion on environmental issues facing their community and their world. While it apparent these climate change issues exit, there is a need for data to be presented through the perspectives of today's youth. This case study does not seek to resolve the climate change concerns of pre-university students in an urban setting. Rather it seeks to create an awareness of the ways environmental education programs can foster awareness in pre-university students. Additionally, the case study seeks to describe and explain the ways environmental education programs can be used as a means for promoting individual as well as community awareness about climate change issues.

Based on these facts, it is reasonable to assume that there is a need to promote climate change awareness through environmental education. Pre-university students have an interest in the ways climate change impacts their lives and are concerned for the future. The problem presented in this study is to describe and explain pre-university student perspectives on climate change issues and environmental education. Findings from this case study will assist teacher, high school administrators, and community members in seeing the need for increased environmental education at the pre-university level.

Purpose of the Study

The purpose of this study was to describe and explain pre-university student perspectives on climate change issues and environmental education. The results may assist teachers, high school administrators, and community members in several areas.

The findings of the study will inform assist teachers, high school administrators, and community members of the perspectives students' have on climate change issues facing the environment and more specifically their community. The findings will also inform teachers, high school administrators, and community members of the need to offer more environmental education for pre-university students. The study will provide the basis for greater understanding of the need to promoting climate change awareness through offering environmental education at the pre-university level.

This immersion into the research process actively involves the qualitative researcher in the quest for gaining a deeper understanding of the social phenomena. A case study was selected for this study. Case studies involve an in-depth study of this bounded system and rely on a number of data collection materials.

According to Yin (2010) "...the distinctive need for case studies arises out of the desire to understand complex social phenomena..." because "...the case study method allows investigators to retain the holistic and meaningful characteristics of real-life events…" (p. 2) such a social problems, for example. In fact, case studies seem to be the preferred strategy when "how or "why" questions are being posed, when the investigator has little control over events, and when the focus is on a contemporary phenomenon within some real-life context (Yin, 2010).

A case study approach was selected for this study to gain an in-depth understanding of undergraduate student perspectives on social issues facing their communities. Willig (2008) asserts, case studies "…are not characterized by the methods used to collect and analyze data, but rather its focus on a particular unit of analysis: a case…" (p. 74). Stake (2005) argues that the topic of the case can be an individual, but not the means by which the individual engages in a particular practice. He writes, "A doctor may be a case. But his or her doctoring probably lacks the specificity, the boundedness to be a case…" (p. 444).

Presentation of the individual cases provides the reader the opportunity to gain understanding of the views, observations, and opinions of the individual pre-university student perspectives on climate change issues. Actual quotes from the student interviews were used in the case studies as an attempt to portray the participant as an individual entity.

Research Question

The exploratory questions that guided the study were:

1. What elements constitute pre-university student perspectives on climate change issues and environmental education?
2. What variables influence this perspective on climate change issues and environmental education?
3. What beliefs do these pre-university students hold which support or negate this perspective?

Methodology

The decision to conduct a qualitative study was influenced by the characteristics of qualitative design discussed by Janesick (2011). She describes research as being alive and active. It is a way of looking at the world and interpreting the world. This study focused on qualitative methods as means to understand the multiple complexities existing in the social world (Janesick, 2011).

Qualitative research involves passion for the work. The qualitative researcher is interactive in the sense used by John Dewey (1934) when writing about artists:

An "expression of the self in and through a medium, constituting the work of art, is itself a prolonged interaction issuing from self with objective conditions, a process in which both of them acquire form and order they did not first possess." (p. 65)

This immersion into the research process actively involves the qualitative researcher in the quest for gaining a deeper understanding of the social phenomena. A case study was selected for this study. Case studies involve an in-depth study of this bounded system and rely on a number of data collection materials. The cases used in this study were four pre-university students from three different public high schools in Chicago, Illinois. The study sought to gain understanding of the perspectives of pre-university students on climate change issues and environmental education.

Presentation of the individual cases provides the reader the opportunity to gain understanding of the views, observations, and opinions of the individual participant perspectives on climate change issues and environmental education. Direct quotes from the participants were used in the case studies as an attempt to portray the participant as an individual entity.

Data Collection

Interviews, researcher reflective journal, observations, researcher field notes, documents, artifacts, and transcripts were collected. At least two in-depth interviews were conducted with the participants. In an attempt to gather the rich, descriptive information required for qualitative research, semi-structured interviews with open-ended questions were used. The first step in the data collection process was conducting interviews with the study participants. The information qualitative researchers seek to gain is rich, thick descriptions of the participants in their social setting. Thus, open-ended questions were used to elicit the most complete and thorough responses from the participants.

The nature of qualitative research is flexible, as participants are being studied in their social setting. While variables in the social world cannot be controlled, the researcher can follow a format to help ensure items such as equipment are functioning.

In addition to interviews, participant observation was used to supplement the data collected in the interviews. Janesick (2011) alludes to observation as the immersion into the social setting which allows the researcher to begin to experience the experiences of the participants. The researcher observed each participant at least one time. Settings for observations included a home setting interaction with other peers in an informal context.

Observations also provided a check as to the credibility of the other data collected. Observations do require a series of planned steps. There are limitations as to the amount of information individuals reveal in the interview. Observations served as a means for verifying that the participants' actions match their words.

Document and artifact analysis, researcher field notes, and a researcher reflective journal also served as other sources of data for this case. The researcher gathered documents and artifacts from all participants as an attempt to further understand selected faculty member perspectives on promoting climate change awareness through environmental education. Field notes consisted of supporting interview and observation notes. Format for the field notes collected during interviews and observations followed suggestions provided by Janesick (2011). A researcher's reflective journal was also kept as another means of data collection. The reflective journal served as means for the researcher to express emotions, ideas, and reactions to the study. During the data analysis stage, the reflective journal provided another resource for identifying emerging themes and sub themes.

The Role of the Researcher

The researcher's role (full participant or observer), issues of entry, reciprocity, personal biography and ethics (i.e., informed consent, privacy, etc.) must be taken in account during the research process. In this case study, the role of the qualitative researcher was to adhere to the possibility of neutrality.

The role of the researcher in this case study follows that of a non-participant observer (Creswell, 2013). This means that the role as researcher and observer was clearly known by all the participants. The researcher did not interact as a participant in the development, delivery, or activities of this class. Researcher presence was kept as passive as possible, except when needed to actively pursue additional information from one of the participants. At times this policy of non-interference restricted observations and abilities to seek out more information which imposed another limitation of the research method, but it also preserved the natural setting for these observations.

Participant Selection

Four pre-university from three different public high schools in Chicago, Illinois were interviewed individually regarding their perspective climate change issues and environmental education. The four pre-university students were from various parts of Chicago. Following the recommendations of Janesick (2011), the researcher relied on collection and analysis of various forms of data. Each pre-university student was interviewed at least two times for a total of eight formal and informal interviews. The pre-university students were audiotaped for the formal interviews. Data was also collected from three observations, thirteen documents and artifacts, and nineteen researcher reflective journal entries.

The pre-university students were selected because of the representation of urban youth, their willingness to talk about their experiences, and their ability to provide different perspectives. Each participant was 16 years of age, came from diverse socio-economic backgrounds, and attended a public high school in Chicago.

Christina was the only male involved in this study. This was not intentional but due to the fact that the other pre-university students willing to participate in the study were female. Alicia was selected because of her interest in environmental education. Kayla and Claire were selected because of their willingness to participate in the study, their interest in photography as a means for promoting awareness of climate change issues, and their connection to the other pre-university students selected for the study. Each participant agreed to use their actual first names in the study. The official names of the high schools were changed to ensure confidentiality of the institution. Table 1 shows the basic profile of each participant.

Each participant discussed his or her perspectives on climate change issues and environmental education. A summary of the responses follows.

Christian: The Influencer

Christian was the only male participant in the study. He attends a Chicago public school on the northwest side of the city and has known the other participants for one year.

When asked: what is the biggest climate change issue facing society or the city of Chicago today, he responded:

Pollution, including noise and light pollution is one of the biggest climate change issues facing the city of Chicago because it's possible to reduce it but people still choose not to. This can be seen in the increase in the number of ozone action days during the summer and the horrible water conditions in the Chicago River.

Table 1. Case ordered matrix, participant characteristics

Pre-University Student	Gender	Age
Christian	Male	16 years
Alicia	Female	16 years
Claire	Female	16years
Kayla	Female	16 years

When asked if he received any environmental education in high school and if there should be more environmental education at the high school level, Christian responded:

No, we do not receive any formal environmental education at our high school. However, I believe that high school students should know about the impact of climate change on their community and the environment as a whole. High school students should also be educated about ways they can take action to improve their neighborhoods and city through projects where students can make a positive difference in the environment. Schools need to do more than just offer a recycling program and offer opportunity to learn about what we can do to improve the environment. High school students also need to learn about climate change and the impact it has on our daily lives and quality of living.

When asked if he thinks climate change is an issue facing society, Christian responded:

Yes, I do because in Chicago we could see all four seasons in one day. This affects activities, the way you dress, and your health. For example, students that have allergies, the increase in flu cases, and your mood. When we have months of really cold weather, like last winter and the polar vortex, you don't feel motivated to go anywhere and that can be depressing.

Christian was asked what high school students can do to promote climate change awareness in his community and responded:

We can influence others to recycle, use products that are more Environmentally friendly and learn the results of climate change on our daily lives and community.

Alicia: The Promoter

Alicia is an athlete that runs middle distance track. She runs outdoors in Chicago and has a deep concern about the quality of air she breathes as well as the need for more green space in the city. She attends a Chicago public school in the South Loop section of the city and has known the other participants for two years.

When asked: what is the biggest climate change issue facing society or the city of Chicago today, she responded:

I feel like the biggest climate change issue is the amount of pollution and trash being produced without any well-known efforts to stop it. This has created a problem with air quality and people's overall health. Look at the large number of people who has asthma for example. While there are other factors contributing to asthma, the amount of emissions from cars and other vehicles has greatly impacted the air we breathe.

When asked if she received any environmental education in high school and if there should be more environmental education at the high school level, Alicia responded:

I have received a very limited amount of environmental education at the high school level, but I know there is an environmental studies class offered that I plan on enrolling in. I feel like there should be more environmental education at the high school level.

When asked if she thinks climate change is an issue facing society, Alicia responded:

Yes, because it affects our everyday lives and hurts our environment So people living here in the future will have a lower quality of life.

Alicia was asked what high school students can do to promote climate change awareness in her community and responded:

We can try and convince schools we need more climate change education or we ourselves can just be more active in researching and learning and then spread our knowledge to our peers using outlets such as social media.

Claire: The Motivator

Claire enjoys photography, especially capturing images of nature. She enjoys spending time by the lake photographing water images and lake creatures. She attends a Chicago public school in the West Loop section of the city and has known the other participants for one year.

When asked: what is the biggest climate change issue facing society or the city of Chicago today, she responded:

Extremely cold winters and cooler summers are the biggest climate change issues facing the city of Chicago. Last winter was very long and produced record-breaking cold temperatures. New terms such as ice caves and polar vortex became common among the people of Chicago. The summers are also shorter with much cooler temperatures. I remember summers when we had several days in the 90's or 100's. Now we rarely have those really warm days. Also, the weather changes from summer to fall literally overnight.

When asked if she received any environmental education in high school and if there should be more environmental education at the high school level, Claire responded:

Yes, we've had conferences in the theater for all of my school's Biology and Advanced Placement Biology to learn more about the environment and environmental issues. This is great but in my opinion there needs to be more than just a presentation. There also needs to be more activities such as an environmental club or program where students can learn more.

When asked if he thinks climate change is an issue facing society, Claire responded:

Yes, because the earth is constantly changing and we will always have to face changes in the atmosphere and the world because of our continuous use and abuse of natural resources.

Claire was asked what high school students can do to promote climate change awareness in her community and responded:

I suppose the biggest component is awareness. Once people know about the impact of climate change, more people will gain knowledge and learn about what they can do to help the environment. This creates a flow of people wanting to help. Word-of-mouth and action by doing can promote climate change education in schools and in communities.

Kayla: The Volunteer

Like Claire, Kayla has an interest in photography, specifically urban setting and images in nature. She enjoys being outdoors, even if the setting is urban. She attends a high school in the West Loop and has known the other participants for two years.

When asked: what is the biggest climate change issue facing society or the city of Chicago today, she responded:

The biggest climate change issues facing society include flooding, drought, and ocean levels rising. While some parts of the world have seen more and more flooding, other parts have seen a reduced amount of rainfall, leading to drought. The polar ice caps are melting because of the rising temperature of the earth. This has caused sea levels to rise. Climate change is an issue for everyone.

When asked if she received any environmental education in high school and if there should be more environmental education at the high school level, Kayla responded:

No, I do not receive that much environmental education at my high school. I believe that should be more.

When asked if he thinks climate change is an issue facing society, Kayla responded:

Yes, because gradually the effects of climate change will dramatically alter the way we live and have a great affect on future generations.

Kayla was asked what high school students can do to promote climate change awareness in her community and responded:

High school students can be charged with learning more about climate change issues by volunteering with organizations or community activities. We have a day of "clean and green" at our local park. In addition to helping clean up the park, students can do their own research on ways each person can help reduce their carbon footprint.

Major Themes

The topics of awareness, research, and education were major themes that emerged from the study. All four high school participants expressed a need for each individual to be aware of the impact of climate change in their lives today and for future generations. Research was another theme the four students discussed. Although there are a limited number of formalized environmental educational programs at their high schools, the participants talked about students doing their own research to learn more. The need for more environmental education was a strong theme that emerged from the study. The participants talked about the lack of environmental education and discussed ways that it could be added into the curriculum. They discussed wanting to do more to help their communities and city.

SOLUTIONS AND RECOMMENDATIONS

Funding is always a challenge when trying to introduce new programs in the public-school setting. For example, despite the overwhelming positive evaluations by teachers of the value and efficacy of an environmental education program in Hawaii, lack of funding and access to equipment were reported as anticipated limitations to implementing their newly gained knowledge in the classroom (Rivera, Malia, Manning, Mackenzie, Krupp, and David, (2013). However, funding is not needed to integrate environmental education and climate change awareness into the high school curriculum. For example, in a science course, field-based research projects could be conducted. Chicago public schools also require that students complete 40 service learning hours prior to graduation. Teachers could help students select environmentally focused projects to complete their service learning hours. Partnerships can also be formed with local businesses and environmental groups to provide environmentally based projects for students as well as presentations on climate change issues and other topics related to the environment.

Based on a 2013 survey conducted by Inverness Associates, interviews, and school site visits, there are ten areas where all independent schools can take steps to become more environmentally sustainable:

- **Organization:** Make environmental education and sustainability a high priority; establish a green council; craft a green mission, goals, and a plan; and report progress on a regular basis.
- **Leadership:** Appoint a compensated sustainability coordinator with responsibilities to develop and direct the school's overall sustainability plan.
- **Resource Efficiency:** Benchmark use of electricity, oil/natural gas, water, and waste disposal, and make systematic plans to reduce usage and to document savings.
- **Facilities:** When renovating or constructing buildings, utilize best practices that conform to green standards.
- **Healthy Operations:** Adopt policies for green purchasing, hazardous waste and pest management, and school wellness.
- **Nutritious Food:** Evaluate and improve the school's food program to focus on good nutrition and health and local, seasonal offerings.
- **Curriculum:** Adopt a definition of environmental literacy, use it to audit the curriculum, and evaluate ways to incorporate environmental education across the academic program.
- **Extracurricular Program:** Enhance opportunities for students to learn about the environment outside the classroom and in nature, and incorporate informal environmental education into the overall school culture.
- **Students:** Include students in meaningful leadership roles in making the school more environmentally sustainable.
- **Partnerships:** Connect with the wider environmental education and sustainability community, provide enhanced professional development, and help create a robust network of green independent schools (Chapman, 2014, p.3).

A review of the literature reveals that students develop a stronger interest in learning when real-world projects are involved. An example can be found in an environmental reef project with Queensland High School (Stepath and Whitehouse, 2006). Following is a quote from a female student who participated in the project:

The good thing about learning here was that we did not have to sit inside. We actually got hands-on experience of the island. I learned a lot about the animals and reef monitoring, about going up and counting everything. I didn't really think with the sea cucumbers, there would be so many. It is a lot better out here. It gets boring in a classroom. I had a really good time (Stepath & Whitehouse, 2006, p. 1).

A high school in Ontario, Canada also found the experiential learning helped foster a sense of passion for environmental issues from the students (Caspell, 2007). Following is a quote from one of the participants in the Roots of Courage, Roots of Change (ROC) Integrated Curriculum Program (ICP):

It was as, to a very large degree, very close to a life-changing experience—partially due to the teachers, the classmates and the curriculum, but also due to the length [of the program]... I feel I came out a different person... [and] a lot of it had to do with how I thought. I think differently [about] my impact on the world, and how I interact with people (Caspell, 2007, p. 20)

The city of Chicago offers many opportunities to volunteer for environmental initiatives. Students can volunteer for the Chicago Conservation Corps. The Chicago Conservation Corps (C3) recruits, trains, and supports a network of volunteers who work together to improve the quality of life in our neighborhoods and schools through environmental service projects. If students have an interest in environmental issues and wish to make a difference in their communities, they can get involved in one of the following C3 programs:

1. Each spring and fall, adult residents of Chicago can apply to C3's Environmental Leadership Training program. In this training, individuals can learn about conservation issues in Chicago and techniques in community organizing in a series of classes, and then receive support from C3 staff to design and implement a conservation service project in their communities.
2. Chicago Public School teachers can apply to start an after-school conservation club for 8th-12th grade students. Students in these clubs assess conservation practices in their schools, focusing on land use, waste management, air quality, energy conservation, water conservation and water quality issues. Based on the results of their assessments, clubs then implement projects to meet the conservation needs identified.

3. C3 Explorer Projects are one-time environmental volunteer opportunities for Chicagoans who want to improve the quality of life in their neighborhoods, but lack the time to commit to C3s Environmental Leadership Training Program. Many of these projects are run in conjunction with C3's partner organizations. Partners include: Chicago's Environmental Fund, Alliance for the Greater Lakes, Department of Water Management, Chicago Public Schools, CNT: Sustainable Communities Attainable Results, Chicago Cares, Chicago Center for Green Technology, Friends of the Chicago River, LISC Chicago, SCA, Shedd Aquarium, Chicago Park District, Faith in Place, Calumet Stewardship Initiative, Green Corps Chicago, Openlands, and Friends of the Forest Preserve (Chicago Conversation Corps, 20140).

The program offers a program called, "The Explorer Track" provides students with the opportunity to get their feet wet, hands dirty and make a real impact on the environment. The Explorer track provides Chicago residents with hands-on volunteer opportunities that protect Chicago's unique natural resources in nature areas and in their own backyard. In addition to building knowledge and skills, the Explorer Track is an opportunity to tap into a network of environmental leaders that are working to create a movement of environmental action across Chicago and beyond.

Several short-term volunteer opportunities with partner organizations and respected C3 leaders are available year-round. By becoming a part of the Explorer Track, students could:

- **Protect Natural Resources Areas:** Take action to protect Chicago's natural resources through stewardship, beach conservation, river conservation, native species conservation and more.
- **Get Involved in Neighborhood Environmental Action:** Join a garden work day, storm water management action, community cleanup day, sustainable community development event, learn about alternative transportation and weatherization opportunities in your neighborhood.
- **Partner With C3 Leaders:** Tap into the grassroots network of C3 environmental leaders. Be a part of the next leader workshop in your neighborhood.
- **Volunteer for Special Events:** Be a part of a team promoting responsible waste management for Mindful Metropolis or Green Fest and your work will be rewarded with the chance to see documentaries, attend music events or eat delicious food (City of Chicago, 2014).

There are also a number of other organizations both nationally and locally that can assist high school students with increasing awareness of climate change issues and enhance their knowledge of environmental education. An example can be found in the Chicago Environmental Network. The purpose of the CEN is to inform the public about opportunities to get involved in environmental issues in the greater Chicago metropolitan area and to promote public participation in environmental activities. Member organizations offer a wide variety of opportunities that let you participate in conservation activities and educational programs. CEN was established with initial support from the Chicago Zoological Society and the Illinois Environmental Protection Agency (Chicago Environmental Network, 2014).

There are also thousands of nationally based environment volunteer organizations available fromVolunteerMatch.org (2014). A comprehensive list of internally based volunteer environment program can be found at GoAbroad.com (2014), the Global Volunteer Network, Conservation International Volunteer Program, GVI, EarthWatch Institute, and Projects Abroad.

SEATRUST INSTITUTE AND AWARE

SeaTrust Institute is a scientific research and educational nonprofit organization. The project teams are actively working on environmental science and policy issues both locally and globally. Through working at both levels at the same time, SeaTrust Institute is in a position to provide grassroots knowledge and information to the international decision arena, providing a voice for local issues and knowledge within international policy processes. The dual scope also allows SeaTrust Institute to engage forefront and timely global science and policy decisions in local projects, education and capacity building.

Through experience, the organization knows that the most effective environmental work is "Local to Global and Back Again℠." AWARE is framed around the Halcyon Adaptation Scenario℠ which integrates knowledge sets and processes through enabling students to interweave science with locally relevant knowledge, engage community stakeholders to develop a policy voice on environmental concerns. Students may work with the science, engineering, social or policy processes that make their community more resilient to environmental change; AWARE involves students with interests ranging from chemistry and engineering to politics and social improvement.

At the international level, climate change and health are key focal issues for the organization's work within the United Nations Framework Convention on Climate Change (UNFCCC). Local projects providing climate change and health adaptation assistance to regions with low adaptive capacity are a link to exchanging knowledge while bringing local voices into the larger conversation.

AWARE (Action Within a Resilient Environment) is a SeaTrust Institute youth project that engages high school and select middle school students in research, education and community action to increase climate change and environmental awareness through experiential learning activities. AWARE allows students to acquire public service credits required for high school graduation by many states and cities while contributing to their local communities - and it provides experiential environmental education that enhances applications to top universities. Projects are framed around the following questions:

- What impact has climate issues does your city/community face?
- What happens as things change?
- What does this mean for the health for the people in the community?
- Why is the work doing important?

Questions Guiding Youth Projects

Each project addresses some aspects of the following questions. It involves three steps: finding the information, discussing what you found and learned with other AWARE participants and global students facing similar issues, and sharing your new knowledge with some group in your own community. This is part of the SeaTrust Institute's charge to work "Local to Global: and Back Again" to help communities prepare for and better react to climate change that is affecting and will continue to affect everyone's lives. Students are provided with information collection templates and ideas as well as guidance and support throughout your project.

AWARE-Chicago River Project Ideas

- Compare Discharge Monitoring Report (DMR) data at wastewater treatment plants (WWTP) to that found at river.
    - Based off permits WWTP have to perform certain tests (BOD, fecal coliform, nitrates, phosphates, etc.) and report them. We will take their data collected in the past and future and compare it among WWTP that perform tertiary treatment (disinfection) to those who do not.
    - Students will sample upstream and downstream specific WWTP to determine the effects of WWTP influent into the river.
    - Collect fecal coliform data from WWTP and compare those who disinfect to those who do not. See how this is reflected in the environmental conditions of the outfalls.
- Green infrastructure to reduce the effects of storm water to the Chicago River. (WWTP has to discharge untreated sewage to the river because they have too

much influent, by using green infrastructure we can reduce the amount of storm water flow to treatment plants and reduce the discharge of untreated sewage)

  - **Rain Gardens:** Store water, can place in medians, brownfields, etc.
  - **Rain Barrels:** Store rainwater runoff from buildings and utilize for irrigation
  - **Permeable Pavement:** Pair with builders to utilize permeable pavement to decrease amount of storm water flow to Chicago River (will have to show them the cost effect of permeable pavement vs. impermeable pavement)
- **Brownfield Restoration:** In industrial areas or low income areas there is often abandoned property and what is present at the space is unknown. Students can locate brownfield sites, test for various contaminants, and turn space into a community garden or rain garden
- **Biotic Inventories:**
  - **Invasive vs. Native Plant Species:** Students can take a set cross section of an area and determine the amount of invasive species compared to native plants. Students can create an event to remove invasive species and compare data over years or can compare effects of invasive species over time
  - **Invasive Fish Species:** Set up an online inventory for amount of native fish species versus invasive fish species to determine how the competition between the two is affecting the diversity within the river (For example, Asian carp. Students can do the same with zebra mussels)

FUTURE RESEARCH DIRECTIONS

The study findings and review of the literature raise interest related to the need to promote climate change awareness among pre-university students. Most importantly, there is a need to increase environmental education programs at the pre-university level and to align with the SDGs. This is a trend that will continue to increase both on a national and international level as climate change issues will continue to impact our everyday lives. The research also shows that youth have a vested interest in improving the quality of their world both today and for future generations. Finally, additional research on using arts-based projects as a way for promoting environmental education and climate change awareness would be beneficial. The example below provides an overview of one such project conducted at an alternative high school in the U.S.:

The environmental education through filmmaking project, a case study at an alternative U.S. public high school, investigates environmental literacies of at-risk students who produced two short documentary films, one on recycling and one on water conservation. The filmmaking project sought to promote students awareness of environmental issues and increase their environmental literacy by exploring the conservation topics on multiple levels: carrying out research for their films, conducting interviews, and operating camera equipment (Harness and Drossman, 2011, p. 1).

There are several opportunities for further research based on themes that emerged from the research and review of the literature. Teachers, researchers, and pre-university administrators may wish to conduct studies from the teachers' perspective. Understanding the teachers' viewpoints on climate change issues and environmental education would assist in promoting awareness within their classrooms. Additional research on the perspectives of pre-university administrators would also call for further study. Specially, researching the ways pre-university administrators can partner with environmental organizations and non-governmental agencies to offer environmental education programs, training, and volunteer opportunities for students. Drawing opinion from international pre-university students would also offer a more diverse perspective on a global scale. This study was limited to three public schools in Chicago, Illinois and it would be beneficial to gain a global perspective.

FUTURE RESEARCH DIRECTIONS

Findings of this study revealed that there is a need to promote environmental education programs to pre-university students. This need for fostering environmental awareness aligns with SDG #Quality Education. In particular developing countries can utilizing the resources of more developed countries to create environmental education programs for pre-university students. Involving youth in sustainability development is critical to gaining awareness of the impact climate change has on local and national resources.

Youth are very concerned about climate change issues and want their voices to be heard. These concerns are very real and are seen in the dramatic climate changes in their own communities. Additionally, students saw the impact of climate change, especially extreme heat and cold in Chicago, on health issues experienced by their own friends and family. For example, each of the participants knew at least one person suffering from asthma.

Additionally, quality of education as a whole is still lacking, as outlined in the progress report by the Secretary-General listed below. These concerns about the overall access to quality education presents a challenge for promoting environmental educational program due to lack of resources. However, educating youth on the impact of environmental issues can assist in preparing their countries for sustainable development.

PROGRESS OF GOAL 4 IN 2017

Achieving inclusive and equitable quality education for all will require increasing efforts, especially in sub-Saharan Africa and Southern Asia and for vulnerable populations, including persons with disabilities, indigenous people, refugee children and poor children in rural areas.

- In 2014, about 2 in 3 children worldwide participated in pre-primary or primary education in the year prior to official entry age for primary school. However, in the least developed countries, the ratio was only 4 in 10.
- Despite considerable gains in education enrolment over the past 15 years, worldwide, the adjusted net enrolment rates were 91 per cent for primary education, 84 per cent for lower secondary education and 63 per cent for upper secondary education in 2014. About 263 million children and youth were out of school, including 61 million children of primary school age. Sub-Saharan Africa and Southern Asia account for over 70 per cent of the global out-ofschool population in primary and secondary education.
- Even though more children than ever are going to school, many do not acquire basic skills in reading and mathematics. Recent learning assessment studies show that in 9 of 24 sub-Saharan African countries and 6 of 15 Latin American countries with data, fewer than half of the students at the end of primary education had attained minimum proficiency levels in mathematics. In 6 of 24 sub-Saharan African countries with data, fewer than half of the students who finished their primary schooling had attained minimum proficiency levels in reading.
- Equity issues constitute a major challenge in education according to a recent assessment. In all countries with data, children from the richest 20 per cent of households achieved greater proficiency in reading at the end of their primary and lower secondary education than children from the poorest 20 per cent of households. In most countries with data, urban children scored higher in reading than rural children.
- The lack of trained teachers and the poor condition of schools in many parts of the world are jeopardizing prospects for quality education for all. Sub

-Saharan Africa has a relatively low percentage of trained teachers in pre -primary, primary and secondary education (44 per cent, 74 per cent and 55 per cent, respectively). Moreover, the majority of schools in the region do not have access to electricity or potable water.

- On the basis of data from 65 developing countries, the average percentage of schools with access to computers and the Internet for teaching purposes is above 60 per cent in both primary and secondary education. However, the share is less than 40 per cent in more than half of sub-Saharan countries with data.
- Official development assistance (ODA) for scholarships amounted to $1 billion in 2015, a decrease from $1.2 billion in 2014. Australia, France and the United Kingdom of Great Britain and Northern Ireland were the largest contributors. Source: Report of the Secretary-General, "Progress towards the Sustainable Development Goals", E/2017/66

Youth-based environmental education programs can be an agent for change. Future research needs to be conducted on ways environmental education programs can be affordable and accessible to all individuals. Research findings related to the ways youth participants use their knowledge to benefit the community is needed. In addition to completing an AWARE project, following up with having students complete a service learning component would be a benefit to the students and the community. Identification of issues is the first step; creating an action plan for improvement would lead to positive change.

EPILOGUE AND LESSONS LEARNED

In summary, the findings of this study and review of the literature conclude that pre-university students are concerned about climate change issues and wish to promote awareness to their peers and community. There is also a need for more formalized environmental education offered in the Chicago public school system at the pre-university level. Although funding is limited and curriculum changes difficult, schools can partner with existing local or international organizations to offer students resources on climate change, environmental education, and volunteer opportunities. Students want to make a difference by taken action within a resilient environment.

The findings of the research questions presented here, if conducted, may help to further promote awareness related to climate change issues on a national and international level, provide pre-university students with resources on volunteer organizations to take action, and increase the level of environmental education offered to students at the pre-university level. As the earth is warming and climate change is more prevalent, youth need to be resilient to an ever-changing environment. Education and action are needed to foster an understanding of ways they can assist their communities to adapting to environment changes.

REFERENCES

Chapman, P. (2014). Environmental Education and Sustainability. *Independent School*, *73*(4), 20.

Chicago Conservation Corps. (2014). Partner organizations. Retrieved from http://chicagoconservationcorps.org/blog/about-this-weblog/c3-partner-organizations/

Chicago Environmental Network. (2014). Contact CEN. Retrieved from http://www.chicagoenvironment.org/contact/index.cfm

City of Chicago. (2014). City service. Retrieved from http://www.cityofchicago.org/city/en/depts/cdot/provdrs/conservation_outreachgreenprograms/svcs/volunteer_for_thechicagoconservationcorps.html

City of Chicago. (2014). Environment: Programs and initiatives. Retrieved from http://www.cityofchicago.org/city/en/progs/env.html

Conservation International Volunteer Program. (2014). Destinations. Retrieved from http://www.conservationvip.org/destinations/

Creswell, J. W. (2013). *Research Design: Qualitative, Quantitative, and Mixed Methods Approaches*. Thousand Oaks, CA: Sage.

Dewey, J. (1934). *Art as Experience*. New York: Minton, Malach, and Co.

Earthwatch Institute. (2014). Expeditions. Retrieved from http://earthwatch.org/expeditions

Global Volunteer Network. (2014). Most popular programs. Retrieved from http://www.globalvolunteernetwork.org/programs/index.php?type=Environmental

GoAbroad.com. (2014). Volunteer abroad in environment. Retrieved from http://www.goabroad.com/volunteer-abroad/search/environment/volunteer-abroad-1

Grayson, J. (2011). Charting a green course. *T.H.E. Journal*, *38*(4), 26.

GVI. (2014). GVI youth programs. Retrieved from: http://www.gviusa.com/find-a-program/search-results/

Harness, H., & Drossman, H. (2011). The environmental education through filmmaking project. *Environmental Education Research, 17*(6), 829–849. doi:10.1080/13504622.2011.618626

High School Students Go to the Reef. (2006). Online Submission. *Paper presented at the New Zealand Association for Environmental Education (NZAEE) Conference*, Auckland, New Zealand, Jan 22-25.

Janesick, V. J. (2011). *Stretching Exercises for Qualitative Researchers* (3rd ed.). Thousand Oaks: Sage.

Priyanto, Y., Fanani, Z., & Sasmitojati, S. (2013). Environmental awareness as an education paradigm for sustainable development in vocational high schools of Kediri East Java. *International Journal of Academic Research, 5*(5), 7–12. doi:10.7813/2075-4124.2013/5-5/B.1

Projects Abroad. (2014). Teen volunteer abroad: High School specials. Retrieved from: http://www.projects-abroad.org/volunteer-projects/high-school-specials/

Rivera, M.A.J., Manning, M.M. & Krupp, D.A. (2013). A Unique Marine and Environmental Science Program for High School Teachers in Hawai'i: Professional Development, Teacher Confidence, and Lessons Learned. *International Journal of Environmental and Science Education, 8*(2), 217-239.

SeaTrust Institute. (2014). AWARE. Retrieved from http://www.seatrustinstitute.org/AWARE.html

Stake, R. E. (2010). *Qualitative Research: Studying How Things Work*. New York, NY: Guilford. Press.

Stepath, C., & Whitehouse, H. (2006). *Encouraging Proximal Relations*. Queensland.

United Nations (2017). Report of the Secretary-General, Progress towards the Sustainable Development Goals, E/2017/66.

United Nations Division for Sustainable Development Goals. (2018). Sustainable development goals. Retrieved from https://sustainabledevelopment.un.org/sdgs

United States Environmental Protection Agency. (2014). Adaptation overview. Retrieved from http://epa.gov/climatechange/impacts-adaptation/adapt-overview.html

VolunteerMatch.org. (2014). Opportunities. Retrieved from https://www.volunteermatch.org/search?l=USA&k=Environment

Willig, J. (Ed.). (2008). Qualitative Research Methods in Education and Educational Technology. Charlotte: Information Age Publishing.

Yin, R. K. (2010). *Qualitative Research from Start to Finish*. New York, NY: Guilford Press.

ADDITIONAL READING

Alewmaw, F. Berhanu and Nako M. Sebusang (2017). Climate change and adaptation-induced engineering design and innovations in water development projects in Africa. *African Journal of Science, Technology, Innovation and Development*, 9(5-6), 1-13, Retrieved from http://www.tandfonline.com/doi/abs/10.1080/20421338.2017.1355601

Allchin, D. (2015, April). Global Warming: Scam, Fraud, or Hoax? [abstract]. *The American Biology Teacher*, *77*(4), 309–313. Retrieved from http://abt.ucpress.edu/content/77/4/39. doi:10.1525/abt.2015.77.4.12

Brügger, A., Thomas, A. M., & Suraje, D. (2016). 'Proximising' climate change reconsidered: A construal level theory perspective. *Journal of Environmental Psychology*, *46*, 125-142. Retrieved from: http://www.sciencedirect.com/science/article/pii/S0272494416300238

Clarke, P. (2012). *Education for Sustainability: Becoming Naturally Smart*. London: Routledge.

Clover, D. E. de O. Jayme, B., Hall, B.L., & Follen, S. (2013). The Nature of Transformation: Environmental Adult Education. Rotterdam: Sense Publishers.

Dickinson, J. L., & Bonney, R. (Eds.). (2012). *Citizen Science: Public Participation in Environmental Research*. Ithaca, New York: Cornell University Press. doi:10.7591/cornell/9780801449116.001.0001

Elder, J. L. (2003). *A Field Guide to Environmental Literacy: Making Strategic Investments in Environmental Education*. Environmental Education Coalition.

Falk, J. H., Heimlich, J. E., & Foutz, S. (Eds.). (2009). *Free-Choice Learning and the Environment*. Lanham, MD: AltaMira Press.

Goleman, D., Bennett, L., & Barlow, Z. (2012). *Ecoliterate: How Educators are Cultivating Emotional, Social, and Ecological Intelligence*. San Francisco, CA: Jossey-Bass.

Gruenewald, D. A., & Smith, G. A. (2008). *Place-Based Education in the Global Age: Local Diversity*. New York: Lawrence Erlbaum Associates.

Hayward, B. M. (2012). *Children, Citizenship and Environment: Nurturing a Democratic Imagination in a Changing World*. London, UK: Routledge. doi:10.4324/9780203106839

Hoot, J. L., & Szente, J. (Eds.). (2010). *The Earth is Our home: Children Caring for the Environment*. Olney, MD: Association for Childhood Education International.

Hungerford, H. R., Bluhm, W. J., Volk, T. L., & Ramsay, K. (Eds.). (2001). *Essential Readings in Environmental Education* (2nd ed.). Champaign, Illinois: Stipes Publishing.

Johnson, E. A., & Mappin, M. J. (Eds.). (2005). *Environmental Education and Advocacy: Changing Perspectives of Ecology and Education*. Cambridge, UK: Cambridge University Press.

Leou, M. J. (2005). *Readings in Environmental Education: An Urban Model*. Dubuque, Iowa: Kendall/Hunt Publishing Company.

Louv, R. (2011). The Nature Principle: Human Restoration and the End of Nature-Deficit Disorder. North Carolina: Chapel Hill.

Meier, D. R., & Sisk-Hilton, S. (Eds.). (2013). *Nature Education with Young Children: Integrating Inquiry and Practice*. New York: Routledge.

NAAEE. (2017). *Guidelines for Excellence: Professional Development of Environmental Educators*. Washington, DC: North American Association for Environmental Education.

Patrick, P. G., & Dale Tunnicliffe, S. (2013). *Zoo Talk*. Dordrecht: Springer. doi:10.1007/978-94-007-4863-7

Rickinson, M., Lundholm, C., & Hopwood, N. (2009). *Environmental Learning: Insights from Research into the Student Experience*. Dordrecht: Springer.

Saylan, C., & Blumstein, D. T. (2011). *The Failure of Environmental Education (and how we can fix it)*. Berkeley, CA: University of California Press.

Schlottmann, C. (2012). *Conceptual Challenges for Environmental Education: Advocacy, Autonomy, Implicit Education and Values*. New York: Peter Lang.

Smith, G. A., & Sobel, D. (2010). *Place- and Community-Based Education in Schools*. New York: Routledge.

Yazan, B. (2015). Three Approaches to Case Study Methods in Education: Yin, Merriam, and Stake. *Qualitative Report*, *20*(2), 134–152. Retrieved from http://nsuworks.nova.edu/tqr/vol20/iss2/12/

KEY TERMS AND DEFINITIONS

Adaptation: Adjustment or preparation of natural or human systems to a new or changing environment which moderates harm or exploits beneficial opportunities.

Adaptive Capacity: The ability of a system to adjust to climate change (including climate variability and extremes) to moderate potential damages, to take advantage of opportunities, or to cope with the consequences.

Alternative Energy: Energy derived from nontraditional sources (e.g., compressed natural gas, solar, hydroelectric, wind).

Climate Change: Climate change refers to any significant change in the measures of climate lasting for an extended period of time. In other words, climate change includes major changes in temperature, precipitation, or wind patterns, among others, that occur over several decades or longer.

Ecosystem: Any natural unit or entity including living and non-living parts that interact to produce a stable system through cyclic exchange of materials.

Emissions: The release of a substance (usually a gas when referring to the subject of climate change) into the atmosphere.

Energy Efficiency: Using less energy to provide the same service.

Fossil Fuel: A general term for organic materials formed from decayed plants and animals that have been converted to crude oil, coal, natural gas, or heavy oils by exposure to heat and pressure in the earth's crust over hundreds of millions of years.

Global Average Temperature: An estimate of Earth's mean surface air temperature averaged over the entire planet.

Global Warming: The recent and ongoing global average increase in temperature near the Earth's surface.

Greenhouse Effect: Trapping and build-up of heat in the atmosphere near the Earth's surface.

Landfill: Land waste disposal site in which waste is generally spread in thin layers, compacted, and covered with a fresh layer of soil each day.

Ozone Layer: The layer of ozone above Earth that shields it from harmful ultraviolet radiation from the sun.

Recycling: Collecting and reprocessing a resource so it can be used again. An example is collecting aluminum cans, melting them down, and using the aluminum to make new cans or other aluminum products.

Renewable Energy: Energy resources that are naturally replenishing such as biomass, hydro, geothermal, solar, wind, ocean thermal, wave action, and tidal action.

Resilience: A capability to anticipate, prepare for, respond to, and recover from significant multi-hazard threats with minimum damage to social well-being, the economy, and the environment.

Sensitivity: The degree to which a system is affected, either adversely or beneficially, by climate variability or change.

Sustainable Development Goals (SDGs): Otherwise known as the Global Goals, the SDGs are a universal call to action to end poverty, protect the planet and ensure that all people enjoy peace and prosperity.

Vulnerability: The degree to which a system is susceptible to, or unable to cope with, adverse effects of climate change, including climate variability and extremes.

Chapter 7
When Things Fall Apart:
Global Weirding, Postnormal Times, and Complexity Limits

Christopher Burr Jones
Walden University, USA

ABSTRACT

The chapter addresses the challenges facing first responders and public administrators due to accelerated warming, global weirding, and the limits to complexity. Similarly, these same challenges are also likely to have an impact on the ability of governments, international organizations, and non-governmental organizations to implement and realize the sustainable development goals and their 169 targets. The chapter focuses on the state of critical infrastructure, primarily in the USA, and the maintenance and sustainability of the physical systems of energy distribution, transportation, communication, and other basic services that support economic development and social systems. The chapter posits the need to explore these themes through the lens of futures studies and the need to envision and create preferred futures.

INTRODUCTION

This chapter was initially conceived as a provocation, a thought experiment, for public administrators to consider the benefits and hazards of considering worst-case scenarios for the future. The original presentation, given at the 2018 annual meeting of the American Society for Public Administration (ASPA), addressed preferred, sustainable futures and the implementation of the Sustainable Development Goals (SDGs) as strategies to help avoid a major societal collapse or build bridges to

DOI: 10.4018/978-1-5225-7727-0.ch007

post-industrial civilization. The researcher's rationale for bringing these issues to ASPA was that, given the times that we live in, public administrators are arguably already among the most likely to deal with the consequences civilizational decline and collapse, with increased attention on the evidence of such decline, such as terrorism and mass shootings, extreme weather events, and increasing chaos in society generally. The questions that the researcher raised for public administrators are no less relevant for broader consideration of the implementation of the SDGs: those people who will implement them are political leaders, public administrators, safety and security agencies, international organizations, nongovernmental organizations, and civil society actors.

BACKGROUND

Broadly, this chapter explores the tension between the forces of positive evolution of our species and planet and the entropic forces of chaos and uncertainty. It draws on the work of futures studies and assessments of the state of play in the building, maintenance, and stability of physical infrastructure. Infrastructure is a key indicator of social commitment to economic and social development in the medium-term future, so it has emerged as a concern in the research literature (American Society of Civil Engineers [ASCE], 2017; Kemp 2017; Marcuson, 2008; Zimmerman, 2009) and it's resilience in the face of climate change (Katz, 2017; Rapaport et al., 2015; Repetto & Easton, 2010). With some few exceptions, futurists have been reluctant to consider the consequences of broader societal collapse (Gidley, 2017; Slaughter, 2004, 2010). While there may be resistance to take a doom and gloom view, it may be time to consider some of the broader consequences of Decline and Collapse futures, if as some have argued, we have passed a tipping point in the Earth's carrying capacity (Caton, 1982; Kolbert, 2014). But I argue that threats to civilization need to be considered in a broader context, not as an acceptance of doom and gloom, but as part of a transition to a desirable, sustainable future. The challenge may be to envision and realize wise, ethical, and good futures particularly in the face of pessimism about growing environmental degradation (Lombardo, 2017).

One central driving force in global weirding is the accelerating warming of the Earth's atmosphere, which may continue to rise until it reaches a new state of dynamic thermal equilibrium, as suggested in Lovelock's (1995, 2009a, 2009b) Gaia theory. The approach of this paper is informed by the futures studies tradition, particularly the alternative futures typologies, i.e., the Four Futures, of Dator (2009a), and shares many of the other assumptions of academic future studies (Bell, 1997; Dator, 1995; Gidley, 2017).

This chapter will also explore the emergence of post-normal conditions—those global trends or dynamics that behave contrary to expectations, and are characterized by complexity, uncertainty, chaos, and contradictions (Sardar, 2017). Global forces of change including technology innovation, climate change, global economics, demography, and social movements drive the growing levels of complexity, chaos, and contradiction in social and economic systems. All of these forces make informed decision-making more challenging, but more urgent. We are now living in a global village, with blurring boundaries, and so much at stake collectively, our decision-making in terms of implementing the SDGs and operationalizing the 169 targets must consider the post-normal times we are in as well as the potential for catastrophe.

MAIN FOCUS OF THE CHAPTER

This chapter explores secondary research on the plausibility and possibility of widespread societal decline and collapse, due to accelerating climate change and *peak complexity*. The 2018 ASPA panel titled Accelerated Weirding: Policy and Administration in Hothouse Futures was informed and inspired by the idea of global weirding first coined by Lovins (Friedman, 2007). In this formulation, global climate change is not solely about heating, although that is a big part of it, but also about weather extremes and other concomitant natural events. So, for example, the double hurricane impact on Puerto Rico resulted in massive electrical system collapse. Abnormal heat waves were measured in the Arctic as this chapter was written. Freak weather increasingly becomes part of the normal background of experience. The physical changes in the world are one outcome of unintentional human experimentation with adding greenhouse gases to the Earth's atmosphere, but we may also need to consider the unintentional experimentation with human social and economic systems.

Peak complexity is an idea suggested by Tainter (1988) that many past civilizations have collapsed because they reach a point of saturation, an inflection point, where social structures are no longer able to achieve marginal efficiency in the functions of social, economic, and political systems. If the inefficiencies in those systems grow and are not addressed, they have a cascading effect, a runaway train, where the problems multiply and social and political systems ultimately collapse. Thus, this chapter focuses both on the impacts of global weirding and peak complexity on infrastructure primarily in the United States. All of these changes are likely to have direct and indirect implications for the implementation and adoption of the SDGs over time.

Issues, Controversies, Problems

Dator's (2009a) four generic future typologies are: continuation, collapse, discipline, transformation, which he initially based on images of the future in popular literature and media, have been used extensively over three decades in strategic planning and futures research. Extensive literature elsewhere has analyzed the practice and experience of using these alternative futures, but it is important to note that they are not conceived as necessarily discrete, but rather that one alternative future could shift or overlap with another. Relevant to this chapter is the tension between two of the typologies: collapse futures on one hand and technological transformation futures, on the other.

Collapse futures have a long tradition in dystopian literature that includes the work of Malthus. The *Population Bomb* and other alarmist literature emerged in the 1970s and 1980s, and the work of the Club of Rome and the Massachusetts Institute of technology on the *Limits to Growth* (Meadows, Randers, & Meadows, 2004) underscored the potential of pollution and resource depletion to threaten the existence of a highly technological civilization. Science fiction and Hollywood have taken post-apocalyptic visions to extremes, there being no small number of late 20th century movies such as *Godzilla*, *Terminator*, *The Day After Tomorrow*, and the *Mad Max* movie series to give nightmares to generations of children growing up. A classic futures studies text in the Collapse genre, Vacca's (1976) *The Coming Dark Age,* makes a strong case for a civilizational path that follows Murphy's Law: whatever can go wrong, will go wrong.

A large and growing literature posits that climate change and industrial growth threaten, not only humans, but the biosphere as a whole (Ehrlich & Ehrlich, 2013; Feffer, 2014; Kunstler, 2005; Wallace-Wells, 2017), although there is ample opposition to doom and gloom views (see Mooney, 2017). Catastrophe studies have become a small cottage industry. A growing list of potential sources of existential risk includes:

- Anthropogenic
    - Artificial intelligence
    - Biotechnology
    - Cyberattack
    - Global warming
    - Earthquakes due to fracking
    - Environmental disaster
    - Mineral resource exhaustion
    - Experimental technology accident
    - Nanotechnology

    - Subsidence from oil/water extraction
    - Warfare and mass destruction
    - World population and agricultural crisis
- Non-Anthropogenic
    - Asteroid/comet impact
    - Extraterrestrial invasion
    - Natural climate change (ice age; ocean circulation stops)
    - Cosmic threats (dust, supernova, black holes)
    - Geomagnetic reversal
    - Global pandemic
    - Megatsunami
    - Solar irregularity/dimming
    - Super volcano

One might ask, what good have these frightening and negative images of the future done for us? Aside from entertainment, are they not impediments to aspiring to better, preferred futures?

Some futurists argue that such bleak images of the future not only desensitize and numb us, but they actually result in angst and depression (Gidley, 2017; Slaughter, 2004, 2010). Similar arguments have been made about some of the classic studies on civilizational collapse, such as Diamond's (2005) *Collapse*. A strong counterargument can be made that we are better off preparing for the worst, even if it never comes, particularly when it comes to life-sustaining functions of the biosphere, such as the atmosphere. If we can learn lessons about how other civilizations have met their demise, wouldn't it be wise for us to be aware of what the potential warning signs might be?

Global civilization faces a double-threat of peak complexity, a kind of inertia dragging down political and economic systems, as well as rapid global heating. These threats collectively create a kind of global weirding that goes beyond the original meaning of stranger and wilder weather events, and potentially includes limits to civilizational and societal complexity (Tainter, 1988) and then on top of that, forecasts of 6° C or more of global heating within the next century (Lovelock, 2015; Lynas, 2008). Acknowledged as the seminal researcher on the various reasons for the collapse of complex societies and civilizations (cited in Diamond, 2005, p. 420), Tainter (1988) posited four concepts that explain the limits to complexity. According to Tainter, these concepts are:

1. Human societies are problem-solving organizations
2. Social political systems require energy for their maintenance

3. Increased complexity carries with it increased costs per capita
4. Investment in social political complexity as a problem-solving response often reaches a point of declining marginal returns (1988, p. 194)

Tainter notes that the first three are the underpinnings of the fourth. It could be argued, given much of the evidence from "tribal" politics, leadership dysfunction, and the US withdrawal from the world since 2016, that we have reached that point of declining marginal returns across a range of sectors and systems in modern society. To be sure, technology and innovation have been the answer to many of these challenges, but at the same time have created their own vulnerabilities and complexities. And the author assumes that there is a connection between infrastructure and the evidence that such systems have reached a point of declining marginal returns, or as seems to be the case, are in actual decline. It must be acknowledged that this notion of limits to complexity is challenged by popular ideas about evolution, where systems tend to respond to entropy by transformation into more complex forms. Technology innovation, such as super machine intelligence, may serve to overcome or transcend the limits that Tainter identified in previous societal collapse. Yet, human civilization arguably is now more chaotic, complex, and contradictory than at previous points in human evolution.

The other part of global weirding is based on some of the worst forecasts for the consequences of carbon dioxide and greenhouse gases in the atmosphere, which may push global temperatures past the point of civilizational sustainability. One of the more compelling narratives is by Lynas (2008) who analyzes the potential consequences of global warming, by degree, by using analogies to previous periods in the earth's geological history. Because Lynas adopts a very straightforward yardstick for the potential consequences of accelerated warming, this is a recommended text for public administrators to get a sense of the scope of the potential changes ahead. Extending the analysis of Lovelock (2006, 2009, 2015) who argued that we already are past the tipping point towards driving the planetary system to a new thermodynamic steady-state, similar to the temperature range in the Carboniferous era, Lynas argued that even at 3 °C above the baseline, the US (and by extension, many other parts of the world) will experience severe drought, dust bowls, and extreme heat events. The UN Paris climate accords had aimed to cap greenhouse gases that would allow no more than 2 °C rise in temperature, but CO_2 projections for the near-term will place us at 3 °C at a minimum. Given the potential for even greater temperature shifts, communities, leaders, and institutions need to prepare for the worst. And these additional stressors and challenges must be reconciled in planning for and implementing the goals and their 169 targets.

The final nail in the civilization coffin may be the inability of governments to adequately respond to the global warming challenge. Dator (2009b) argued that the coming problems of peak oil, environmental catastrophe, and global economic and fiscal collapse are all bad enough, but that government intervention has been, and will be, insufficiently focused on the longer term to respond in time. To raise consciousness about the need to prepare for global weirding, so that we can work our way through it, hopefully not just to survive, but to flourish as a species.

The purpose of this chapter is to consider the implications of global weirding and constraints on complexity on the SDGs and critical infrastructure to help better inform public administrators and first responders. For example, how will the likelihood of more frequent catastrophic weather events, drought, wildfires, dust storms, and other consequences of thermal stress and the implications of the limitations on complexity have for implementation of the SDGs? This is an attempt to raise consciousness about the interconnected forces weaving together environmental challenges and human society. A great deal of our resilience in the face of such catastrophes will depend on our ability to prepare and plan for the worst and to imagine and envision desirable futures. Communities arguably need to engage in "what if" exercises to explore the topography of futures where remediation may be too late, and adaptation the only solution to accelerated global heating. Because our collective ability to move social justice and climate change initiatives forward depends on the external environment, we need to consider that milieu as we prioritize and operationalize the SDG targets.

The theoretical framework for this study arises from both the traditions of hermeneutics and post-structuralism, particularly critical theory, in deconstructing structures of power within contemporary society. These theories form the bedrock of much of the work of the last few decades on critical futures, integral futures, and causal layered analysis. This research also adopts the assumptions of academic futures studies including the ideas: that a single future cannot be predicted, that there are multiple possible futures, that the images in people's heads have a role in determining which futures become the present, and that the future is not predetermined (Dator, 1995; Gidley, 2017).

The academic discipline of futures studies posits that there are probable, possible and preferred futures, and that trends are not destiny (Bell, 1997; Gidley, 2017). However, the Gaia theory argues that solar evolution over billions of years is destiny, our understanding of the physical evolution of our sun is that it has increased its intensity over billions of years, and will continue to increase its radiation output into the far distant future. In spite of solar evolution and output that is now 30% more than when life began, the Earth's biosphere has managed to maintain the optimal temperature range for the existence of life. Over the last 10 million years or so, that thermal balance has been maintained during cycles of glaciation that have stabilized the earth's temperature between cold and warm cycles entrained to orbital mechanics, the Milankovitch (orbital forcing) effect. Unfortunately, the key regulatory system, the CO_2 geological cycle and the addition of other greenhouse gases will likely disrupt that homeostasis. Lovelock argued that the most likely planetary system response will be to find a higher state of thermodynamic equilibrium, possibly the global temperatures that were common during the Carboniferous era, 300+ million years ago.

The average planetary temperature is currently 16 °C; it was 20 °C during the Carboniferous. One challenge is that we are pushing CO_2 levels in the atmosphere even higher than they were during the Carboniferous, so Lynas (2008) speculates that even 6 °C is possible–painting a gloomy picture of the consequences of much more than 3-5 °C increase in average global temperature. And one of the major problems is the lag time, the nature of feedback loops that means that once the changes begin, they can accelerate in a vicious negative feedback loop. The classic example in global weirding is the potential release of methane in the Arctic, melting methane solids in the ocean, and greater rates of decomposition in a warming world. All of those release potent greenhouse gases that could potentially pose existential threats if accelerating warming becomes "a runaway train."

The post-normal theoretical framework holds that our prospective futures are likely to be characterized by: complexity, chaos, and contradiction (Sardar & Sweeney, 2015; Sardar, 2017). This framework is congruent with a critical theory standpoint, acknowledging the complexity of a highly technological global civilization, and making problematic the structures of power that extract resources for the rich, plunder the earth, and fail to share these riches equitably. The critical theory tradition appears to have had an influence in post-colonial social movements, intentional communities, slow growth, no growth, and regrowth discourses. The systems view of the Gaia theory, and the challenge of maintaining marginal returns both align with the suggestion from the post-normal times that we have to understand phenomenon as complex objects within a complex web of relationships.

Another key concept in addressing apocalyptic images of the future is also the suggestion of the work of Polak (1973) that images of the future of individuals have a collective impact on the direction that societies take. Societies with apocalyptic eschatology tend to be less successful, and those with more open-ended or optimistic End Times had greater longevity. There is the danger of negative wishful thinking, to acknowledge an obvious danger. How might this be reflected in the state of the built environment and physical infrastructure that supports basic functions of transportation, communication, and the provision of basic goods and services?

It is widely acknowledged that the US has aging industrial infrastructure that is in need of repair or replacement. In other words, even before we consider the threats of accelerating warming, we already face significant challenges providing adequate infrastructure for the coming generations. It has been estimated that simply to upgrade Highway, seaport, and airport basic infrastructure will cost in the neighborhood of $6 billion. Some parts of the basic infrastructure, such as water and plumbing in older urban areas are in a state of crisis or collapse, such as in Detroit. Demographic changes, such as flight to the suburbs have exacerbated funding and scale challenges. Neglect and weather impacts have also taken their toll on infrastructure. Other structural shifts have reduced the power of unions, corporate and financial power have grown, and governments face growing opposition from conservatives and a limited government ideology that has altered spending priorities.

What are the major aspects of infrastructure that public managers and administrators should consider, when planning for crisis events? Those include various parts of the transportation system, planes, trains, and automobiles and their platforms-- airports, rails, highways and bridges. Basic infrastructure also obviously includes energy distribution, water, sewerage, solid waste disposal, and communications networks. It helps to see each of those systems within the context of levels of analysis — how they serve individuals, families, neighborhoods, communities, states, countries, and regions. Ultimately, understanding how those systems can be optimally nested to protect resilience and sustainability could help offset many of the challenges ahead in adapting to climate change and civilization in decline. Will we continue to grow community, counter trends that appear to fragment social institutions and build greater complexity into local structures, or revert to tribalism? That question may only be answered over the course of time, over the next century or so.

How can public administrators and political leaders prepare for 6° C of change, civilizational stress, and post-normal conditions driven by complexity, chaos, and contradiction? The threats to infrastructure have been fairly clearly identified, the types of events, and driving forces behind the threats. Hurricanes, as we have learned again in the past hurricane season, can produce extreme flooding, storm surge, and wind damage. Three major storms to hit the United States was a typical this last season, but it is argued they will be more common and severe in the future. The frequency of extreme events becomes more likely. Serious, and widespread drought would be expected above 3° C, accompanied by dust storms, worse than the dust bowl of the 1930s in the Plains. The consequences for agriculture are likely to be serious. Extreme heat events will likely become more common, and as the overall heat index continues to climb, there will be adverse effects for workers, already in evidence in Texas and Florida. Maintaining healthy temperature environments indoors will also be challenging, and greater use of air conditioning will only add to energy consumption and CO_2 in energy production.

The ASCE's (2017) report on the state of US infrastructure was a gloomy report. Virtually all of the 26 elements were worse than the previous 2013 report. Additional literature documents the specific threats faced by elements of the infrastructure menu. Major studies have addressed some specific geographic and political regions, for example Alaska (Larsen et al., 2008), South Africa (Chinowsky et al., 2012); threats to specific demographics and communities (Rapaport, Manuel, Krawchenko, &

Keefe, 2015), and community/agency resilience (NRC, 2012) in the face of disaster and climate change related events.

Aviation received a D (ASCE, 2017) among the worst scores given out. Recent research has addressed the consequences of sea level rise, weather issues, runway damage, and the likely need for lighter loads as temperatures rise (Burbidge, 2017; Ives, 2017; Larsen, 2015). Bridges received a C+ (ASCE, 2017). Larsen et al. (2008) addressed the particular needs of Alaska, for example, with bridges challenged not only by sea level rise and coastal erosion, but also a decline in permafrost. Permafrost soil failure also impacts pipeline footings as well. Dams earned a D (ASCE, 2017) and received special attention, along with levees (earning a D) in the National Research Council (2012) and ASCE reports. Potable drinking Water received a D from ASCE (2017) and the continuing problems of lead contamination in Detroit and other Rust Belt cities underscore the challenges of urban replacement of aging infrastructure, compounded by the challenges to freshwater due to drought, pollution, and declining sources of aquifer water (Vorosmarty, Green, Salisbury, & Lammers 2000). Energy infrastructure was graded D+ (ASCE, 2017). Burke (2015) recently addressed challenges to the aging energy grid, but noted the emergence of smart grid technology that could help transform that sector. Hazardous waste also earned a D+ (ASCE, 2017) and Inland Waterways a D (ASCE, 2017).

Parks and Recreation earned a D+ (ASCE, 2017) and national parks are already facing damage and challenges directly related to changing climate (source). Ports scored a relatively good C+ (ASCE, 2017) but are among the facilities most at risk from sea level rise and hurricanes and possibly terrorist attacks. Rail improved from the previous assessment and earned a B grade (ASCE, 2017). Roads, earning a D grade (ASCE, 2017), are particularly important given the dependence of communities, particularly in rural areas, on roads for many aspects of life and are particularly vulnerable to climate change and whether damage (Chinowsky et al., 2012; Chinowsky, Price, & Neumann, 2013, Larsen et al., 2008). In the same vein, the facilities and infrastructure that round out the list have similarly poor scores: Schools—D; Solid Waste—C+; Transit—D; and, Wastewater—D + (ASCE, 2017).

Former Department of Homeland Security director Jeh Johnson recently stated (MSNBC, January 15, 2018) that one of the biggest threats to national defense and homeland security was aging infrastructure and climate change. The ASCE 2017 report puts that claim in sharp focus, and provides a map of how to improve those elements of our national infrastructure. The estimated gap to bring US infrastructure up to par is estimated at $2 trillion. And to maintain all of it in optimal condition will require annual investments by federal and state governments to the tune of $206 billion per year (ASCE, 2017). Until those investments are made, our infrastructure will continue to deteriorate, and it could cost more in the long run to fail to maintain what we have.

Post-normal times may very well mean the collapse of the federal government's ability or willingness to respond adequately to infrastructure funding needs, and the responsibility may fall increasingly on states and local government. This trend of decentralization, might be positive, if it can enhance local resilience and adaptation (Larsen et al., 2008). State and county leaders and administrators should consider the relative vulnerability of consumers and elements of infrastructure (Cutter, 1996). Similarly, those leaders can share and pool knowledge with each other and with federal leaders in response to disaster and recovery (Cutter et al., 2015). A large body of knowledge and cross-sector approaches exist in risk management and preparedness that can inform decision-making in the aftermath of events stimulated by accelerating climate change and civilizational stress and decline. Insurance pools, and other funding strategies, such as crowdsourcing, and new fees may need to be considered to support infrastructure projects.

SOLUTIONS AND RECOMMENDATIONS

In terms of facing the future, political leaders and public administrators need to plan for the worst but also engage communities to envision their preferred futures, and vision the sustainable cities and communities they would like to inhabit. The best way to build resilience is to raise consciousness and build shared understanding of the challenges, risks, and responses to accelerating change. Administrators, planners, and first responders should practice for multiple catastrophes and collapse scenarios. The year 2017 provided a good example of multiple major hurricanes, widespread wildfires, and continuing drought, much more likely to be the norm in post-normal times. Experience shows that preparedness for emergencies pays off.

The role of technology has been understated in this paper, but also has the potential to be transformative. One of the very weaknesses of civilization, following Tainter (1988), may well be solved by the artificial intelligence, or what Gidley (2017) calls super machine intelligence. The dilemma of maintaining complex systems in the face of the limits to complexity may well be mitigated by neural networks, super machine intelligence, and expert systems. A rapid transition into a solar and renewable energy economy could also have significant positive impacts on infrastructure. For example, automated control systems and smart grid technologies, efficacious decentralization, and tax incentives could empower individuals, families, and communities. It has also been argued that widespread environmental catastrophes could be the wake-up call to galvanize widespread social movements to counter the fossil fuel economy and end the destructive behavior of consumer capitalism. In any case, it looks like we are in for a dramatic ride into the future, whether in the form of autonomous electric cars or horse and buggies.

In the face of Collapse, potential limits two complexity, and post-normal times, why not just "party until the End," bury our heads in the sand, or give up envisioning and realizing sustainable futures? One answer is that sustainability is not a lifestyle choice, but rather an inevitable requirement for co-evolution with our planet. If the projections of the Club of Rome and *Limits to Growth* are reasonable forecasts based on the assumptions of growth, growing pollution, and resource depletion, then arguably at some point we will reach a steady state. The models argue that the current macro economic assumptions are unsustainable, so at some point, collective behavior changes will be necessary, but not necessarily sufficient to avoid major human catastrophe.

One key to transformation into a more stable civilization or bridge through the decline and/or collapse is envisioning preferred, sustainable futures (Jones, 1998; 2018). Defeatism and escapism are inevitable facing the challenges of infrastructure and global weirding, but as Macy (1983) noted hope and optimism are tools to cope with the depersonalization and anomie that come from the dangers that confront us. While it may not be an immediate solution to the crisis in infrastructure, the empowerment of individuals and communities to envision and create better futures is one way to instill hope and optimism for the future. There is an abundant literature and examples of such futures planning across the globe (Gidley et al., 2009). The World Futures Studies Federation and Teach the Future are both organizations that are promoting futures education at all levels, from professionals to kids. Teach the Future offers free resources, for example, to K–12 educators for foresight education. Projects across the globe, from favelas in Brazil to children in Africa and futures labs in Europe are deliberately trying to reach the youth of today to build more responsible leaders for the future (Jones, 2018). As Lombardo (2017) argued, we have a responsibility to envision and create good futures empowered by wisdom, a more evolved human and planetary ethics.

Engaging in the SDGs is clearly another way to model that wisdom by addressing environmental and social equity and justice. The SDGs do not currently have high visibility, but there is obviously a direct relationship between the support of physical infrastructure, both in the USA and the world at large, and society's ability to implement the goals and their targets. Without adequate infrastructure, the likelihood of attaining those goals diminishes. Conversely, an understanding of humankind's aspirational goals beyond the struggle for infrastructure is also a key to sustaining hope and optimism for the future. Setting higher goals reminds us that infrastructure is a means to higher ends.

Implementing the SDGs and envisioning preferred futures converge as parallel processes of seeing the potential for a more equitable and just society, and helping provide the means to help manifest those values.

CONCLUSION

Post-normal times is by definition a transitional period of time between normal times, so part of the project of seriously addressing climate change mitigation, and not just adaptation, in the global climate regime is a fundamental change in worldview. The project of the SDGs and their targets it is to begin moving some of the values embedded in social equity and justice to the forefront, ahead of the exigencies of scarcity economics and the current system of growth at all costs. Certainly, political leaders and public administrators will be required to deal with the unfolding crises that will likely peak during the mid-century, but must not lose sight of the longer-term goals of establishing sustainable development values and practices as a baseline for the social and economic goals for humanity and planetary health. However, it just may be the case that the aphorism "things will get worse before they get better" is true, but we cannot neglect or disregard our collective responsibility for future generations. Infrastructure, particularly, represents that kind of investment in the long-term future that will either make it harder or easier for future generations to envision and build preferred, sustainable futures. We need to engage policy-makers, communities, and individuals in envisioning those good futures (Lombardo, 2017) at the same time that we promote and implement the SDGs. It is especially important to engage young people in the visioning of those good futures, to give them hope and ownership of the future that they will inhabit. Realizing the SDGs by 2030 and beyond will require not only financial and human resources, but also a commitment to creativity, imagination, and the will to make better, good futures.

REFERENCES

American Society of Civil Engineers. (2017). 2017 Infrastructure Report Card.

Bell, W. (1997). *Foundations of Futures Studies*. New Brunswick, N.J. and London UK: Transaction Publishers.

Burbidge, R. (2016). Climate-proofing the airport of the future. *Journal of Airport Management*, *11*(2), 114–128.

Burke, S. E. (2015). Enemy Number One for the Electric Grid: Mother Nature. *SAIS Review*, *35*(1).

Caton, W. (1982). *Overshoot: The Ecological Basis of Revolutionary Change*. University of Illinois.

Chinowsky, P. S., Price, J. C., & Neumann, J. E. (2013). Assessment of climate change adaptation costs for the U.S. road network. *Global Environmental Change, 23*(4), 764–773. doi:10.1016/j.gloenvcha.2013.03.004

Cutter, S. (1996). Vulnerability to environmental hazards. *Progress in Human Geography, 20*(4), 529–539. doi:10.1177/030913259602000407

Cutter, S., Ismail-Zadeh, A., Alcántara-Ayala, I., Altan, O., Baker, D. N., Briceño, S., ... Wu, G. (2015, June). Pool knowledge to stem losses from disasters. *Nature, 522*(18), 277–279. doi:10.1038/522277a PMID:26085255

Dator, J. (1995). *What futures studies is, and is not*. Honolulu: Hawaii Research Center for Future Studies.

Dator, J. (2009a). Alternative Futures at the Manoa School. *Journal of Futures Studies, 14*(2), 1–18.

Dator, J. (2009b). The Unholy Trinity, Plus One. *Journal of Futures Studies, 13*(3), 33–48.

Diamond, J. (2005). *Collapse*. New York, NY: Penguin Group.

Ehrlich, P., & Ehrlich, A. (2013). Can a collapse of global civilization be avoided? In *Proceedings of the Royal Society*.

Feffer, J. (2014). *Earth: Game Over?* Foreign Policy in Focus.

Friedman, T. (2007). The People We Have Been Waiting For. *New York Times*.

Gidley, J. M. (2017). *The Future: A Very Short Introduction*. Oxford: Oxford University Press. doi:10.1093/actrade/9780198735281.001.0001

Gidley, J. M., Fien, J., Smith, J., Thomsen, D. C., & Smith, T. F. (2009). Participatory Futures Methods: Towards Adaptability and Resilience in Climate-Vulnerable Communities. *Environmental Policy and Governance, 19*(6), 427–440. doi:10.1002/eet.524

Ives, M. (2017). Climate Change Lands at the Airport. *New York Times*. Retrieved from https://www.nytimes.com/2017/09/30/business/airports-climate-change-global-warming.html

Jones, C. (1998). Envisioning Sustainable Futures. In D. Hicks & R. Slaughter (Eds.), *World Yearbook of Education 1998: Futures Education*. London: Kogan Page.

Jones, C. (2018). Sustainable futures. In W. Leal (Ed.), *The Encyclopedia of Sustainability in Higher Education.* Springer.

Katz, J. M. (2017) Florida: We Don't Discuss the Cause. *Weather*. Retrieved from https://features.weather.com/us-climate-change/florida/

Kemp, R. (2017) America's Infrastructure: Is Our Country on the Road to Ruin?! *PA Times*. Retrieved from https://patimes.org/americas-infrastructure-country-road-ruin/

Kolbert, E. (2014). *The Sixth Extinction: An Unnatural History*. New York: Henry Holt and Co.

Kunstler, J. H. (2005). *The Long Emergency: Surviving the Catastrophes of the Twenty-First Century*. New York: Atlantic Monthly Press.

Larsen, K., & Gunnarsson, U. (2009). Climate change scenarios and citizen-participation: Mitigation and adaptation perspectives in constructing sustainable futures. *Habitat International*, *33*(3), 260–266. doi:10.1016/j.habitatint.2008.10.007

Larsen, O. M. (2015). Climate change is here to stay: Reviewing the impact of climate change on airport infrastructure. *Airport Management*, *9*(3), 264–269.

Larsen, P., Goldsmith, S., Smith, O., Wilson, M., Strzepek, K., Chinowsky, P., & Saylor, B. (2008). Estimating future costs for Alaska public infrastructure at risk from climate change. *Global Environmental Change*, *18*(3), 442–457. doi:10.1016/j.gloenvcha.2008.03.005

Lovelock, J. (2006). *The Revenge of Gaia*. London: Penguin Books.

Lovelock, J. (2009). *The Vanishing Face of Gaia*. New York: Basic Books.

Lovelock, J. (2015). *A Rough Ride to the Future*. New York: The Overlook Press.

Lynas, M. (2008). *Six degrees. Our future on a hotter planet*. Washington, D.C.: National Geographic Society.

Macy, J. R. (1983). *Despair and personal power in the nuclear age*. Philadelphia, PA: New Society Publishers.

Marcuson, W. F. III. (2008, January). Fixing America's Crumbling Infrastructure, A Call to Action for All. *Public Works Management & Policy*, *12*(3), 473–475. doi:10.1177/1087724X07312145

Meadows, D. H., Randers, J., & Meadows, D. (2004). *Limits to Growth. The 30-year Update*. White River Junction, VT: Chelsea Green Publishing Co.

Mooney, C. (2017). Scientists challenge magazine story about 'uninhabitable Earth.' *The Washington Post.* Retrieved from https://www.washingtonpost.com/news/energy-environment/wp/2017/07/12/scientists-challenge-magazine-story-about-uninhabitable-earth/?noredirect=on&utm_term=.019e1445a7d4

National Research Council. (2012). Dam and Levee Safety and Community Resilience: A Vision for Future Practice. Washington. D.C., The National Academies Press.

Polak, F. (1973). *The Image of the Future*. New York: Jossey-Bass.

Rapaport, E., Manuel, P., Krawchenko, T., & Keefe, J. (2015). How Can Aging Communities Adapt to Coastal Climate Change? Planning for Both Social and Place Vulnerability. *Canadian Public Policy*, *41*(2), 166–177. doi:10.3138/cpp.2014-055

Repetto, R., & Easton, R. (2010). Changing Climate, More Damaging Weather. *Issues in Science and Technology*, *26*(2).

Sardar, Z. (2017). *The Postnormal Times Reader*. Great Britain: Centre for Postnormal Politics and Futures Studies.

Sardar, Z. & Sweeney, J. (2015). The Three Tomorrows of Postnormal Times. *Futures*, *75*, 1-13.

Schmidt, C. W. (2009). Beyond Mitigation. *Environmental Health Perspectives*, *117*(7), A306–A309. doi:10.1289/ehp.117-a306 PMID:19654902

Slaughter, R. A. (2004). *Futures Beyond Dystopia*. London: Routledge Falmer.

Slaughter, R. A. (2010). *The Biggest Wake up Call in History*. Indooroopilly, Australia: Foresight International.

Tainter, J. A. (1988). *The Collapse of Complex Societies*. Cambridge, UK: Cambridge University Press.

Vacca, R. (1973). *The Coming Dark Age*. Garden City, N.Y: Doubleday.

Vorosmarty, C. J., Green, P., Salisbury, J., & Lammers, R. B. (2000, July 14). Global water resources: Vulnerability from climate change and population growth. *Science*, *289*(5477), 284–288. doi:10.1126cience.289.5477.284 PMID:10894773

Wallace-Wells, D. (2017). The Uninhabitable Earth. *New York Magazine*. Retrieved from http://nymag.com/daily/intelligencer/2017/07/climate-change-earth-too-hot-for-humans.html

Zimmerman, R. (2009). Making Infrastructure Competitive in an Urban World. *The Annals of the American Academy of Political and Social Science*, *226*.

KEY TERMS AND DEFINITIONS

Accelerated Warming: The idea that global heating is a "runaway train," based on the idea that atmospheric levels of CO_2 and other greenhouse gases will continue rise, even with aggressive climate change mitigation. It also posits that other feedback mechanisms, such as ocean thermal expansion, methane releases, and albedo changes will exacerbate positive feedback processes, further warming the planet.

Four Futures: The futures studies methodology that uses futures typologies (Business As Usual, Collapse, Disciplined, and Transformational societies) as tools to explore possible alternative futures, developed by Jim Dator, the Institute for Alternative Futures, the RAND Corporation, and SRI International to capture the dominant images of the future in literature and popular culture.

Futures Studies: A cross-disciplinary field that evolved from state planning, technological forecasting, normative futures, and post-colonialism that addresses possible, probable, and preferred futures. Futures studies posits that there is no single future, but that many possible alternative futures exist, that the future cannot be predicted, and that visioning preferred futures play a role in creating alternative futures.

Gaia Theory: The idea that Earth's biosphere is a cybernetic, homeostatic system that tends to create and sustain conditions favorable for life. Gaia theory describes a process of balance that has been in place for millennia and that maintains a thermodynamic equilibrium and long-term geochemical cycles (i.e., oxygen, CO_2).

Global Weirding: The idea that global warming is not simply about rising global temperatures, but about the concomitant emergence of variable and freakish weather, including extreme weather events, drought, dust storms, ice storms, hurricanes, and other atypical weather and geophysical events.

Peak Complexity: The idea that all societies and civilizations have natural limits to complexity and diversity. The idea is based on the observations of past societal collapse based on the inability of those societies to adapt to environmental, economic, and demographic growth. At some point, marginal improvements decline and energy costs grow resulting in decline and collapse of social and/or economic systems.

Post-Normal Times/Theory: A theory that there is a transition period between "normal" societal states, characterized by complexity, chaos, and contradiction.

Sustainable Futures: The idea that there are preferred, alternative futures, that are based on different normative values and assumptions that are not based on the assumptions of growth, economic development, and materialism. These futures are based on other values, such as spirituality, personal and emotional development, and other non-material measures of "growth" and "progress."

Chapter 8
Active Learning and Disaster Risk Reduction:
Playing the Game of Your Life

LaMesha Lashal Craft
Coastal Carolina University, USA

ABSTRACT

The author provides a robust discussion of an ethnographic case study to facilitate creative thinking about how to use communications and social media technology to build resilience and improve citizen disaster preparedness through a "Be Ready" trivia campaign. This research can inform strategies to achieve several of the United Nations Sustainable Development Goals (SDGs) as well as the United Nations Office for Disaster Risk Reduction's Sendai Framework for Disaster Risk Reduction (SFDRR). Future research directions include a new community resilience index that measures citizens' use of communications and social media technology. Implications for social change include raising the level of public awareness and facilitating a means to improve personal responsibility for disaster preparedness through low cost education programs. This could improve efforts by government and non-government organizations to improve disaster risk reduction; increase access to information and communication technology; increase disaster emergency planning and response; and build resilient communities.

INTRODUCTION

Education is the foundation upon which we build our future. (Christine Gregoire)

It is with the future in mind that the author submits education is the key to international efforts to sustain mother earth, build a preferable future, and to grow as a human race.

DOI: 10.4018/978-1-5225-7727-0.ch008

The future is plural (Bell, 1997; Dator, 2009; Sardar & Sweeney, 2016; Wallace-Wells, 2017) and given the pending state of climate and environmental change, it is imperative to provide tangible and achievable education tools that are applicable in local and global societies to demonstrate how individuals can prepare for large-scale man-made or natural disasters. As a career military intelligence professional, the author is trained to analyze the potential for future events to impact or threaten the current environment. Some threats are known yet poorly mitigated (black elephants) while others have not been considered as plausible threats to stability (*black swans*). Such preparation requires and eye towards the future and a pulse on real-world and relevant threats. In essence citizens, organizations, and government entities must think outside of the proverbial box to implement the most effective disaster risk reduction programs and ultimately prevent social disorganization and anomie before, during, or after disasters.

The objectives of this chapter are three-fold. First, the author discusses findings from an ethnographic case study titled *Perceived Threats to Food Security and Possible Responses Following an Agro-terrorist Attack*. The findings demonstrated how even in a community with high social capital and perceived community resilience, the lack of knowledge about emergency preparedness procedures, mismanaged expectations regarding relief, and failure to plan for unexpected yet plausible disasters can lead to social disorganization and anomie. Secondly, this research can facilitate more creative thinking about how to use communications and social media technology to build resilience and improve citizen disaster preparedness through a "Be Ready" trivia campaign. Thirdly, this research can inform strategies to achieve several of the United Nations Sustainable Development Goals (SDGs) as well as the United Nations Office for Disaster Risk Reduction's Sendai Framework for Disaster Risk Reduction (SFDRR). Specifically, the author submits that all of the 17 Sustainable Development Goals (SDGs) are linked to SDG #4 (quality education), because without it improving society, building resilience to climate and environmental changeand creating a preferable future are impossible. The author's recommendations include using an education program to achieve targets of SDG #3 (healthy lives and well-being); SDG #9 (building resilient infrastructure); SDG #11 (safe, inclusive, and resilient cities and human settlements), and SFDRR Priorities 2 and 3 (strengthen and invest in disaster risk reduction). Moreover, future research directions include the development of a new community resilience index and the development of questionnaires to leverage local community knowledge of community and faith-based organizations in developing countries where maintaining 72-hour emergency kits are not economically feasible.

Implications for social change include raising the level of public awareness and facilitating a means to improve personal responsibility for disaster preparedness through low cost education programs. This could improve efforts by government and non-government organizations to improve disaster risk reduction; increase access to information and communication technology; increase disaster emergency planning and response; and build resilient communities.

BACKGROUND

Understanding and Challenging Human Nature

The 2030 United Nations Agenda for Sustainable Development acknowledges the dire need to reduce the risk of disasters and has set forth the goals to do so. However, it is important to note that the very nature of preparedness for future events, the state of the ever-changing climate, and understanding the butterfly effect of how actions today will impact the world in 10 or 20 years runs counter to human nature. It is human nature to discount events that have yet to occur in one's lifetime because of how humans process information (Holdeman, 2011). According to Richards Heuer (1999), humans tend to perceive what we expect to perceive. Subconsciously, our patterns of expectations tell us what to look for, what is important, and how to interpret what is seen (Heuer, 1999). In that same vein, the mind is poorly wired to deal effectively with inherent uncertainty (Heuer, 1999). However, to accomplish the Sustainable Development Goals and maintain a proactive edge it is imperative to consider things, events, and conditions that are plausible and that could directly or indirectly dissolve the very fabric of humanity.

Case Study Discussion: Perceived Threats and Possible Responses

In 2017, the author conducted an ethnographic case study to examine, describe, and analyze the perceived threats to food security and the possible responses to food shortages in an agricultural community in Yuma, Arizona. The study was based on a hypothetical agro-terrorist attack on the U.S. food supply. Yuma's agricultural significance lies in its title as the "winter lettuce capitol of the world." Yuma produces approximately ninety percent of all leafy vegetables grown in the US between November and March. The author used the following research questions to frame the study:

Research Question: What are the perceived threats to food security in an agricultural community in Arizona and the possible responses to food shortages to a hypothetical agro-terrorist attack on the U.S. food supply?

Sub-Questions:

1. To what extent do the citizens of Yuma, Arizona believe their local government can provide assistance when a threat to food security exist?
2. To what extent does community resilience in a community within Yuma, Arizona affect possible responses to food shortages?

In her research, the author suggested that citizens of Yuma, Arizona who experienced an agro-terrorist attack and were subsequently disappointed in the government's response could become rebellious if the community experienced low social capital or if social capital was nonexistent. In a mixed methods research study of two resource-dependent rural communities that had experienced wildfires, Kulig, Reimer, Townshend, Edge, and Lightfoot (2011) found that the Index of Perceived Community Resilience (IPCR) provided a useful composite of perceived community resilience. This is the basis for which the author designed the interview questions to answer the aforementioned research questions pertaining to community resilience (see Appendix 1). During the study, the author interviewed nine residents using 18 open-ended questions. Through the field study and data analysis, the author explored four themes: the perception of fear following a threat to food security; how residents felt about their local government's ability to provide assistance when threats to food security exist; the levels of social capital in the community; and the existence of community resilience. The author also interviewed six experts from the Fire Department, the Sheriff's Office, the County Emergency Management Office, Yuma County Public Health Services, and the Police Department. The author used 10 open-ended questions (see Appendix 2) to examine the following themes: their organization's involvement in preparations against threats to food security; their responsibility to the citizens of Yuma, Arizona following a disaster that threatens food security; their perceptions of social cohesion among the citizens; and the perceived functionality of community resilience (Craft, 2017).

The research demonstrated that Yuma residents exhibited characteristics of high social capital such as trust in their neighbors, trust of government officials and responders, and community solidarity. Furthermore, eight out of nine residents trusted their neighbors to varying degrees and believed that the government leaders, law enforcement officials, and professionals within emergency management services (EMS) would assist them during a disaster. All of the residents and the experts interviewed believed that the community of Yuma, Arizona was "small but strong" and resilient. Yuma is three hours west of Phoenix, Arizona and three hours east of California and several experts stated citizens have a frontier mentality. However, the residents also demonstrated a lack of knowledge, mismanaged expectations, and confusion that if left unaddressed could cause a breakdown in society during a disaster (Craft, 2017).

Notably, the residents' level of trust in neighbors changed once asked to posit an agro-terrorist attack occurred and a food shortage ensued; only two neighbors maintained their original level of trust (see Figure 1). None of the residents knew if a plan to respond to an agro-terrorist attack existed. Nor did they demonstrate a clear understanding of which organizations were responsible for food security (see Figure 2).

Additionally, none of the residents mentioned the 72-hour emergency preparedness kit when asked how they would protect themselves and their family following an agro-terrorist attack and the threat of a food shortage. The lack of discussion about the kit contradicted initiatives by the Fire Department, the Sheriff's Office, and the Emergency Management Office to educate the public on how to care for themselves during the first 72 hours of an emergency or disaster (Craft, 2017).

Moreover, when discussing emergency management, six out of nine residents expected that emergency support would be delayed because EMS personnel would

Figure 1. Residents' degree of trust in neighbors after hypothetical disaster (Craft, 2017)

Figure 2. Organizations that residents believe are responsible for food security (Craft, 2017)

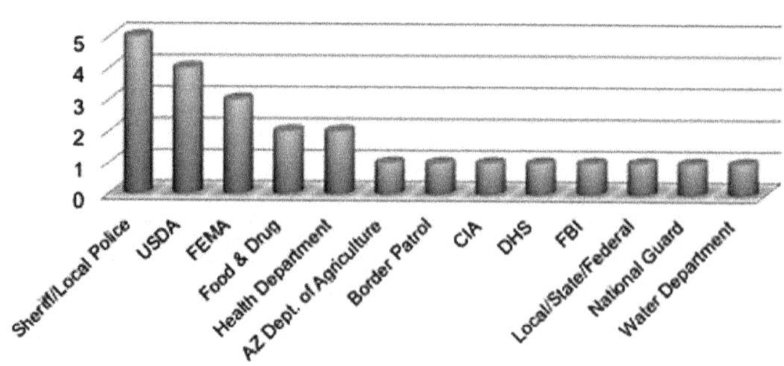

assist their own family prior to assisting the citizens. This suggests the residents were unaware of the "Prepare to Serve" program. Experts from Yuma learned a valuable lesson from the Hurricane Katrina disaster when some EMS personnel did not come to work because of their responsibility to their families. Experts from the Yuma Fire Department teach the "Prepare to Serve" program to EMS personnel, employees of Yuma City, Yuma County employees, and the Chamber of Commerce. It emphasizes the necessity of public servants training their families for disasters, so they can fulfill their responsibilities of restoring basic services for the citizens of Yuma (Craft, 2017).

The absence of discussion about the 72-hour kit, and the expectation that support would be delayed suggests the residents lacked knowledge about emergency management procedures and resources, which could indicate a deficit in education and community competence. The lack of education could negatively impact the residents' decision making before, during, and after a natural or man-made disaster (Craft, 2017).

MAIN FOCUS OF THE CHAPTER

The main focus of this chapter is to provide a dynamic examination of anomie theory as a cautionary tale to the futuristic plight of society in an environment where citizens are not resilient and lack social capital. The author also offers discussion on community resilience and its role as the cornerstone of the whole community approach (WCA) and the global implications of WCA. The author also facilitates creative thinking through an innovative use of communications and social media technology to build resilience and improve citizen disaster preparedness through a "Be Ready" trivia campaign. In that same vein, the author demonstrates how a proactive approach to the quality education of citizens can empower them with the requisite knowledge to improve community competence and community resilience, thus applying the fundamentals of the whole community approach. As illustrated in Figure 3, the author's research and the subsequent recommendation to establish a trivia campaign using the current information and communication technology can aid in the achievement of some sustainable development goals (SDGs). Specific SDG targets and SFDRR priorities include: SDG #3D (strengthened capacity for risk reduction); SDG #9C (increased access to information and communications technology and access to the internet); SDG #11B (increased disaster and emergency planning and response in line with SFDRR), and SFDRR Priorities 2 and 3 (strengthen and invest in disaster risk reduction).

Figure 3. Links between author's research and United Nations' initiatives (Craft, 2018)

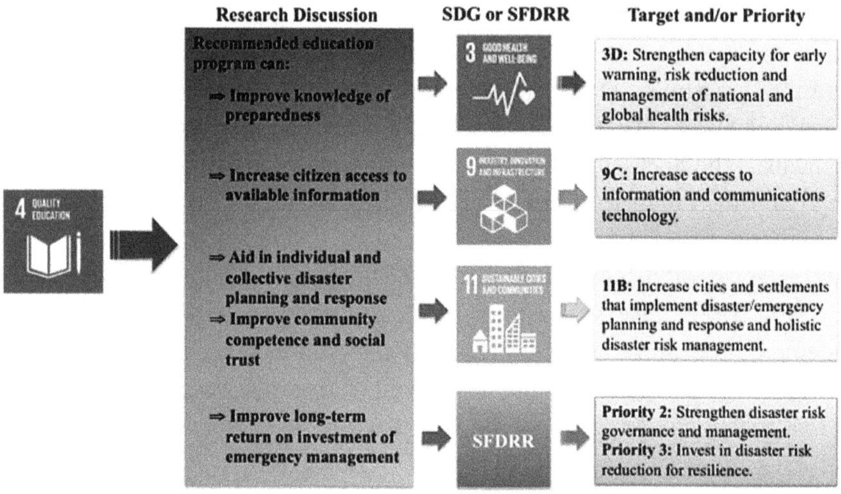

Issues, Controversies, Problems

In Theory: Are We One Event From Anomie?

The research of theorists Durkheim (1984) and Merton (1968) are salient when analyzing the potential impacts that a real or perceived loss of basic needs could have on a community. The subsequent affect on the degree of social cohesion and relative deprivation also become a factor. In 1897, Durkheim introduced the concept of anomie, which is the occurrence of social instability after standards and values break down in society (Durkheim, 1984). Merton (1968) later expanded on Durkheim's findings (Zembroski, 2011); when he discovered that most cases of anomie occur when the population does not have a sense of cultural meaning and societal integration. This is also known as relative deprivation (Kawachi, Kennedy, & Wilkinson, 1999; Zembroski, 2011). The threat of social instability can be aligned within the context of social (i.e., networks, norms, and trust) and cultural (i.e., values and beliefs) capital (Smith & Lawrence, 2014). The combination of social and cultural capital constitutes social capital theory, which facilitates the examination of a community's capacity to respond to and recover from an adverse event or the likelihood of a perilous decline towards anomie.

Relative deprivation refers to group reference theory in which we compare ourselves (and our social situations) to individuals that belong to other groups. This system of comparison forms the basis for the rationalization of situation and behavior

(Merton, 1968). One's sense of relative deprivation leads to resentment--the sense that others' gains are somehow based upon or result in our losses. Resentment could, in turn, result in the formation of a counter-cultural orientation based upon reversal of normal cultural goals and social institutionalized means (Merton, 1968). Kawachi et al. (1999) argued that social characteristics, including the existence of relative deprivation, coupled with the level of social cohesiveness among citizens, could influence the amount of criminal activity. Citizens who live in communities with limited social networks were less likely to contribute to community development and were susceptible to potentially high levels of crime in an environment where social institutions have begun to fail (Sampson & Groves, 1989).

Community Resilience, Migration, and Disaster Mitigation

In contrast, resilient communities are capable of responding and adapting to adversity or a change in conditions (Bromley et al., 2017; Hyvarinen & Vos, 2015; Kulig, Edge, Townshend, Lightfoot, & Reimer, 2013; Leykin, Lahad, Cohen, Goldberg, & Aharonson-Daniel, 2016; Lichtveld, 2018; Linnell, 2014; Plodinec, Edwards, & White, 2014; Smith & Lawrence, 2014; U.S. Department of Health and Human Services [U.S. HHS], 2012). It also enables effective mitigation of vulnerabilities, which in turn facilitate the timely restoration of a community's functionality after a disaster (Bromley et al., 2017; FEMA, 2009; Kulig et al., 2013; Lee, 2010; Plodenic, Edwards, & White, 2014; Smith & Lawrence, 2014; U.S. HHS, 2012).

Additionally, community resilience is one of the cornerstones to the whole community approach cited in the 2014 Quadrennial Homeland Security Review report and the National Preparedness Framework (U.S. DHS, 2013, 2014). The whole community approach enables citizens, emergency management organizations, government officials, and stakeholders to develop a greater understanding of the needs of the respective community (FEMA, 2011). The benefits of this approach include shared understanding, citizen empowerment, resource integration, improved social infrastructure, preparedness, risk reduction, and greater resiliency (Chandra et al., 2013; Plodenic, Edwards, & White, 2014).

For over a decade scholars have analyzed the correlation between community resilience and the mitigation of disasters (Aldrich & Meyer, 2015; Bromley et al., 2017; Chan, Yue, Lee, & Wang, 2016; Chandra et al., 2013; Chen, Chen, Vertinsky, Yumagulova, & Park, 2013; Faulkner, Brown, & Quinn, 2018; Kulig et al., 2013; Kulig et al., 2011; Lichtveld, 2018; Plodenic et al., 2014; Smith & Lawrence, 2014). The United States Presidential Policy Directive-8 (National Preparedness) noted that improving resilience against disasters require improving public awareness and motivating ordinary people to assume individual responsibility for preparedness.

However, given the urgency of the 2030 Agenda for Sustainable Development to reduce the risk of disasters, the author submits that the whole community approach is also an international endeavor.

Data from the majority of studies on community resilience suggest the community is established and the residents have lived in the same neighborhood over a period of time. However, human mobility is a complex phenomenon that can impact the location, size, composition, and environmental stability of societies and communities (International Organization for Migration [IOM], 2014; Martine, Schensul, & Guzman, 2013). Human mobility can also generate risks to disaster preparedness such as the incapacity to relocate before, during, and after disasters (IOM, 2014). For example, between November and March the population in Yuma, Arizona increases by 90,000 (nearly doubling in size). The warm winters and harvest season attract vacationers and migrant workers respectively. The inclusion of visitors with disparate perspectives, experiences, and backgrounds creates a disaster preparedness challenge. Half of the experts interviewed during the author's research acknowledged that the increased population created a level of uncertainty because visitors may not share the same perceptions about community relations and a sense of belonging in Yuma (Craft, 2017).

The 2030 Agenda for Sustainable Development identifies the significance of international migration and the need for a comprehensive plan of action. The number of international migrants has continued to grow to approximately 258 million in 2017, which is an increase of 38 million in less than 8 years. Concerns include the safe and responsible migration and mobility of people under the construct of well-managed policies (United Nations, 2017). Imagine the complexity of attempting to accomplish this goal under the pressure of a large-scale disaster. Now imagine the inclusion of the whole community approach as an international endeavor, where the essential information of how to prepare for and respond to a disaster was globally available via an education program.

SOLUTIONS AND RECOMMENDATIONS

Education Is the Lynchpin to a Sustainable Future

Norris, Stevens, Pfefferbaum, Wyche, and Pfefferbaum (2008) examined the utility and applicability of community resilience to disaster readiness. They used a set of capacities to describe conditions that exist in a community that is resilient: economic development, social capital, information and communication, and community competence. The emphasis on information, communication, and competence is imperative in accomplishing the United Nations 2030 goal to ensure all persons have the knowledge and skills to promote sustainable development and sustainable lifestyles (United Nations, 2015).

Over the years the importance of education to avoid catastrophic incidents has been acknowledged. However, most literature highlights the importance of educating and training first responders, emergency management professionals, and scientists (Chen, Chen, Vertinsky, Yumagulova, & Park, 2013; Crutchley, Rodgers, Whiteside, Vanier, & Terndrup, 2007). Educating the average citizen is equally, if not more, important because the state of this earth depends on those that are literally responsible for its demise or improvement. Investing in education is a means of investing in prevention as well as disaster risk reduction. Disaster risk reduction is an invaluable part of economic and social development. It is also paramount in ensuring sustainability for the future.

As a proponent of social change, the author submits that a partial solution to addressing most domestic vulnerabilities and effectively managing disasters rests with education, community networking, and social activism. Part of establishing a comprehensive plan for emergency management and preparedness involves setting conditions prior to an emergency through education and training (Bromley et al., 2017; Chan, Yue, Lee, & Wang, 2016; Chandra et al., 2013; Chen, Chen, Vertinsky, Yumagulova, & Park, 2013; Cutter et al., 2008; Muttarak & Pothisiri, 2013). Educating citizens on disaster preparedness should focus more on the real-world learning experience (practical application) rather than providing information on a portal. To be clear, the entities responsible for homeland security, homeland defense, emergency management, and emergency response have done an honorable job providing information on the requisite websites regarding what to do when preparing for emergencies, the kinds of natural disasters that occur in the respective region, and general information. However, if this remains the primary means of providing information, it is fair to assume that most Americans (and international communities) will likely wait until an emergency or disaster has occurred to access the information, which means they were not prepared to respond prior to the incident.

Establishing a "Be Ready" Trivia Campaign

The author recommends that governments leverage the current lines of communication (e.g., Twitter and Facebook pages of local government departments, as well as the local television and radio channels) to create a "Be Ready" trivia campaign as a vehicle to disseminate information about emergency management programs and procedures (see Figure 4). The campaign should leverage current trends of using social media technology and data as part of disaster warning, response, and recovery (Bromley et al., 2017; Reuter, Hughes, & Kaufhold, 2018). In the United States, the current

Figure 4. Concept of "Be Ready" trivia campaign
(Craft, 2018)

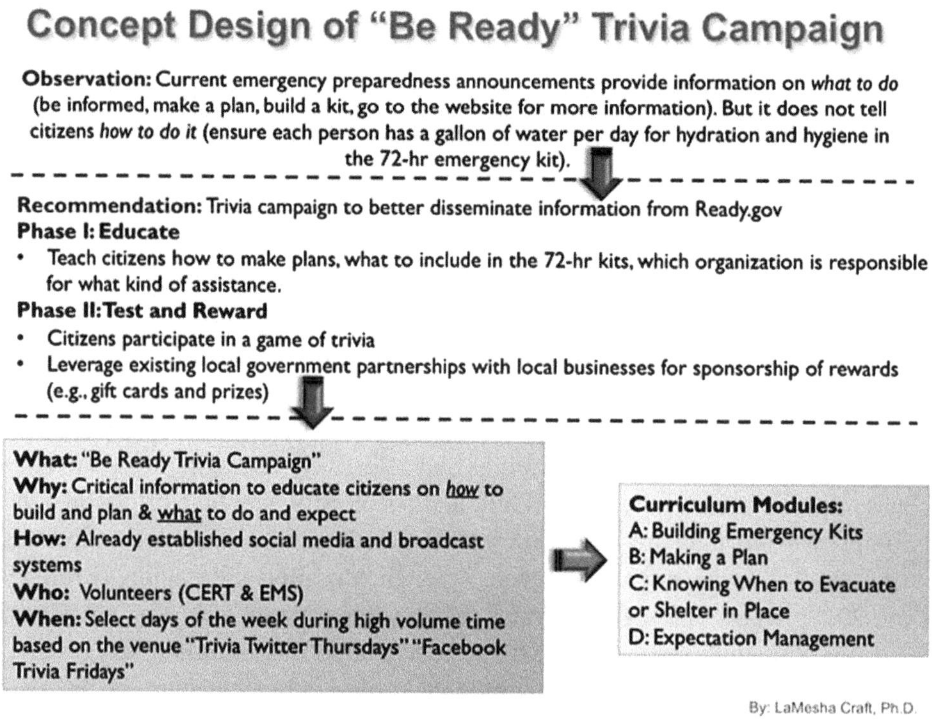

public service announcements about emergency preparedness that are broadcasted via the Ad Council provide information on *what to do* (e.g., be informed, make a plan, build a kit, go to the website for more information) but it does not tell citizens how to do it (e.g., ensure each person has a gallon of water per day for hydration and hygiene in the 72-hr emergency kit (Craft, 2018).

The recommended first phase of the campaign involves the use of local social media (such as Twitter and/or Facebook) and broadcasting platforms (such as the radio) that are readily available to teach citizens how to make plans and build 72-hr kits. In most cases, the local fire department, police department, and emergency management office have an official Twitter or Facebook page. As noted in Figure 5, in order to make the trivia question, the author simply created a trivia question in a text box, included the answer to the question in the bottom right corner, and saved the text box as a picture. This procedure enabled the timely upload via the respective media forum. The author also created an example of a script disseminating the same information during a 30 second radio broadcast sponsored by a local business. The significance is the partnerships between local government and local radio stations to address community-specific challenges and considerations. Moreover, previously established partnerships with local businesses and organizations could facilitate the sponsoring of 20 or 30 second advertisement spots to provide the information during the trivia campaign. In the example (see Figure 5), the sponsor is Wegman's food store. The duration of the campaign would depend on the topic and the needs of the community such as preparedness, natural hazards, and emergency information.

During the second phase citizens get an opportunity to demonstrate their knowledge of the information through a game of trivia. Ideally, this event could occur in conjunction with the local community fair or festival. In this case, rather than emergency management or emergency preparedness organizations handing out pamphlets during the community event, citizens could participate in the trivia to answer questions regarding the topics that have been broadcasted. Another option is hosting the trivia game via local radio stations. The theme of the trivia campaign could be in the form of popular radio contests that require participants to provide the name of a song after listening to a portion of it. However, in this scenario,

Figure 5. Examples of how to broadcast trivia questions
(Craft, 2018)

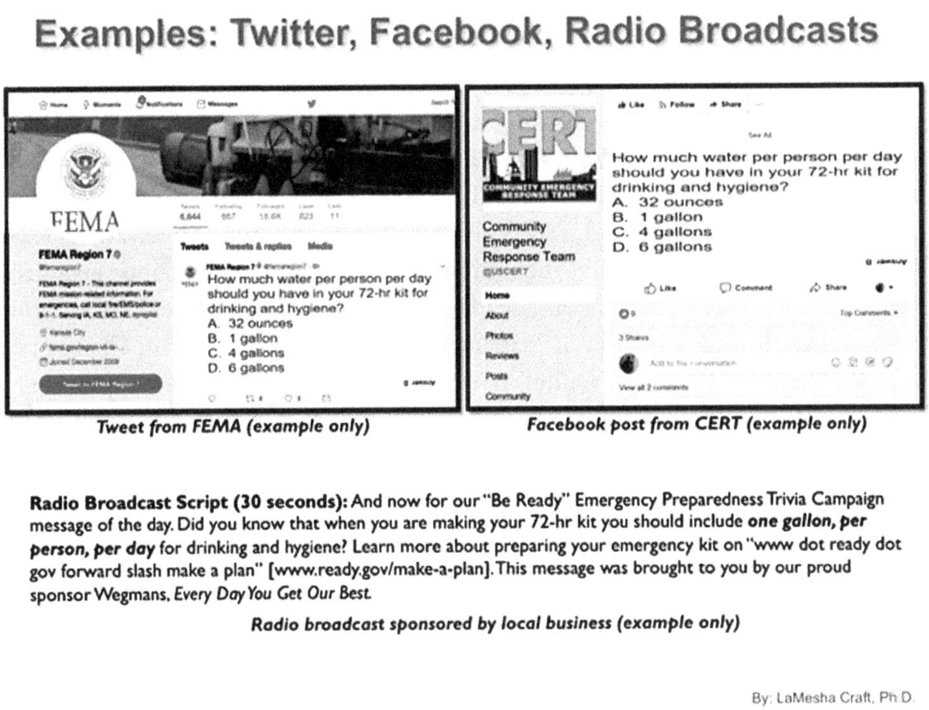

citizens would receive a short list of questions to answer correctly for a prize. Local organizations or businesses could sponsor the reward for those winning first, second, and third place (Craft, 2018).

Complacency Is the Crux to Preparedness

Reinforcing the necessity of collaboration across echelons is important. As noted in the author's research, while Yuma's frontier mentality served as a forcing function of collaboration and high social capital among local entities and communities, the citizens and government officials were not prepared for an agro-terrorist ("black swan") event. In communities where government entities have established procedures for the most common incidents introducing occasional black swan scenarios into the emergency management tabletop exercises could limit complacency, emphasize the importance of coordination across echelons of governance, and promote the urgency to prepare for what has yet to occur (Craft, 2017). Emergency management experts could also explain the concept of a primary, alternate, contingency, and emergency (PACE) program to citizens to help them establish a flexible individual and/or family response plan (Craft, 2017).

Education Is a Source of Empowerment

Communities tend to perform better before, during, and after disasters if they share a sense of empowerment, are socially cohesive, and believe in the good of the community (Bromley et al., 2017; Brown, Haun, & Peterson, 2014; Chandra et al., 2013; Faulkner, Brown, & Quinn, 2018; FEMA, 2011; McCrea et al., 2014; Muttarak & Pothisiri, 2013; Norris et al., 2008; Plodenic, Edwards, & White, 2014; U.S. DHS, 2010, 2014). Moreover, at the individual level, social connections establish informal and formal ties that encourage the development of generalized trust within a community (Bromley et al., 2017; Brown, Haun, & Peterson, 2014; Chandra et al., 2013; Faulkner, Brown, & Quinn, 2018; George & Stark, 2016; Glanville et al., 2013; Reininger et al., 2013; Rothstein, 2013). Social scientists have hypothesized that informal and formal ties will foster generalized trust because relationships enable individuals to form expectations of goodwill (Bromley et al., 2017; Faulkner, Brown, & Quinn, 2018; Glanville et al., 2013). Therefore, setting conditions such as active learning techniques for ordinary people to become more involved in their community and knowledgeable about preparedness is paramount to improving individual and government capacity to respond to and recover from disasters.

FUTURE RESEARCH DIRECTIONS

Adapting Community Resilience Index to Measure Social Media Interactions

Buikstra et al. (2010) acknowledged that a synergistic relationship existed between individual and community resilience. Studies between the 1990s and 2010 resulted in social models to describe resiliency (Brown & Kulig, 1996/7; Kulig, 1999; Kulig, Edge, & Joyce, 2008). Scientists acknowledged the difficulty in identifying and measuring levels of individual and community resilience because it is dynamic and it changes over time (Brown & Kulig, 1996/7). Therefore, studying individual or community resilience requires the study of an entity's response to stress over time. Studies focused on economic downturns, loss of natural resources in resource dependent communities, and the resulting effects on the resilience of the respective communities.

Scientists also discovered the difficulty of identifying when and how indicators influence the level of community resilience. Norris et al. (2008) argued that resilience is enhanced by resources that include three properties: the ability to withstand stress without suffering degradation; the extent to which elements are substitutable in the event of disruption or degradation; and the capacity to achieve goals in a timely manner to minimize losses and avoid disruption (p.134).

The model proposed by Norris et al. (2008) improved the identification of factors that contribute to community resilience. However, measuring it remained problematic. Kulig, Edge, Townshend, Lightfoot, and Reimer (2013) created a revised scale titled the Index of Perceived Community Resilience (IPCR). The IPCR incorporated elements of the Community Resilience Scale (Kulig, Edge, & Joyce, 2008; Kulig, Reimer, Townshend, Edge, & Lightfoot, 2011). The Community Resilience Scale included community members' perceptions of resilience-related items (the scale consisted of 15 items). While creating the IPCR, Kulig et al. (2013) narrowed the scope and created an 11-item index that incorporated the Norris et al. (2008) set of capacities, elements of the Community Resilience Scale, as well as Cutter et al.'s (2008) Community and Regional Resilience Initiative model and Cutter's (2008) Disaster Resilience of Place model.

As noted, the author used the IPCR during her field study. However, during data analysis the author discovered that the current measurement of interaction between people in a community does not account for the non-face to face forms of communication and interaction via social media platforms. For example, some residents noted they were too busy with work to attend community events. However, during the interviews some residents stated they used the social media websites of the local fire and police departments to stay informed. The inclusion of additional items to measure non-face to face communication, education, and information sharing could improve the identification of all factors that contribute to community resilience.

Trivia Is Trivial in Developing Countries and Communities

The author acknowledges that the aforementioned trivia campaign may not be feasible in developing countries and among vulnerable communities where the struggle to meet daily nutritional needs far surpasses concerns of having a 72-hour emergency kit (Lichtveld, 2018). However, the integration of community local knowledge into disaster preparedness and risk reduction remains valuable elements of community competence (Adams, Prelip, Glik, Donatello, & Eisenman, 2018; Bromley et al., 2017; Brown, Haun, & Peterson, 2014; Chan, Yue, Lee, and Wang, 2016; Chandra et al., 2013; Faulkner, Brown, & Quinn, 2018; George & Stark, 2016; Johnson, Edwards, Gardner, & Diduck, 2018). A modified application of the concept of active learning and practical application with an emphasis on developing questionnaires could aid entities in strengthening the capacity of developing countries for early warning, risk reduction, and management of national and global health risks. Further research could develop a central education campaign for disaster risk reduction.

CONCLUSION

The best way to build resilience is to raise consciousness and develop a shared understanding of the challenges, risks, and responses to climate and environmental change. Such consciousness requires and eye towards the future and a pulse on current and futuristic threats to mother earth and her inhabitants.

In this chapter the author used findings from an ethnographic case study to demonstrate how vulnerable society remains when there is a lack of knowledge, mismanaged expectations, and a systemic lack of planning for the unexpected disaster. The author also offered a creative means of using readily available communications and social media technology to advance resilience and citizen disaster preparedness through a "Be Ready" trivia campaign. Moreover, this chapter provided a comprehensive illustration of how this data can inform strategies to achieve several of the United Nations Sustainable Development Goals (SDGs) as well as the United Nations Office for Disaster Risk Reduction's Sendai Framework for Disaster Risk Reduction (SFDRR). Specifically, this chapter discussed SDG #3 (healthy lives and well-being); SDG #4 (quality education); SDG #9 (building resilient infrastructure); SDG #11 (safe, inclusive, and resilient cities and human settlements), and SFDRR Priorities 2 and 3 (strengthen and invest in disaster risk reduction).

Future research directions include the development of a new index to measure community resilience to include citizens' use of communications and social media technology. The author also noted the necessity for additional research regarding the development of questionnaires to leverage local community knowledge in developing countries where maintaining 72-hour emergency kits are not economically feasible.

Raising the level of public awareness and facilitating a means to improve individual participation in disaster preparedness through low cost education programs could assist government and non-government efforts to improve disaster risk reduction; increase access to and use of information and communication technology; increase disaster emergency planning and response; and build resilient communities.

REFERENCES

Adams, R. M., Prelip, M. L., Glik, D. C., Donatello, I., & Eisenman, D. P. (2018). Facilitating partnerships with community-and faith-based organizations for disaster preparedness and response: Results of a national survey of public health departments. *Disaster Medicine and Public Health Preparedness*, *12*(1), 57–66. doi:10.1017/dmp.2017.36 PMID:28735594

Aldrich, D. P., & Meyer, M. A. (2015). Social capital and community resilience. *The American Behavioral Scientist*, *59*(2), 254–269. doi:10.1177/0002764214550299

Bell, W. (1997). *Foundations of futures studies*. New Brunswick, NJ: Transaction Publishers.

Bromley, E., McCreary, M., Wells, K. B., Eisenman, D. P., Williams, M., Chandra, A., ... Kim, B. (2017). How do communities use a participatory public health approach to build resilience? The Los Angeles County Community Disaster Resilience Project. *International Journal of Environmental Research and Public Health*, *14*(10), 1267. doi:10.3390/ijerph14101267 PMID:29065491

Brown, D., & Kulig, J. (1996). 1997). The concept of resiliency: Theoretical lessons from community research. *Health and Canadian Society (Winnipeg, Man.)*, *4*(1), 29–52.

Brown, L. M., Haun, J. N., & Peterson, L. (2014). A proposed disaster literacy model. *Disaster Medicine and Public Health Preparedness*, *8*(03), 267–275. doi:10.1017/dmp.2014.43 PMID:24992944

Buikstra, E., Ross, H., King, C., Baker, P., Hegney, D., McLachlan, K., & Rogers-Clark, C. (2010). The components of resilience: Perceptions of an Australian rural community. *Journal of Community Psychology*, *38*(8), 975–991. doi:10.1002/jcop.20409

Chan, E. Y., Yue, J., Lee, P., & Wang, S. S. (2016). Socio-demographic predictors for urban community disaster health risk perception and household based preparedness in a Chinese urban city. *PLoS Currents*. doi:10.1371/currents.dis.287fb7fee6f9f45 21af441a236c2d519 PMID:28856059

Chandra, A., Williams, M., Plough, A., Stayton, A., Wells, K. B., Horta, M., & Tang, J. (2013). Getting actionable about community resilience: The Los Angeles County community disaster resilience project. *American Journal of Public Health Association*, *103*(7), 1181–1189. doi:10.2105/AJPH.2013.301270 PMID:23678906

Chen, J., Chen, T. Y., Vertinsky, I., Yumagulova, L., & Park, C. (2013). Public-private partnerships for the development of disaster resilient communities. *Journal of Contingencies and Crisis Management*, *21*(3), 130–143. doi:10.1111/1468-5973.12021

Craft, L. (2017). *Perceived Threats to Food Security and Possible Responses Following an Agro-Terrorist Attack* [Doctoral dissertation]. Walden University.

Craft, L. (2018, March 13). Accelerated Weirding: Policy Administration in Hothouse Futures [Panel]. In *American Society for Public Administration Annual Conference.* Denver, CO.

Crutchley, T. M., Rodgers, J. B., Whiteside, H. P. Jr, Vanier, M., & Terndrup, T. E. (2007). Agroterrorism: Where are we in the ongoing war on terrorism? *Journal of Food Protection*, *70*(3), 791–804. doi:10.4315/0362-028X-70.3.791 PMID:17388078

Cutter, S. L., Barnes, L., Berry, M., Burton, C. G., Evans, E., Tate, E. C., & Webb, J. (2008). *Community and regional resilience: Perspectives form hazards, disasters and emergency management. CARRI Research Report 1*. Oak Ridge: Community and Regional Resilience Institute. Retrieved from http://www.resilientus.org/wp-content/uploads/2013/03/FINAL_CUTTER_9-25-08_1223482309.pdf

Dator, J. (2009). Alternative futures at the Manoa School. *Journal of Futures Studies.*, *14*(2), 1–18.

Faulkner, L., Katrina, B., & Tara, Q. (2018). Analyzing community resilience as an emergent property of dynamic social-ecological systems. *Ecology and Society*, *23*(1), 24. doi:10.5751/ES-09784-230124

Federal Emergency Management Agency. (2009). Mitigation. Retrieved from http://www.fema.gov/what-mitigation

Federal Emergency Management Agency. (2011). A whole community approach to emergency management: Principles, themes, and pathways for action. Retrieved from https://s3-us-gov-west-1.amazonaws.com/dam-production/uploads/20130726-1813-25045-0649/whole_community_dec2011_2_.pdf

George, N., & Stark, A. (2016). Community resilience and crisis management: Policy lessons from the ground. *Policy and Politics*, *44*(4), 591–607. doi:10.1332/030557315X14351420409945

Glanville, J. L., Andersson, M. A., & Paxton, P. M. (2013). Do social connections create trust? An examination using new longitudinal data. *Social Forces*, *2*(545). doi:10.1093fot079

Heuer, Richards J. & Center for the Study of Intelligence (U.S.). (1999). *Psychology of intelligence analysis*. Center for the Study of Intelligence, Central Intelligence Agency. Retrieved from http://www.loc.gov/catdir/enhancements/fy0634/99489116-t.html

Holdeman, E. (2011). Black swan events. *Emergency Management*, *6*(3), 10.

Hyvärinen, J., & Vos, M. (2015). Developing a conceptual framework for investigating communication supporting community resilience. *Societies (Basel, Switzerland)*, *5*(3), 583–597. doi:10.3390oc5030583

International Organization for Migration. (2014). Human mobility: Shaping vulnerability and resilience to disasters. Retrieved from https://www.iom.int/files/live/sites/iom/files/What-We-Do/docs/background-paper-for-HFA2-light.pdf

Johnson, R.M., Edwards, E., Gardner, J.S., Diduck, A.P. (2018). Community vulnerability and resilience in disaster risk reduction: An example from Phojal Nalla, Himachal Pradesh, India. doi:10.100710113-018-1326-6

Kawachi, I., Kennedy, B. P., & Wilkinson, R. G. (1999). Crime: Social disorganization and relative deprivation. *Social Science & Medicine*, *48*(6), 719–731. doi:10.1016/S0277-9536(98)00400-6 PMID:10190635

Kulig, J. (1999). Sensing collectivity & building skills: Rural communities & community resiliency. In W. Ramp, J. Kulig, I. Townshend et al. (Eds.), Health in rural settings: Contexts for action (pp. 223-244). Lethbridge: University of Lethbridge.

Kulig, J., Edge, D., & Joyce, B. (2008). Understanding community resiliency in rural communities through multimethod research. *Journal of Rural and Community Development*, *3*(3), 77–94.

Kulig, J., Reimer, W., Townshend, I., Edge, D., & Lightfoot, N. (2011). Understanding links between wildfires and community resiliency: Lessons learned for disaster preparation and mitigation. Retrieved from http://www.ruralwildfire.ca/files/Final

Kulig, J. C., Edge, D. S., Townshend, I., Lightfoot, N., & Reimer, W. (2013). Community resiliency: Emerging theoretical insights. *Journal of Community Psychology*, *41*(6), 758–775. doi:10.1002/jcop.21569

Lee, J. V. (2010). Policing after 9/11: Community policing in an age of homeland security. *Police Quarterly*, *13*(4), 347–366. doi:10.1177/1098611110384083

Leykin, D., Lahad, M., Cohen, R., Goldberg, A., & Aharonson-Daniel, L. (2016). The dynamics of community resilience between routine and emergency situations. *International Journal of Disaster Risk Reduction*, *15*, 125–131. doi:10.1016/j.ijdrr.2016.01.008

Lichtveld, M. (2018). Disasters through the lens of disparities: Elevate community resilience as an essential public health service. *American Journal of Public Health*, *108*(1), 28–30. doi:10.2105/AJPH.2017.304193 PMID:29211534

Linnell, M. (2014). Citizen response in crisis: Individual and collective efforts to enhance community resilience. *Human Technology*, *10*(2), 68–94. doi:10.17011/ht/urn.201411203311

Martine, G., Schensul, D., & Guzmán, J. M. (2013). *The demography of adaptation to climate change*. New York, NY: United Nations Population Fund, International Institute for Environment and Development, and El Colegio de México.

McCrea, R., Walton, A., & Leonard, R. (2014). A conceptual framework for investigating community wellbeing and resilience. *Rural Society*, *23*(3), 270–282. doi:10.1080/10371656.2014.11082070

Merton, R. K. (1968). Social theory and social structure (Enl. Ed.). Free Press.

Muttarak, R., & Pothisiri, W. (2013). The role of education on disaster preparedness: Case study of 2012 Indian Ocean earthquakes on Thailand's Andaman Coast. *Ecology and Society*, *18*(4), art51. doi:10.5751/ES-06101-180451

Norris, F. H., Stevens, S. P., Pfefferbaum, B., Wyche, K. F., & Pfefferbaum, R. L. (2008). Community resilience as a metaphor, theory, set of capacities, and strategy for disaster readiness. *American Journal of Community Psychology*, *41*(1-2), 127–150. doi:10.100710464-007-9156-6 PMID:18157631

Plodinec, M. J., Edwards, W. C., & White, R. K. (2014). Applications of a "Whole Community" Framework for Enhancing Community or Campus Resilience. *Procedia Economics and Finance*, *18*, 9–16. doi:10.1016/S2212-5671(14)00907-1

Reininger, B. M., Rahbar, M. H., Lee, M., Chen, Z., Alam, S. R., Pope, J., & Adams, B. (2013). Social capital and disaster preparedness among low income Mexican Americans in a disaster prone area. *Social Science & Medicine*, 8350–8360. doi:10.1016/j.socscimed.2013.01.037 PMID:23465204

Reuter, C., Hughes, A. L., & Kaufhold, M. A. (2018). Social media in crisis management: An evaluation and analysis of crisis informatics research. *International Journal of Human-Computer Interaction*, *34*(4), 280–294. doi:10.1080/10447318.2018.1427832

Rothstein, B. (2013). Corruption and social trust: Why the fish rots from the head down. *Social Research*, *80*(4), 1009–1032.

Sampson, R. J., & Groves, W. B. (1989). Community structure and crime: Testing social-disorganization theory. *American Journal of Sociology*, *94*(4), 774–802. doi:10.1086/229068

Sardar, Z., & Sweeney, J. A. (2016). The three tomorrows of postnormal times. *Futures*, *75*, 1–13. doi:10.1016/j.futures.2015.10.004

Smith, K., & Lawrence, G. (2014). Flooding and food security: A case study of community resilience in Rockhampton. *Rural Society*, *23*(3), 216–228. doi:10.1080/10371656.2014.11082066

United Nations. (2015). Transforming our world: The 2030 agenda for sustainable development. Retrieved from https://sustainabledevelopment.un.org/post2015/transformingourworld

United Nations. (2017). International migration report 2017. Retrieved from http://www.un.org/en/development/desa/population/migration/publications/migrationreport/docs/MigrationReport2017.pdf

United Nations Office for Disaster Risk Reduction (UNISDR). (2015). Chart for the Sendai Framework for Disaster Risk Reduction 2015-2030. Retrieved from https://www.preventionweb.net/publications/view/44983

U.S. Department of Health and Human Services. (2012). Community resilience fact sheet. Retrieved from http://www.phe.gov/Preparedness/planning/abc/Documents/community-resilience.pdf

U.S. Department of Homeland Security. (2010). Quadrennial homeland security report review: A strategic framework for a secure homeland. Retrieved from http://www.dhs.gov/xlibrary/assets/qhsr_report.pdf

U.S. Department of Homeland Security. (2013). National Response Framework (2nd ed.). Retrieved from http://www.fema.gov/national-response-framework

U.S. Department of Homeland Security. (2014). The 2014 quadrennial homeland security review. Retrieved from http://www.dhs.gov/sites/default/files/publications/qhsr/2014-QHSR.pdf

Wallace-Wells, D. (2017). The uninhabitable earth. The New Yorker. Retrieved from http://nymag.com/daily/intelligencer/2017/07/climate-change-earth-too-hot-for-humans.html

Zembroski, D. (2011). Sociological theories of crime and delinquency. *Journal of Human Behavior in the Social Environment*, *21*(3), 240–254. doi:10.1080/10911359.2011.564553

ADDITIONAL READING

Atkinson, C. L. (2014). *Toward resilient communities: Examining the impacts of local governments in disasters*. New York, NY: Routledge.

Blackburn, F. (2017). Community engagement, cultural competence and two Australian public libraries and Indigenous communities. *International Federation of Library Association and Institutions*, *43*(3), 288–301. doi:10.1177/0340035217696320

Kawalerowicz, J., & Biggs, M. (2015). Anarchy in the UK: Economic deprivation, social disorganization, and political grievances in the London Riot of 2011. *Social Forces*, *94*(2), 673–698. doi:10.1093fov052

Moats, J. B. (2007). *Agroterrorism: A guide for first responders* (1st ed.). Texas A&M University Press.

Monke, J. (2007, March). Agroterrorism: Threats and preparedness (Congressional Research Service Report No. RL32521). Federation of American Scientists. Retrieved from http://www.fas.org/sgp/crs/terror/RL32521.pdf

Schaible, L. M., & Altheimer, I. (2016). Social structure, anomie, and national levels of homicide. *International Journal of Offender Therapy and Comparative Criminology*, *60*(8), 936–963. doi:10.1177/0306624X15595420 PMID:26216918

Seebeck, L. (2007). Responding to systemic crisis: The case of agroterrorism. *Studies in Conflict and Terrorism, 30*(8), 691–721. doi:10.1080/10576100701200165

KEY TERMS AND DEFINITIONS

Agro-Terrorism: The deliberate attack of food and water supplies for the purpose of generating fear, causing economic losses or undermining social stability.

Anomie: Social instability that occurs after societal bonds, common rules, standards, and values break down in society.

Black Swan: An event (positive or negative) that is considered outside the realm of expectation but remains plausible and will have an extreme impact.

Community Competence: The collective aptitude of individuals to learn about their social environment and use the information to identify problems and establish consensus to collectively address the problems to meet the needs of the community.

Community Resilience: The collective efforts of individuals within a community to resist, absorb, recover from, or successfully adapt to, adversity or change in conditions.

Food Security: A state that exists when people, at all times, have the requisite physical and economic access to safe and nutritious food to meet their dietary needs for a healthy life.

PACE Plan: In this chapter, an acronym and methodology to build well-constructed emergency/disaster preparedness plans that account for the unknowns and unexpected. PACE stands Primary, Alternate, Contingency, and Emergency. When establishing this plan, individuals account for additional courses of action when the primary is no longer an option.

APPENDIX 1: INTERVIEW QUESTIONS FOR RESIDENTS

Perception of Fear

Question 1: What do you perceive as the greatest threat to you and your family after an agro-terrorist attack that threatens the food security?

Follow-up Question: How would you protect yourself and your family from this threat?

Perception of Trust in Government Assistance/Agencies

Question 2: In the event of an agro-terrorist attack that causes a food shortage what is your local government's emergency management plan?

Follow-up Question: If you do not know, how would you find out following an agro-terrorist attack?

Question 3: Which local and state agencies do you believe are responsible for food security during a disaster in your area?

Follow-up Question: What do you think they are supposed to do?

Follow-up Question: If you do not know, how would you find out following an agro-terrorist attack?

Question 4: To what extent do you trust your local and state law enforcement to aid you and your immediate family following an agro-terrorist attack? Why or why not?

Perceptions of Social Capital

Question 5: To what degree do you believe your neighbors are friendly?

Question 6: How often do you visit or get together with your neighbors to chat or for social visits?

Question 7: How often do you and your neighbors do favors for each other such as watching each other's children, lending tools, or helping with daily requirements/chores?

Question 8: To what degree do you believe there is a feeling of belonging in your community?

Question 9: To what degree do you trust your neighbors in the event of an agro-terrorist attack and a food shortage?

Question 10: How would you protect yourself and your immediate family following an agro-terrorist attack and the threat of a food shortage?

Perceptions of Community Resilience

Question 11: How does the physical environment contribute to your overall health?
Question 12: Who do you identify as the leaders within your community?
Question 13: How well do you believe the leaders in your community listen to the residents?
Question 14: How frequently do you participate in community events?
Question 15: Do you think your community would be able to endure a large-scale disaster like an agro-terrorist attack that threatened the food supply?

APPENDIX 2: INTERVIEW QUESTIONS FOR EXPERTS

Question 1: How is your organization involved in the following: food security, emergency management, and community services?
Follow-up Question: How frequently is your organization involved?
Question 2: In the event of an agro-terrorist attack that threatens food security what are the responsibilities of your organization?
Question 3: What are your organization's responsibility to the citizens of Yuma, Arizona following an agro-terrorist attack and a threat to food security?
Question 4: How does your organization work with community and the citizens of Yuma, Arizona (e.g., meet and greets and educational classes on community topics of interests)?
Follow-up Question: Please provide specifics on your organization's work with the community.
Follow-up Question: How frequently does your organization work with the community?
Question 5: To your knowledge, what kind of social networks are available for the citizens of Yuma that could assist in their preparedness and/or response to a disaster such as an agro-terrorist attack?
Question 6: In your opinion, do you think the citizens of the community function as a community?
Follow-up Question: Do you think they would pull together and support and protect one another if a threat to food security exists?

Chapter 9
Science Communication for Climate Change Disaster Risk Management and Environmental Education in Africa

Innocent Chirisa
University of Zimbabwe, Zimbabwe & University of the Free State, South Africa

Abraham Rajab Matamanda
University of the Free State, South Africa

ABSTRACT

This chapter describes and explains the role of science communication to advance environmental education on climate change with a special reference on Africa being one of the regions suffering from the effects of climate-induced disasters and risks in the increased anthropogenic effects of modern development. It is argued that scientists are poor communicators on what they do, and hence, are often misunderstood by the media and society. This then calls for attention to be paid with regards to science communication, which has to be packaged in ways that make it easier for the generality of citizens to interpret and understand. Efforts in this regard are made possible through environmental education, which has proved to be useful in the discourse of disaster risk management in different parts of the world. The buttressing methodological philosophy to this chapter is applied systems approach. Critical areas of reference are health and diseases, resilient communities, coastal adaptation, and farming practices and technologies.

DOI: 10.4018/978-1-5225-7727-0.ch009

INTRODUCTION AND BACKGROUND

There exists a very strong connection between science communication and environmental education. In southeast Mexico, for example, a multifaceted, comprehensive teacher-training project was implemented and code-named Education for Sustainable Development. The project included training workshops, seminars, congresses, forums and distance education. New teaching units were added to textbooks, including topics on biodiversity, pollution, and deforestation. Strategies to raise awareness include media campaigns, messaging through youth radio programs, theatre and music, conferences, environment day celebrations, and other events and programs within and outside of school can complement formal curricula and significantly strengthen learning (Iltus, 2013). In all these cases, there is an aspect of disseminating scientific knowledge through other means, and at the same time raising awareness on environmental issues. Overall, in both formal and informal arrangements of education, the communication on the definition and science, characterization, policy, and advocacy on climate change, a lacuna exists on how to do this in a defined and compelling manner. As would be expected by any fuzzy yet real problem, there are many misconceptions, myths and isolated and piecemeal approaches to the whole debate on climate change. The major cause of disagreement oscillates around priority and action. The major questions become what needs to be done now, by whom and for whom? This is a policy concern and governments and key actors are rather concerned with immediate pressing issues like poverty, unemployment and managing politics around resources. To make matters worse, the climate science and environmental scientists seem to be talking only to themselves with their decorated models and deep-thought-out studies. The overall impression globally and regionally is the scientists are poor communicators on what they do hence are often misunderstood by the media and society over the work they spend their life and effort doing. In addition, they seem not to be getting their message to the policymakers and the communities themselves, those being the tipping centers that decide what actions have to be made practically and as implementing agents. It behooves, therefore, the authors of this chapter to explain how Africa can pass on the important message of innovative climate change education that will result in an improvement of how communities have to adapt and cope with the crisis across sectors - agriculture, water and sanitation, housing, to name these few. The present chapter makes a case for Africa in light with science communication, environmental education, and dissemination of climate change information. This is more or less a broad analysis of the issue of science communication as a mechanism of advancing environmental education on climate change. The dimensions are taken in this analysis range from the focus on public awareness, public understanding, climate change literacy, media coverage and how this relates to climate change education in the African context.

The aim of this chapter to describe and explain the role of science communication as a method to advance environmental education on climate change with a special reference on Africa being one of the regions suffering from the effects of climate-induced disasters and risks in the increased anthropogenic effects of modern development. We argue for a strong possibility of science communication as a tool for advancing environmental education on climate change, focusing on Africa. Africa has been witnessing serious climate-induced disasters. The issue under discussion is not to debate on whether Africa is able to cope with climate change or not. Rather, the debate centers on the mechanisms that can be used to make people understand what climate change is and how they can mitigate its negative impacts. Moreover, we also explore the possibilities of harnessing environmental education as a tool for the effective communication of this science information and data. Banda (2013) brings in the aspect of media coverage as one of the most important mechanisms for spreading climate change related information. Everyone needs to know about climate change.

Climate change is a science and it is this science that ordinary people need to fully understand so that they can adequately deal with its consequences. The reporting of information on climate change can be done in many forms of media including radios, televisions, and magazines among others. Banda (2013) points out that the most important and significant research appears in journals such as Nature Climate Change, Geophysical Research Letters, Nature, Science, PNAS, and Climatic Change. However, Chirisa et al. (2018) are of the opinion that indigenous knowledge systems are commendable when it comes to disseminating climate change information. It is not an issue of where climate-related information is disseminated, but the biggest question is whether the victims of climate-induced disasters really grasp the message being conveyed. Radio journalists are more able than most to interact directly with their audiences and bring audience-contributions into their stories. According to Goh et al. (2008:8):

effective science communication requires attention to the messages that are to be conveyed that depends on the "art" of communication, allowing adequate time to produce science communication products that include time for feedback and revision. Good science communication requires attention to both the science and the presentation. In general practice, the vast majority of the effort by scientists is in the collection and analysis of data, with little time or resources devoted to the communication of science. Rather than science communication being an afterthought, factoring in the time and resources that are needed for developing a quality communication product is recommended.

Climate Change as a Subject in Environmental Education

Climate change is more than an environmental concern. It has emerged as the biggest developmental challenge for the planet. Its economic impacts, particularly on the poor, make it a major governance issue as well (United Nations Development Programme, 2009). Climate change is indeed a very complex subject that needs to be explored from many different perspectives. The subject of climate change can be understood from social, economic and gender perspectives among others (UNESCO, 2009). climate change relates not solely to the science domain but deals with all aspects of society. From the impacts of climate change on women and community life, climate change is not gender neutral that in fact, gender inequality is the most pervasive of all inequalities. Gender equality itself is shaped by other social differentiators such as race-ethnicity, class, and age. Climate change is also about the economy. Industrialized countries have managed to de-link sulfur dioxide emissions from economic growth. In other words, emissions have fallen even as national income has risen. But they have failed to do the same with carbon dioxide (CO_2) emissions. Per capita, CO_2 emissions remain closely related to a country's level of economic development, and thus standard of living. It is evident that as long as the world economy is carbon-based – driven by energy from coal, oil, and natural gas – growth cannot be de-linked substantially from CO_2 emissions. Climate change has taken a centre stage in environmental education in recent years. The need to include climate change in environmental education has been extended to the school curriculum. In the last two decades, Climate Change and Environmental Education (CCEE) and Education for Sustainable Development (ESD) have emerged as critical tools for protecting the environment and fostering sustainable development. The Rio Declaration on Environment and Development of 1992 confirmed the indispensability of children in achieving sustainable development. The United Nations Framework Convention on Climate Change (Article 6) and the Kyoto Protocol (Article 10) both encourage governments to educate, empower and engage all stakeholders and major groups on climate change policies. In 2002, the United Nations General Assembly proclaimed the United Nations Decade of Education for Sustainable Development (2005–2014), underscoring the indispensable role of education in achieving sustainable development. Facilitating access to information on climate change is critical to winning public support for climate-related policies (Iltus, 2013).

The buttressing methodological philosophy to this chapter is applied systems approach that views actors, processes, inputs, and outputs of the climate change, policy and science as intricately related hence requiring a trans-disciplinary, interdisciplinary and multi-disciplinary approach. Environmental education that fails to approach the environment this way is not only flawed but faces outright difficulties when juxtaposed to the complex reality of this world. . The chapter is built on systematic reviews of work done by scientists in Africa through a review of recent communications, scholarly views and media reports on climate issues in Africa. Reviews cover aspects of the science of mitigation and adaptation across sectors. A sampling of the studies material is random and systematic. However, critical areas of reference will be health and diseases, resilient communities, coastal adaptation and farming practices and technologies. Textual, content and discourse analyses will form the pedestal of data analysis for the chapter.

CONCEPTS AND DEBATES

Climate Change

Climate change is one of the most important issues on the global political and economic agenda, yet it has taken at least 20 years to reach the international priority hierarchy (British Council, 2004). Mostly, this is because climate change was originally communicated as a scientific problem, using technical jargon. Complex, confusing, and at times contested scientific information resulted in a slow public and political response to the climate crisis (Chanza and de Wit, 2015; Chirisa et al. 2018). The climate change debate has also taken place in industrialized nations, among a public largely safe from its worst effects. Yet, for many, especially those who are constantly at risk from the climate-induced disasters, climate change remains an abstract concept. In Africa, climate change is far from a new phenomenon – it is already determining the course of people's lives (Chanza & de Wit, 2015).

Various climate-induced disasters overwhelm Africa and in recent years these seem to be escalating and increasing in severity. The under-developed nature of the continent is also attributed for exacerbating the vulnerability of Africa to climate-induced disasters. Climate change has thus emerged as a disaster on the African continent and it is increasingly becoming an agenda for most governments. For example, in Zimbabwe since 2014 the country broadened the scope of the then Ministry of Environment and Natural Resources Management to Ministry of Environment, Water and Climate (MoEWC) owing to the climate-disasters which plague the country. By definition, a disaster is described as a specific phase when there is a sudden or slow natural or human-caused occurrence which results in widespread damage far exceeding the affected community's ability to cope with using their own resources (UNISDR, 2009). From another perspective, a disaster is considered to be an occurrence that disrupts the normal conditions of existence and causes a level of suffering that exceeds the capacity of adjustment of the affect community (Federal Ministry for Economic Cooperation and Development, 2015; WHO/EHA, 2002). On the whole, a disaster is a product of a hazard which results in society failing to cope when a particular event occurs. Considering the catastrophy of climate change which is associated with hazards such as floods, droughts, cyclones and heat waves, there is a need for understanding the complexity of climate change so as to mitigate the negative effects of the climate-induced disasters.

From the foregoing definitions of disasters, it emerges that there are both human-induced and natural disasters. There are various climate-induced disasters which are increasingly overwhelming communities across the world. Although some proponents attempt to distinguish between natural and human-induced causes of climate change, the bottom line is all are categorized as clmate-induced disasters. Droughts are among the climate-induced hazards which plague most nations across the world. Masih et al. (2014) outlines that drought is a recurrent global phenomenon which affects human in different ways. According to Guha-Sapir et al. (2014) a total of 642 drought events were reported across the world between 1900 and 2013. Drought hazards result in various challenges which include food insecurity, death and loss of incomes. For example, the reduction of rural income distribution in Spain has been attributed to the occurrence of climate induced droughts in the country (Quiroga and Suarez, 2016). Rajsekhar and Gorelick (2017) outline how climate change-induced droughts escalates water scarcity woes which ultimately exacerbates the conflicts in Jordan.

Extreme weather events and greater unpredictability in weather patterns are having serious consequences for people who rely on land, lakes and seas to feed themselves and to earn a living. As a result, Africa's engagement with the issue is evolving rapidly, presenting an opportunity to leapfrog the slow evolution of western public opinion and political action. African citizens' response to climate change is hampered by a fundamental shortage of relevant, useful information for African audiences. The intensive media coverage and public awareness campaigns prevalent in much of the industrialized world have been largely absent in Africa, particularly outside major urban centers such as rural areas of Africa. Too often, African voices are absent from the international climate debate. The need for scientists to communicate more effectively about climate change is urgent.

For people to take climate change seriously and support appropriate responses, they need to feel sure it is happening and is caused primarily by humans. But while the rise in global temperature is a fact (see, e.g., Intergovernmental Panel on Climate Change (IPCC) ((2007)), which calls the warming "unequivocal"), 56% of Americans believe there is a lot of disagreement among scientists about whether global warming is even occurring. While the most authoritative scientific body of knowledge (e.g. IPCC, 2007; American Association for the Advancement of Science, 2006) attributes most of the warming of the past 50 years to human activity 0, only 41% of Americans believe that humanity is the dominant cause. Forty-two percent (42% believe it is due about equally to natural and human causes. This is according to an April 2007 poll by ABC News, The Washington Post, and Stanford University. Yet, in recent years the global attention to climate change discourse is astounding. Surprisingly, at a time when most world leaders are finding ways to combat climate change and related disasters, the U.S. President Donald Trump does not subscribe to the notion that climate change is a reality. Rather, earlier in 2018 when six of the G7 committee on Climate Action were engaged on climate change comminuqie, Trump decided not even to join the conversation (Inside Climate News, 2018). Such actions show the diverging views on climate change even among heads of states. Ironically the developed world is fingered to have the capacity to champion the combating of climate change considering the resources and expertise they have (Chirisa et al. 2018). Why is there an understanding gap? There is plenty of blame to go around, from general scientific illiteracy to the media's failings, to a disinformation campaign (e.g., see Union of Concerned Scientists, 2007) designed to sow doubt.

Science Communication and Environmental Education Nexus

Scientific communication is a multi-faceted subject that is going through a conceptual transformation. In the current public discussion of scientific communication, most of the attention is focused on journals, especially on the "journal crisis," in which libraries cannot afford the rapidly increasing subscription rates and the growing ranks of journals. Electronic publishing, using the Internet, is often seen as a possible way to relieve the cost pressure. This view is too limited, though, as it concentrates on a small part of scientific communication, and it does not provide a full picture of the evolution that is currently taking place in as far as science communication is concerned (Odlyzko, 2004).

There is no single definition and clear explanation of what science communication is all about or what it is intended to do. Goh et al (2008) explain science communication. According to the authors, science communication is the development of content-rich, jargon-free, communication-based materials. Content-rich entails communication which is replete with data and ideas. Jargon-free refers to the elimination of shorthand notation that scientists use to communicate within their peer groups—this means removing acronyms and maintaining a common language basis for an explanation of concepts. In addition, Communication-based refers to focusing on the intended audience and providing an even broader base of accessibility for a wider audience. Effective communication is an important part of doing science.

Goh et al (2008) identifies three key elements of science communication, which are visualization, context, and synthesis. Raw data do not guarantee anyone's understanding except perhaps the investigator collecting the data. Rather, data that have been analyzed, interpreted and synthesized are needed for meaningful science communication outcomes. Visualization is key, as the audience must be able to see the who, why, what, where, when, and how of the data that is used to support the ideas. Making a point with data visualized is very powerful, but the audience needs to be able to see and interpret the data themselves. For example, images showing the severity of damage caused by climate-induced disasters are instrumental in environmental education as they help to show the extent of climate change damage. In this way, geospatial information may be generated which may help to visualize the nature and pattern of, for instance, hurricanes and tornadoes as well as the damage they cause; tsunamis in Haiti and Hurricane Katrina in New Orleans are both climate-induced disasters. Context provides answers to the important questions "Why should we care?" or more simply, "So what?" and can include using comparative data so that specific examples can be characterized as 'high' or 'low' relative to regional or global extremes. With respect to climate-induced disasters which are context-specific, there is a need to disseminate information which relates to a particular context. For example, in the Arctic and coastal regions, rising sea-levels, hurricane, cyclones and flooding are prominent. In the sub-Sahara African and Asian context, the nature of the disasters include heatwaves, droughts, and flooding (Chanza and de Wit, 2015; Chirisa et al. 2018). Considering these contextual differences in climate-induced disasters environmental education must then be packaged such that it relates to that particular context. In this way, it will make sense to the recipients who will then relate with the kind of information being conveyed to them. Lastly, synthesis is another aspect which needs to be considered as it refers to the combination of components to make a whole. This synthesis is also evident in the disaster management cycle where there is an overlap of processes which in as much as they may appear to be isolated yet in actual essence they overlap and complement each other.

Burns et al (2003) have proffered their perspective on the concept of science communication. The authors try to raise questions on how the concept differs from public awareness of science, public understanding of science, scientific culture, and scientific literacy. Science communication is not about encouraging scientists to talk more about their work, nor is it an offshoot of the discipline of communications. Although people may use the term "science communication" as a synonym for public awareness of science (PAS), public understanding of science (PUS), scientific culture (SC), or scientific literacy (SL), to date, science communication has not been clearly defined. "Science communication is typically thought of as the activities of professional communicators (journalists, public information officers, scientists themselves)" or simply as ". . . the promotion of the public understanding of science . . ." The 2000 report '*Science and the Public: A Review of Science Communication and Public Attitudes to Science in Britain*, defines science communication as a term that "encompasses communication between:

- Groups within the scientific community, including those in academia and industry
- The scientific community and the media
- The scientific community and the public
- The scientific community and government, or others in positions of power and/or authority
- The scientific community and government, or others who influence policy
- Industry and the public
- The media (including museums and science centers) and the public
- The government and the public."

This definition is useful in that it identifies the important participants in science communication. However, it is lacking in that it is only descriptive and does not address the how or why of science communication. Science communication, thus, simply depicts the spread of information from a scientist or other practitioner to the ordinary person to the extent that the latter clearly understands the message being communicated. Although, the definition may seem to be descriptive it is critical in disaster management as there is a component of mitigation which occurs when a community is aware of the risks and they are able to take necessary action.

Science Communication can be, indeed, instrumental as a driver of environmental education on climate change-induced disasters. Science communication is critical in disaster management where it becomes a panacea for dealing with and avoiding the risks posed by climate-induced disasters. Disaster management is a task which involves multiple stakeholders who deal with issues such as planning, coordination, communication and risk management of disasters (Guha-Sapir et al. 2014). The logic and rationale is on the management of resources and information as far as a disastorous event is concerned and seamlssly one coordinates these resources (Modh, 2010). Disaster management is meant to increase the resilience of cities and places such that they become safe and promote human well-being. Hence, there is great need to consider how the science communication will be packaged such that the audience will share the same vision with decision-makers and understand the need for having communities which are resilient and regenerative to climate-induced disasters (Bizzotto, 2018).

As climate change takes hold, people will demand information about what is happening and what they and their governments can do about it. Wise and responsible media managers see that climate change presents an opportunity to grow and better serve these audiences. Three of the media's traditional roles — informing audiences, acting as watchdogs and campaigning on social issues — are especially relevant.

Media coverage also provides a vital link between the science and the service providers upon which much will depend. For African journalists, coverage of climate change means several things. At home, it can save lives, change policy and empower people to make informed choices. Through informed reporting, journalists can shine a light on the wealth of activities that people are already undertaking to prepare for climate change. Internationally, it can bring African stories to global audiences and help encourage the rich and powerful countries, their citizens and the companies based there, to act in solidarity with climate vulnerable communities (Banda, 2013).

Science communication serves a very critical tool for raising climate change literacy among people. This was the case in southeast Mexico, where a multidimensional, comprehensive teacher training project was launched on Education for Sustainable Development (ESD). The project comprised of training workshops, seminars, congresses, forums and distance education. New teaching units were added to textbooks, including topics on biodiversity, pollution, and deforestation. Strategies to raise awareness include media campaigns, messaging through youth radio programmes, theatre and music, conferences, environment day celebrations, and other events and programmes within and outside of school can complement formal curricula and significantly strengthen learning. These strategies require collaboration with non-governmental organizations (NGOs) and the private sector. In Albania, for example, the Child-Led Environmental Education Initiative created a partnership with local companies that introduced schools to recycling, curricular reform, teacher training and a community-wide media campaign. As a result, nearly half of the children in selected schools adopted at least one environmentally-friendly behavior at home (Iltus,2013). This is a testimony that science communication, is well used by scientists and policymakers can serve as a vehicle of improving knowledge on climate change-related issues.

Science communication has also been applied in coastal areas in the American context, so as to disseminate climate change related information. Coastal areas can be thought of as "people magnets," drawing ever-increasing numbers to the shores to live, work and play. "It is estimated that 53% of the U. S. population lives on the 17% of land in the coastal zone, and these areas become more crowded every year" (U.S. Global Change Research Program, 2001). With the ambiance, however, comes risk. Coastal areas are subject to chronic environmental threats such as sea level rise and natural flooding, as well as the more dramatic danger from the wind, surge, and inundation associated with hurricanes and tsunamis. Thus, continued population growth in coastal areas is exposing people to serious climate change-related impacts. Scientists are constantly seeking new and better ways to deal with the negative impacts of natural forces, such as climate change and tropical cyclones, as well as human-produced impacts related to the coastal development and use. It has become increasingly clear that, while many solutions lie within the realm of physical science, it is not enough to know what risk-aversive actions are effective. There must be a public will to take these actions. Decision-makers at all levels, including the general public, need to understand the issues, including the risks, and to become motivated toward appropriate action (Morrow, 2009). The Coastal Services Center (CSC) has been a leader in this respect through its Human Dimensions Program and its emphasis on providing tools and methods to assist coastal managers and planners in acquiring "people-related" information.

The CSC focuses on educating local officials, planners, resource managers, and other local decision-makers, as intermediary channels with the general public, on the linkages between hazard impacts, community vulnerabilities and policy alternatives. The goal of this risk communication project is to contribute to coastal resilience through better knowledge of the processes involved in risk perception, how these are influenced by experience, and what strategies might be effective in promoting better citizen understanding of coastal hazards and more effective mitigation and response. This project is designed to inform National Oceanic and Atmospheric Administration (NOAA) coastal managers and their partners on how they might most effectively tailor risk-related education and outreach to target audiences – that employs social marketing techniques to promote coastal resilience (Morrow, 2009). Central to both processes is knowledge about the targeted constituency and how it currently understands and feels about the hazard in question (Plattner, 2005).

The problems associated with public lack of understanding is well illustrated in the case of climate change (Sea Grant News, 2008). Experts often provide information without knowing what lay people understand. If they over-estimate they will be talking over their heads and if they under-estimate they will appear to be talking down to them. Information on the educational level is important when designing written warnings and materials; most tend to be written at too high a reading level for much of the intended population. It cannot be assumed that the targeted group has access to or knows how to use the latest technology. As an example, in one study it was determined that most of the public could not interpret a GIS map (Zarcadoolas et al. 2007). Further, in today's diverse communities there are likely to be many cultures that need to be understood.

A community-based approach is more likely to be successful in changing risk-related behavior. Risks are shared and experienced collectively. "The way risks are perceived within the communities influence the range of actions undertaken to reduce them" (Flint and Luloff 2005, p. 408). People look to their social networks for information and guidance, particularly their trusted sources. An effective way to change risk behavior is by facilitating community interaction to address the issue. Experts, usually from a government agency, are often required to inform the public about a particular risk, such as nuclear power. The typical mode is through community meetings where they stand in front of members of the community and "give them the facts." Scientists are not trained to be communicators and they quickly run into trouble. It's not unusual for an expert to get annoyed and overact to the public's concerns (Lanard 2003). Refuting false statements can serve to reinforce them (Maibach, Roser-Renouf and Leiserowitz 2008). On the other hand, it is better to publicly acknowledge their beliefs in an understanding way before starting to carefully lead them toward realistic attitudes and actions related to their safety and wellbeing. It is important to treat audiences with fairness, honesty, and respect. The risk communication literature offers an abundance of ideas and judgments for making an impact with messages.

Effectiveness depends on the development of arguments based on the values, interests, and needs of the targeted audience. Thus, it is important to target separate messages to specific stakeholders – not the general public (Sandman 2003). An effective paradigm might be the audience segmentation analysis used by Maibach, Roser-Renouf, and Leiserowitz (2009) to address the issue of climate change. Research on public attitudes led to the identification of six different audiences that they labeled the Alarmed, the Concerned, the Cautious, the Disengaged, the Doubtful, and the Dismissive. Reaching each segment calls for tailored communication and educational program, and some are likely unreachable. Using the concepts of hazards and outrage, Sandman (2008) suggests a paradigm for developing risk management strategies according to levels of each. In the case of a risk with high hazard potential but limited public concern (outrage), a typical situation related to hurricane risk, he advocates precaution advocacy.

Educational materials for the public have been developed for many natural, environmental and technological hazards, as well as health issues. They can be found on a multitude of websites, including those of the National Weather Service, the Federal Emergency Management Agency, the American Red Cross, and the Centers for Disease Control. It is important that materials be carefully matched with the audience related to reading level, values, and knowledge. This is often not the case. For example, the materials on county emergency management websites in one state were found to be unsuitable for use with low-income minorities. One suggestion was that materials be packaged not by hazards, but by users – a public-specific view (Mileti, 2006). It is well acknowledged that information alone does not lead to action. In a study of campaigns promoting household hurricane preparedness information alone was insufficient (Mileti and Peek 2002). The more successful campaigns had simple clear messages that specified what was at risk, how severe and probably was the risk and what can be done to reduce the risk or losses. These messages needed to be communicated often using a variety of media and interpersonal channels and trusted messengers. Guidelines for developing warning messages and other risk communication materials are abundant. Form the foregoing review, it can be concluded that science communication takes the center stage in spreading information on climate change. The approaches used to disseminate such information are different depending on the context in question.

The concept of Environmental Education (EE) has been explained differently in the academia. Environmental Education (EE) is a debatable concept that means different things to different people. Environmental education as an interdisciplinary and holistic form of education that is geared towards action and change, which promotes the use of participatory learning, learning by doing and action based methodologies (Mukoni, 2013). Environmental education is thus a range of educational processes and programs through which human beings respond to environmental issues in order to foster change in the direction of community life in a healthy environment. The two definitions above show, that, EE is a process, an unending series of activities or interactions between learners, educators and the larger community, which provides opportunities for learners to find out and share environmental information. The discourse of environmental education entails the dissemination of knowledge and information about the natural systems, ecological, economic and political factors about the environment (Mukoni, 2013).

The definitions put forward stem from the International Conferences conveyed by the United Nations Education, Scientific and Cultural Organization (UNESCO). There are two definitions of environmental education developed during the Belgrade Working Conference on Education (1975) and the 1977 Tbilisi Intergovernmental Conference on Environmental Education. The working definitions were published in the Journal of Environmental Education. Firstly, environmental education should be the integral part of the education process, aimed at practical problems of an interdisciplinary character, build a sense of values and contribute to public well-being (Stapp, 1997). Secondly, environmental education is a process aimed at developing a world population that is aware and concerned about the environment and its associated problems and has the attitudes, motivations, knowledge, commitment, and skills to work individually and collectively toward solutions of current problems and the prevention of new ones. The Wisconsin Environmental Education Board (WEEB) defines environmental education as a lifelong learning process that leads to an informed and involved citizenry having the creative problem-solving skills, scientific and social literacy, ethical awareness and sensitivity for the relationship between humans and the environment, and commitment to engage in responsible individual and cooperative actions. By these actions, environmentally literate citizens will help ensure an ecologically and economically sustainable environment ((Fortier et al, 1998: 11).

The purpose of environmental education is to produce a citizenry that is knowledgeable about the biophysical environment, its associated problems, aware of how to solve them as well (Stapp, 1997). It can be noted that what environmental education seeks to promote is environmental literacy among the world population.

The major objectives of environmental education include: helping an individual gain a clear understanding that man is an inseparable part of the system, consisting of man, culture and the biophysical environment, and that man has the ability to alter this interrelationship, and exposing an individual to environmental concerns. Environmental education has arisen due to the realization that it is crucial to educate people in ways that build their capacities to conserve their natural resources and to construct viable communities (Johnson-Pynn and Johnson, 2010). The role of an environmental education program in an aquarium is an example of improving understanding of the human relationship with the non-human world (Kilian, 2012).

AFRICA AND CLIMATE CHANGE

Climate change has been an issue of global concern, with Africa suffering too many of the impacts. Climate change is already a reality in Africa (Chirisa et al. 2018). Some of the signs that Africa is under the threats of climate change include prolonged and intensified droughts in eastern Africa; unprecedented floods in western Africa; depletion of rainforests in equatorial Africa; and an increase in ocean acidity around Africa's southern coast (De Wit and Jacek 2006; Brown et al. 2007). The Global Humanitarian Forum Human Impact Report (2009) reports that 15 out of the 20 countries in the world that are most vulnerable to climate change are from Africa. Vastly altered weather patterns and climate extremes threaten agricultural production and food security, health, water, and energy security, which in turn undermine Africa's ability to grow and develop. Climate and environmentally related disasters which threaten human security can induce forced migration and produce competition among communities and nations for water and basic needs resources, with potentially negative consequences for political stability and conflict resolution (Besada and Sewakambo, 2009).

The effects of climate change in Africa have been experienced in both coastal and inland areas. For instance, informal settlements across the globe are exposed to extreme impacts of climate change. They are characterized by the low resilience and low adaptive capacity (UNHABITAT, 2012). Various climate-induced disasters have plagued Africa in recent years showing the vulnerability of the continent to these disasters. Floods resulting from heavy rains are among the climate-induced disasters in Africa. For example, in 2001 close to 900 people were killed by floods in Algeria with an additional 45,000 being adversely affected (Lukamba, 2010). In the following year, that is 2002, heavy rains again brought floods and landslides in eastern Africa forcing many people in Tanzania, Uganda, Kenya, Burundi and Rwanda to evacuate their homes (Huq et al. 2007). In August 2018, floods and mudslides occurred in Sierra Leone, close to the capital city Freetown and an approximately

1,141 people are reported to have lost their lives or as missing while thousands were left homeless (Bizzotto, 2018). This is considered to be the deadliest event of 2018 considering the large number of deadths which occured and the impact on infrastructure. Lukamba (2010) provides a detailed analysis of the occurrence of climate-induced disasters in Africa and the severity of damage which occurs on infrastructure, human lives, livelihoods and health. Matamanda et al (2017) outlines how climate-induced disasters in Zimbabwe have been responsible for damage of infrastructure in the urban areas.

There is growing recognition that urban dwellers living in slum environs need to be educated so as to raise their awareness in as far as climate change issues are concerned (Bizzotto, 2018). Consultation with local people is used to create expectations which must be managed throughout implementation (Chanza and de Wit, 2015). This is done through awareness raising and transparency, and adapting methods to a public with a lower education level or illiteracy; through mass media, seminars (Chirisa et al.2018). Such initiatives are aimed at strengthening the response to climate-induced disasters such that communities are even aware of the timing of the implications of certain disasters and the responsive measure to be taken in order to prevent and or mitigate the effects. Such is a testimonyfor environmental education on climate change to be tailor-made so that the ordinary people can understand what it is, what are the impacts, and how it can be mitigated. The question that needs to be addressed is how is environmental education practiced in African slum settlements. There are generally low levels of knowledge about climate change in African informal settlements.

By its nature, it seems Africa is not fully equipped with regards climate-induced disaster management. This is evident from the severity of the negative impacts which usually result in the death of many people, while homelessness and famine are rampant (Lukamba, 2010). Disaster management in African context is mainly reactive and people are nearly always caught unaware and baffled by the events which even makes it difficult for rescue and assistance to be brought to the victims (Hinchberger, 2017). Communication is the major obstacle which limits the prevention or mitigation of the climate-induced disasters. Through environmental education, citizens at risk of disasters have the chance to take preventive measures or adopt coping strategies in time before disaster strikes.

Science communication and environmental education have become areas of interest in African rural development today. This is due to the realization that climate change is not only confined to urban areas but also its impacts and implications are trans-boundary. In Sub-Saharan Africa, mobile phone networks have become instrumental in the dissemination of seasonal climate information to rural farmers. One effort to aid farmers in this intricate decision-making process is through seasonal climate prediction (hereafter "forecast"). Having advance information regarding seasonal precipitation is applicable to decision-making in the agricultural domain, particularly for crop selection, which influences decision-making across all domains. For example, a reliable forecast of the high probability of severe drought (conveyed in an appropriate manner) may allow the farmer ample time to migrate elsewhere to live with relatives in order to avoid investing in a poor harvest. Currently, in Africa, the World Meteorological Organization has helped organize three Regional Climate Outlook Fora to facilitate collaboration among leading scientists and sector representatives in the creation of a consensus forecast for their respective regions (Chavas, 2009). These initiatives are critical in that farmers and rural inhabitants can take proactive measures to deal with predictable climate-induced disasters such as droughts (Hinchberger, 2017).

The rural farmers continue to suffer from the impacts of climate change despite the dissemination of information from the Meteorological Organization. This can be attributed to that the largest demographic—the rural farmer— remains largely cut off from the benefits of forecast information despite the designation of illiterate rural farmers as among the "key targets" of the Regional Climate Outlook Forum process (Chavas, 2009). There are several barriers that restrict rural farmers from accessing all these benefits, including illiteracy, lack of communication infrastructure and misunderstanding of forecast message being disseminated. Since rural development in Africa is mainly driven by agriculture; science communication has been tried in all facets to make sure that farmers have information regarding climate change. Satellite-based radios are used in villages in order to disseminate agricultural information including forecast products. Many countries also disseminate via agro meteorological bulletins, national radio, and television broadcasts and newspaper articles, and some offer agricultural meetings led by journalists. The biggest question that can be raised from all these mechanisms is to what extent they promote full understanding of climate prediction information.

DISCUSSION AND POLICY OPTIONS

There is an increasing focus on science being linked to providing practical solutions to environmental problems. Although excellent science can be done, of itself this will not cause change mainly because channels to use this information and create change are poorly developed: A study conducted by the British council in 2004, on the Africa Talks Climate Change suggests that one of the biggest challenges in Africa is the issue of climate change 'terminology'. Climate change is often explained as a scientific discipline, hence the failure by the ordinary people to fully understand it. Climate' is a word that is difficult to translate into many local African languages. As special advisor to the Senegalese Prime Minister Youssoupha Diallo points out, "first of all, it is a conceptual problem... how would we say climate in our languages?" In some languages, there is no translation for climate change. In Lingala, for example, the translated term literally means 'weather change'. In other languages, the word for 'weather' itself may have multiple connotations which introduce a barrier to understanding. The Amharic term, for example, can also be translated as 'change in the air'. Sometimes the only way to convey weather is to list the different elements of weather; in Bari, the only way to say climate change is 'changes in clouds, rainfall, wind, and temperature, and seasons'. In many of the countries, most people do not recognize terms for 'climate change' or 'global warming' in either local languages or widely spoken international languages, such as English, French, and Arabic. The exception is South Africa, where there is near-universal recognition of the terms in English (British Council, 2004). Generally, recognition of climate change terminology is higher in urban areas. Citizens from the capital cities and larger urban centers in many of the countries tend to give the most detailed descriptions of climate change. Yet most people, even the scientists themselves find the terms difficult to explain. Certain policy measures and actions need to be taken into account as a way of improving science communication and environmental education on climate change.

This requires broader and more effective communication of the scientific insights being gained. Even where the solutions to environmental problems are clear, management, political, and ultimately public support are needed to institute the solutions. Therefore, utilizing current research effectively requires tools to facilitate effective communication, to not only scientific peers, but also managers, government, and the public. Though science communication is viewed to be a critical strategy, there is still something missing in the African context. When journalists or newspaper editors or story writers do their work, they should not only think about themselves but also have to worry about who is going to read the message or listen to the story. There is a need to be familiar with the audience's level of knowledge about climate change.

Most scientists, planners, and policymakers find it difficult to communicate effectively with the ordinary public. As such, it is important to make use of several strategies. One measure is to stop speaking in code. Words that seem perfectly common to scientists are still jargon to the wider world and always have simpler and easily understandable substitutes. Rather than "anthropogenic," it can be appropriate to use the term "human-caused." Instead of "spatial" and "temporal," try "space" and "time." When scientists talk about trends in degrees per decade, they will provoke people to do calculations. Instead, try giving the total change over the full period of time. In order to reach apathetic audiences, the following actions are put forward. These actions or measures include:

- Knowing one's audience,
- Personalizing the message,
- Having clear, concise, short messages;
- Appealing to people's values, needs, social norms,
- Appealing to their emotions (including their fears),
- Providing action plans regarding what they can do,
- Recognizing their barriers,
- Stressing the benefits,
- Promoting gradual buy-in,
- Starting with the easiest and
- Using trusted messengers

Indeed, building on Africa's regional technology capacity and proficiency — especially in research and development — will work against the current siphoning out of money that takes place as imported technology vendors and foreign service providers earn a profit from African customers and send it overseas. Setting forth with a mandate to invest in education and training opportunities will go a long way in developing regionally relevant technologies and their application to climate change coping strategies. In order to communicate effectively on climate change and appropriate strategies for responding to it, it is important to understand and acknowledge how differently situated individuals and communities think about, interpret, and discuss its drivers and impacts (Harvey, 2012).

CONCLUSION

For science communication to be effective in disseminating information on climate change, there is urgent need combine appropriate language, metaphor, and analogy; combining science with narrative storytelling; using vivid visual imagery and experiential scenarios, and delivering by trusted messengers (CRED 2009). More localized communication initiatives and platforms abound, including community theatre and centers, and discussions within local religious facilities. More recently, community radio, web forums, participatory web 2.0 platforms, mobile phones, and various other information communication technologies (ICTs) are being utilized as channels to advance climate change information dissemination. The issue of ITCs as tools for climate change communication is often debatable since Africa has not yet fully developed in this sector. This explains the differences between Africa and the developed nations in as far as science communication is concerned. Therefore, for Africa, the use of localized channels can help, especially in rural communities where access to ICTs in the biggest challenge. The chapter has demonstrated that science communication is indeed a very critical instrument in advancing environmental education on climate change. Climate Change has become one of the biggest threats to the African continent. It has social, economic and environmental implications. What matters now is how climate change-related information can be disseminated to the general public so that people can easily grasp what it is, how it affects their lives and how they can cope. Information on climate change can be spread through various platforms such as media, newspapers, stories, dramas, and poetry among others. In Africa, there is generally deficiency in as far as knowledge of climate change is concerned. Climate change has become a worldwide concern, increasingly affecting the livelihoods of individuals in Africa. The need to develop effective adaptation and mitigation strategies in Africa has become crucial to securing livelihoods and community development. A critical element in promoting effective and successful adaptation and mitigation strategies is science communication. Originally presented as a complex and abstract scientific problem, climate change information is increasingly being shared and discussed across disciplines and stakeholder groups at a range of scales. Effective science communication among stakeholders can help to identify problems, raise awareness, encourage dialogue, and influence behavioral change.

REFERENCES

Banda, F. (2013). *Climate Change in Africa: A Guidebook for Journalists. United Nations Educational Scientific and Cultural Organization.* UNESCO.

Besada, H., & Sewakambo, N. (2009). *Climate Change in Africa: Adaptation, Mitigation and Governance Challenges. The Centre for International Governance Innovation.* CIGI.

Bizzotto, M. (2018). *Resilient Cities Report 2018.* Bonn: ICLEI World Secretariat.

British Council. (2004). *Africa Talks Climate.The public understanding of Climate change in ten countries.* World Service Trust.

Brown, O., Hammill, A., & McLeman, R. (2007). Climate change as the 'new' security threat: Implications for Africa International Affairs. *Roy Inst Int Aff, 83*(6), 1141–1154. doi:10.1111/j.1468-2346.2007.00678.x

Burns, T. W., Connor, D. J., & Stocklmayer, S. M. (2003). Science communication: A contemporary definition. *Public Understanding of Science (Bristol, England), 12*(2), 183–202. doi:10.1177/09636625030122004

Chanza, N., & de Wit, A. (2015). Enhancing climate governance through indigenous knowledge. *South African Journal of Science, 112*(1/2), 1–7.

Chavas, D. R. (2009). *Seasonal climate prediction dissemination to rural farmers in sub-Saharan Africa: a "bottom-up" perspective and the emergence of the mobile phone*. Cambridge, MA: Massachusetts Institute of Technology.

Chirisa, I., Matamanda, A., & Mutambwa, J. (2018) Africa's Dilemmas in Climate Change Communication: Universalistic Science Versus Indigenous Technical Knowledge. In Handbook of Climate Change Communication: Vol. 1. Climate Change Management. Springer.

De Wit, M., & Jacek, S. (2006). *Changes in surface water supply across Africa with predicted climate change—AEON—Africa Earth Observatory Network. University of Cape Town, Science Express Report, Cape Town Federal Ministry for Economic Cooperation and Development. 2015. Disaster risk management: Approach and Contribution of German development Cooperation*. Berlin: Federal Ministry for Economic Cooperation and Development.

Flint, C. G., & Lutoff, A. E. (2005). Natural Resource-Based Communities, Risk, And Disaster: An Intersection Of Theories. *Society & Natural Resources, 18*(5), 399–412. doi:10.1080/08941920590924747

Fortier, J. D., Grady, S. M., Lee, S. A., & Marinac, P. A. (1998). *WisconsinÕs Model Academic Standards for Environmental Education*. Wisconsin Department of Public Instruction.

Global Humanitarian Forum Human Impact Report. (2009). *Climate change, the anatomy of a silent crisis*. Author.

Goh, B., Pomsagun, A., Le Tissier, M., Dennison, W.C., Kremer, H.H., & Weichselgartner, J. (Eds.). (2008). Science Communication in Theory and Practice. LOICZ Reports and Studies No. 31. LOICZ, Geesthacht, Germany.

Guha-Sapir, D., Hoyois, Ph., & Below, R. (2014). *Annual Disaster Statistical Review 2013: The Numbers and Trends*. Brussels: CRED.

Harvey, B., Ensor, J., Carlile, L., Garside, B., Patterson, Z., & Naess, L. O. (2012). *Climate change communication and social learning–Review and strategy development for CCAFS*. CCAFS Working Paper No. 22. CGIAR Research Program on Climate Change, Agriculture and Food Security (CCAFS), Copenhagen, Denmark. Available online at www.ccafs.cgiar.org

Hinchberger, B. (2017). *Disaster insurance against climate change attracts African countries*. Available from: https://www.un.org/africarenewal/magazine/may-july-2017/disaster-insurance-against-climate-change-attracts-african-countries

Huq, S., Kovats, S., Reid, H., & Satterthwaite, D. (2007). Editorial: Reducing risks to cities from disasters and climate change. *Environment and Urbanisation, 19*(1), 3-15.

Iltus, S. (2013). *Climate Change and Environmental Education*. UNICEF.

Inside Climate News. (2018). *Six of the G7 commit to climate action. Trump wouldn't even join conversation*. Available from: https://insideclimatenews.org/news/10062018/g7-summit-climate-change-communique-trump-allies-estranged-germany-france-canada

Johnson-Pynn, J. S., & Johnson, L. R. (2010). Exploring Environmental Education for East African Youth: Do Program Contexts Matter? *Children, Youth and Environments, 20*(1), 123–151.

Kilian, H. (2012). *Environmental Education Programs: Can Their Effectiveness be Influenced by the Teaching Method of the Visiting School?* IZE Journal.

Lanard, J. (2003). *Scientists And Risk Communication*. GreenFacts.

Lukamba, M. T. (2010). Natural disasters in African countries: What can we learn about them? *TD The Journal for Transdisciplinary Research in Southern Africa, 6*(2), 478–495.

Masih, I., Maskey, S., Mussa, F. E. F., & Trambauer, P. (2014). A review of droughts on the African continent: A geospatial and long-term perspective. *Hydrology and Earth System Sciences, 18*(9), 3635–3649. doi:10.5194/hess-18-3635-2014

Matamanda, A. R., Dzvimbo, M. A., & Kadebu, R. T. (2017). Climate Change and Infrastructure Planning: Implications for Sustainable Urban Management in Harare, Zimbabwe. *Journal of Public Administration and Development Alternatives, 2*(11), 95–111.

Mileti, D. (2006). Promoting Risk Wise Behavior. *Natural Hazards Workshop*.

Modh, S. (2010). *Introduction to disaster management*. New Delhi: Macmillan Publishers.

Morrow, B. H. (2009). *Risk Behaviour and Risk Communication: Synthesis and Expert Interviews*. Final Report for the NOAA Coastal Services Center.

Mukoni, H. (2013). Environmental education in Zimbabwean secondary schools:' greening or transformative social change? *International Journal of Asian Social Science*, *3*(4), 971–991.

Odlyzko, A. (2004). *The future of scientific communication*. AT&T Labs - Research.

Plattner, T. (2005). Modelling Public Risk Evaluation Of Natural Hazards: A Conceptual Approach. *Natural Hazards and Earth System Sciences*, *5*(3), 357–266. doi:10.5194/nhess-5-357-2005

Quirog, S., & Suarez, C. (2016). Climate change and drought effects on rural income distribution in the Mediterranean: A case study for Spain. *Natural Hazards and Earth System Sciences*, *16*(6), 1369–1385. doi:10.5194/nhess-16-1369-2016

Rajsekhar, D., & Gorelick, S. M. (2017). Increasing drought in Jordan: Climate change and cascading Syrian land-use impacts on reducing transboundary flow. *Science Advances*, *3*(8), 1–15. doi:10.1126ciadv.1700581 PMID:28875164

Stapp, B. W. (1997). The Concept of Environmental Education. *The Journal of Environmental Education*, *1*(1), 30–31. doi:10.1080/00139254.1969.10801479

UN-HABITAT. (2012). *Climate Change and Pro-Poor Urban Governance Capacity Development Workshop: Promoting Sustainable Human Development in Asian Cities*. Workshop Summary Report. Bangkok: UNDP.

UNESCO. (2009). *UNESCO International Seminar on Climate Change Education*. Paris: UNESCO.

UNISDR. (2017). *Sendai Framework for Disaster Risk Reduction*. Available from: www.unisdr.org/we/coordinate/sendai-framework

United Nations Development Programme. (2009). *Climate Change Perspectives from India*. UNDP.

U.S. Global Change Research Program. (2001). *Climate Change Impacts On The United States: The Potential Consequences Of Climate Variability And Change*. Cambridge, UK: Cambridge University Press.

WHO/EHA. (2002). *Disasters & Emergencies: Definitions. Training Package*. WHO/EHA Panafrican Emergency Training Center.

Yohe, G. W., Lasco, R. D., Ahmad, Q. K., Arnell, N. W., Cohen, S. J., Hope, C., . . . Perez, R. T. (2007). *Perspectives on climate change and sustainability. In Climate Change 2007: Impacts, Adaptation and Vulnerability. Contribution of Working Group II to the Fourth Assessment Report of the Intergovernmental Panel on Climate Change.* Cambridge University Press.

Zarcadoolas, Boyer, Krishnaswami, & Rothenberg. (2007). GIS Maps To Communicate Emergency Preparedness: How Useable Are They For Inner City Residents? *Journal of Homeland Security and Emergency Management*, 4.

Chapter 10
Revisiting Zanzibari Seaweed:
Global Climate Change – Mitigation and Adaptation

Nadra O. Hashim
Independent Researcher, USA

ABSTRACT

Well before island nations began to consider rising ocean levels, a feature of global climate change, they have been concerned with the allocation of water resources. The purpose of this chapter is to revisit the efforts of Zanzibar's academic, as well as private and public institutions, as they promote environmentally responsible entrepreneurial projects, while advancing women's economic empowerment. Analysis will examine the history of seaweed production and consider how Zanzibar's seaweed farmers have recently responded to the dislocations associated with global climate change. This discussion will also consider to what extent Zanzibari seaweed production reflects the norms enshrined in the United Nation's Rio + 20 platform, and the language of the UN's 2030 sustainable development goals.

INTRODUCTION

As early as 2001, Zanzibar's seaweed farmers reported that their plants, which thrive in cool waters, were 'dying-off,' (Ott 2018). Initially, the Zanzibari government listened with concern, but not much action was taken because, on balance, Zanzibari Seaweed exports were still thriving. According to National Bureau of Statistics, "in 2009, seaweed contributed 7.6 percent of Zanzibar GDP, 2nd after the [historically important] clove cash crop, (at 4.7 percent)," (FAO, 2014 and NBS, 2011)

DOI: 10.4018/978-1-5225-7727-0.ch010

However, the following year, reports emerged suggesting that Zanzibari seaweed was dying off. In subsequent years, as consistent measurement of both atmospheric and sea surface temperature revealed record-levels of heat, the matter of dying-off garnered increased attention.

Earlier research, first published first in 2015, sought to answer whether global climate change, (GCC), might be responsible for the decline in Zanzibar's commercial grade seaweed. Verifying this claim was initially, very difficult due to wide fluctuations in weather, and the size of the seaweed harvest. For example, despite a general trend towards rising temperatures, and declining seaweed harvests, some years were better than others. For example, 2012, exports reached an all-time high of 15,000 tons. During the following year, yields were low, as a result, exports fell by approximately 25%, to 11, 000 tons. Throughout those years, the general success of Zanzibar's seaweed export market, meant that even in low yield years, Tanzania "was ranked as the 9th largest exporter of aquatic plants in the world - at 0.66 percent of the global share [in 2011] (FAO, 2014)." As such, the government showed concern, if not urgency, regarding the matter of rising GCC, and waning aquaculture.

However, during the 2015-16 year, research revealed that dying-off was accelerating, and that there was a 94 percent decline in the harvest of *Kappaphycus alvarezii,* commonly called Cottonii. Zanzibar's most favored seaweed was in grave danger. Government officials began to reconsider the severity of the problem (Ott, 2018). Eliminating other plausible explanations, Dr. Flower Msuya, the lead researcher/director, promoter and coordinator of women's seaweed production at Zanzibar's Maritime Institute, insisted that GCC was having a definitive, and negative, effect on Zanzibar's export economy. Further, she also argued the decline in Seaweed production is having a demonstrable cultural impact.

Seaweed harvesting employs over 20,000 persons, of which almost ninety percent, are women. According to Dr. Msuya, "women losing [are] their crop, losing their income." As lead coordinator of the Institute's Seaweed Cluster, Dr. Msuya suggests that helping the sisterhood of seaweed farmers adapt to GCC has become a serious fight 'for gender equality' (Ott, 2018). As GCC continues, checked only at the margins, the effort to sustain seaweed production may also highlight a larger, more fundamental threat. Namely, the struggle to adapt to GCC, is rapidly becoming an existential struggle. This is especially true for world's island populations who seek to maintain their distinct island livelihoods, and unique cultural heritage, despite an increasingly inhospitable environment. According to the Goddard Institute for Space Studies:

With the exception of 1998, the 10 warmest years in the 134-year record have all occurred since 2000, with 2010 and 2005 ranking as the warmest years on record (McCarthy, 2014).

Thus, while the trend reflected a general rise in atmospheric temperatures, 2013 may not have been the hottest year on record. Despite this fact, that year, Zanzibari Cottoniii, experienced a significant crop failure. What Dr. Msuya's research ultimately revealed, was that the measure of temperature which more accurately explains the reversal in the bounty, and health, of Zanzibari seaweed, is the measure of sea surface temperature, or SST.

SST is distinct from atmospheric temperature, and as a discrete measure, it can be used as a separate index of climate change. According to the National Oceanic Atmospheric Agency, or NOAA, in 2013, global Sea Surface Temperature was 'among the 10 highest on record," (NOAA, 2014; Xue et al., 2013). Thus, despite moderately warm atmospheric temperatures, SST was something altogether different, and for those involved in aquaculture, 2013 was a very hot year.

Considering a trend toward rising temperatures in both SST and atmospheric weather, Zanzibari seaweed is under increasing and severe threat. The recent reversal has gone beyond expert and governmental concern and has now captured national attention. In 2016, Zanzibar's President Ali Mohamed Shein, used his second inaugural address to commit to assisting Seaweed famers.

To measure the value of Zanzibari seaweed production toward the success of the national economy, it is necessary to begin with a brief history of the decline in agriculture. The study will begin with the era of foreign colonialism, and continue with the emergence of aquaculture, following national independence. Following this discussion and the subsequent examination of the cultural value of seaweed production, research will turn to exploring the differences between atmospheric and sea surface temperatures. Then analysis will turn to a deeper consideration of SST with a view to understanding how rising sea temperature influences a range of seaweed varieties, boosting some, and killing others, most notably harming Cottonii, the export variety so much in demand. In addition to considering what new strategies could help mitigate climate change, analysis will now also explore which strategies, already adopted as of 2015, have been most successful.

Global Climate Change, which once had an incremental effect on Cottonii, is increasingly becoming persistent and pernicious. Records indicate that aside from creating an environment hostile to cold-seeking commercial seaweed, it is encouraging the over-growth of competing varieties of sea vegetation. According to Dr. Msuya, Cottonii is experiencing a kind of heat-exhaustion, and dying off, is exacerbated by the crowding of an encroaching invasive plant, one which loves the heat. (Msuya, 2011a; BBC, 2014). The heat-seeking invasive epiphyte, which is killing Cottoniii seaweed, may be a key example of the cascading impact GCC can play beyond the water. As agricultural production is threatened by heat-seeking invasive plants, analysis of the plight of Cottonii seaweed farmers may help inform adaptive strategies. Like many land-based crops, Cottoniii is vital to the commercial economy of its home nation. However, it is also important to other economies which import the crop, for its cosmetic use, as a food, and for medicines. The current situation is ripe for analysis and must begin by discovering how Zanzibar's government, its universities, as well as local NGOs/large multinationals, have worked together to assist the Institute of Marine Science, and the Zanzibari women responsible for harvesting seaweed, to 'adapt' seaweed farming to GCC. This imperative, which has concerned the Institute's Seaweed Cluster for more than a decade, reflects many of the numerous goals enshrined in the U.N.'s platform for Sustainable Development.

The Platform first emerged in 2012, as a result of The United Nations Conference on Sustainable Development. Rio+20, took place in Rio de Janeiro, Brazil when member states announced a set of Sustainable Development Goals (SDGs). The UN issues these SDGs to augment the eight Millennium Development Goals declared in 2000, (MDG 2015). The MDG had a rapidly approaching target date of 2015 and member states encouraged the expansion in the number and breadth of development goals. Perhaps one of the most notable efforts of the Rio + 20 conferees was an desire to help member states promote green economy policies such as those long-standing efforts promoted by Zanzibar's Seaweed Cluster.

The United Nations Platform for Sustainable Development, an enhanced formulation of goals set out at the 2012 Rio+20 meeting, enumerates seventeen goals, it seeks to implement by 2030. In many ways, the first six goals have been the life work of researchers of Zanzibar's Maritime Institute and the life-blood of Seaweed Cluster. These include ending poverty, ending hunger/promoting sustainable agriculture, ensuring inter-generational health, ensuring inclusive and life-long educational opportunities, achieving gender equality/promoting female empowerment, and ensuring availability and sustainable management of water/ sanitation. The Seaweed Cluster has nearly achieved the first six of these seventeen goals merely by encouraging the production of a sustainable export crop, grown largely by 'poor' women, which has become its nation's leading agricultural export product.

In the case of Zanzibari Seaweed, many of these goals may be achievable so long as the Cluster is able to advance GCC mitigation, while promoting the kinds of positive educational, income and cultural opportunities which have earned such high praise.

Generally, projects that promote women's advancement or environmental protection tend to garner, and share, a broad network of technical, political and financial support. As such, these projects are committing to the UN's call for a green economy by promoting progressive social development, as outlined in the seventeen imperatives of the 2030 SDG platform, and should not be controversial. Committing to the goals is one thing; promoting action, is quite another. In order to address the matter achieving an urgent commitment, and in the process of articulating many of the goals at the Rio +20 meeting, the UN created a High Level Political Forum (HLPF). The role of the HLPF is to promote commitment to the SDGs and Agenda 2030 by advancing a program of sponsored social networking. According to the UN language regarding the HLPF, it must sponsor:

Regular reviews of the 2030 Agenda, [which] are to be voluntary, state-led, undertaken by both developed and developing countries, and shall provide a platform for partnerships, including through the participation of major groups and other relevant stakeholders'. (UNGAR 67/290)

Currently, the HLPF is promoting the fourth sustainable development goal, education Among other imperative, the fourth goal seeks:

By 2030, [to] substantially increase the number of youth and adults who have relevant skills, including technical and vocational skills, for employment, decent jobs and entrepreneurship. (Zhu/SDUN 2018) and (SDG4, 2018)

Defining, promoting and teaching sustainable development, speaks to and converges with the Seaweed Cluster's on-going/emerging efforts to expand farmer knowledge beyond the practical, or the procedural. Now that mitigating GCC is a factor in aquaculture, acquiring the knowledge needed to maintain seaweed production requires, a broader sense, and a deeper range, of "technical and vocational skills," (United Nations Department of Economic and Social Affairs,2018). Consistent with its own norms, and those of the UN's 2030 agenda, the Cluster is now experimenting with the best ways to promote a deeper understanding of the scientific knowledge required to help its farmers to adapt to new realities, and to further, to encourage them to innovate, (Kyewalyanga et al., 2011; Hashim, 2015).

From its earliest years, the basic philosophy and governing structure of the Seaweed Cluster, sought to address the most pressing need of Zanzibari women. This meant teaching them learn a trade, creating an independent source of income, and to do so within the strictures of a conservative society. In doing so, the Cluster achieved the first three of the UN's 2030 SD goals. These included ending poverty, hunger and ensuring healthy lives -- ideals that can be grouped together as basic quality of life indicators. Now that GCC is a factor, and the dying off of commercial seaweed defines a new reality, Zanzibar's Seaweed Cluster has a new burden, or opportunity. Namely, it finds that it must expand efforts to help Zanzibari women 'leap-frog,' from remedial skills, towards a rather advanced, technical education. Here the spirit of the 4th sustainable goal, as well as the thirteen other 2030 sustainable development goals, means that development must promote profound commitment to expanding the ideal of quality of life. Measures must move beyond the notion of meeting an individual's basic needs, towards a more nuanced, and inclusive, ideal of life quality.

As such, the 2030 SDG platform promotes a commitment towards enhancing quality of life that naturally mentions the individual, but largely promotes the community, the nation, and the region. Quality of Life indicators, once solely the language of the United Nations Development Program, are now used throughout the UN system and beyond. The notion of quality of life clearly informs the language of the seventeen 2030 SDGs. What also informs the 2030 SDGs is the reality of climate change, and the need for sustainable economic development, that advance quality of life, as the nature of weather is becoming increasingly perilous. The Seaweed Cluster, long committed to these ideals, gained international acclaim by advancing this agenda, well before, but perhaps, in anticipation of both the 2000 MGD and the 2030 SDG were articulated.

Beyond improving individual and community quality of life/sustainable development, seaweed farming has had an impact on tangible measures, such as national GDP. As Seaweed became the most important export crop, aquatic farming, "stimulated the demand for other goods and services,' (Soekawa, 2008). Meanwhile, the positive social, spill-over effects produced by this industry meant other Zanzibaris have sought, and found, employment. Further, women, who have been particularly successful at farming seaweed, hire farm assistants. They have also joined "saving" or micro-credit clubs, and most have sent some, or all, of their children to school. Some private seaweed producing companies which have emerged because of the work of the Seaweed Cluster, have even established schools to attract Zanzibari women to join their firms (Degenhardt, 2015, p. 52).

Ultimately, the economic clout of women, who participate in seaweed farming has given them a voice in government policy, which in turn, has advanced other environmentally-friendly policies. Now that seaweed farmers, have to contend with rising and warming waters, the national conversation about environmental protection, and GCC adaptation/mitigation will require the attention of other 'stake-holders' committed to sustainable development. This includes the broader community of fishing villages, and dock workers whose live will change as SST and rising water becomes a factor.

Though Zanzibar is not threatened with immediate flooding, sea water is, is indeed, encroaching. Encroaching water translates into a number of conditions that can lead to physical, as well as economic, resource and food insecurity. In Zanzibar creeping seawater has been found seeping into fresh groundwater, making water wells brackish. Zanzibari women, and children, have had to walk farther and farther to collect freshwater, and sometimes from sources that are polluted (Kabendra, 2013) The years just before Rio + 20, the United Nations Development Program's Africa Adaptation Programme (AAP), constructed pipelines from village to village. These were connected to a series of central water tanks to ameliorate the rising water, brackish/polluted water situation. The AAP programme continues to work with Tanzania's island communities, on the mainland Tanzania, as well as twenty other African countries, to promote 'climate smart policies' (Kabendra, 2013).

In 2014, as the Zanzibari government began to publicly express concerned about GCC, and seaweed disease, that has followed, it 'commissioned more research,' in a search for a cure to this ailing crop (BBC, 2014). Thought the export market for Zanzibari seaweed remains significant, warm water, and competition from Asian producers, is cutting into profits. Thus by 2016, and as mentioned, Zanzibar's president suggested specific interventions the government would make, in an effort to protect seaweed farmers. These included finding new markets for Zanibari seaweed and finding structural mitigation to revive declining production.

This effort, like many other aqua-culture initiatives, will have to be an urgent and collaborative effort. So far, the Cluster is promoting the planting/harvesting of seaweed in deeper, cooler waters. The problem is Zanzibari women have not typically, either learned how to swim, or to sail boats. Therefore, this new strategy is enjoying, mixed results. Ultimately, whether cold water farming becomes success, or not, may not be as vital for scientists beyond Zanzibar, as for what it signifies. This first matter captures the broad spirit of SDG 2030, namely, that sustainable development is only possible, in the context of economic and social stability, now threatened by changes in the weather. Addressing public policy in the era of GCC requires a focus on creating proactive mitigation/adaptation strategies, no matter how experimental or uncertain they seem.

The second development is derived from the first, and includes practical mitigation experiments, such as the efforts of the Seaweed Cluster. Such 'real-world' experiments bolster the findings of theoretical researchers whose GCC projections are conceived, tested and analyzed, largely in laboratories or simulated on computers. Real world observations regarding the impact of GCC on aquaculture, and in this case Zanzibar, is vital to a larger understanding of sustainable development, and more specifically, food and water security.

Developing new entrepreneurial skills among a sector, which has suffered the effects of chronic under-development, can have material benefits, however. In the case of Zanzibari seaweed, it means making women's work, and agricultural or raw products export, generally under-paid industries, a leading a component of Zanzibar's engine of economic growth.

Ultimately, the continued study of seaweed farming in Zanzibar is also vital because this 'red algae,' *Kappaphycus alvarezii,* or *Cottonii* is often used to produce Carrageenans. Carrageenans, are sulfated polysaccharides, extracted from this edible seaweed, and used throughout the vast processed food industry for thickening and stabilizing.

Now that the climate is change is a factor, the resulting decline in seaweed production, may come to reflect how food supply and food security, will define not only science, but also politics. As far as Zanzibar is concerned this vegetative 'dying off' may ultimately lead to a slow, but permanent, extinction – all of which has political implications. The last time Zanzibaris experienced this degree of crop blight, the ruling government, would, or could not, intervene. Ultimately, plant blight and crop failure, led to political instability, revolt and finally a military coup which routed the sitting monarchical government.

For rather practical reasons mitigation strategies that are small and focus on cultivating adaptive strategies addressing both vulnerable vegetation, and in the 'farmers' which harvest these plants, may be where time, energy and resources are increasingly - and best - spent. As this research has discovered, larger mitigation strategies, those involving infrastructure are often expensive and may not be politically viable or sustainable for vulnerable nations. The study of Zanzibar' Seaweed Cluster and its discovery of GCC, is an effort to find a small, but workable, mitigation/ adaptation strategy.

Today, the very practical and prescient approach of the Institute of Marine Sciences Seaweed Cluster, suggests that mitigation, directed adaptation and other interventions will go a long way towards promoting long term conservation. In order to properly examine how the Seaweed Cluster is navigating "risk-related decision making' it is necessary to examine Zanzibar's history in the arena of cash-crop exporting, and what role, if any, Zanzibari women played in this industry (NRC, 1996, 2010; Webler, 2014). Ultimately the history of Zanzibar women, long before the initiative of women in the Seaweed Cluster suggests they are accustomed to fostering directed adaptation and mitigation strategies, into many areas their economic and political lives.

BACKGROUND: EXPORT CROPS DURING AND AFTER COLONIAL RULE

The history of cash crop production in Zanzibar reflects the existence of a rich, dynamic and complex identity. By definition, this often-excluded women. However, where women's labor may not have been formally recognized in the political process, it was still, a very important factor in the nation's economic development. The participation of women in the Zanzibari commons, during and after colonialism, reflects the larger narrative in East Africa. While there were indigenous traditions that promoted progressivism, there were also other local traditions, which were conservative. Colonial rule tended to reinforce conservative norms, both in the economic and in the cultural realm. Nineteenth century colonial consolidation of lands garnered for clove production reinforced a sharp distinction between cash crop and subsistence production. Men farmed for export produce and women produced for local consumption (Keshodkar, 2013, p. 33)

Broadly speaking, the colonial plantation project, has often reinforced cultural conservativism, and gendered norms, regarding paid work. In the case of Zanzibar, this began well before British rule, with the influx of immigrants from southern Arabia, Persia and the Levant. These 'conservative' norms became the law under colonial governance. According to David Killingray, British customary law often upheld local hierarchies which furthered the administration of colonial rule (Killingray, 1986, pp. 412-413, 418, 426-428). As such, discovering instances where colonial governments provided a progressive intervention in the lives of African women, and men are rare, but they do exist. In Tanzania the impact of Fabian intervention in pre-independence political developments, an effort which made political transition successful and peaceful, had positive spillover effects. This was true especially, where Unguja, or Zanzibar island, was concerned. Ultimately, Tanzania's Ujaama revolution, led by Julius Nyerere, would improve the social welfare, education and economic lives of women, but the impact would occur later, and take time.

Despite the fact that Zanzibari women resisted being drawn into the mono-crop export economy, as late as the mid-20th century Zanzibari women, like many of their class in sub-Saharan Africa, continued to focus on subsistence farming, and rely on their husband wages from producing cloves for export. One of the earliest and most demonstrative reactions to colonial crop production occurred in the 1940s, when Zanzibari women led a boycott of the colonial cash crop tax scheme. They argued that the monocrop export- regime prevented their 'men' from providing for their families (Berger, 1999 p. 47; Papart and Staudt, 1988, p. 7). All of this is relevant to study of current climate change adaptation/mitigation because beginning with the boycott in the 1940s, and continuing with decades of quiet activism, Zanzibari women have helped shape projects which have led to its most progressive environmental policies.

LIBERAL ECONOMICS AND PROGRESS: DEVELOPMENT IN FITS AND STARTS

After independence Tanzanian government supported women in communal organizations such as the UWT (Umoja wa Wanawake wa Tanzania), and in the national aspiration of communal self-sufficiency, or *ujaama*, promoted by President Nyerere, but this assistance still focused largely on food, not cash-crop production (Killingray, 1986, p. 59). However, while Nyerere continued to embrace ujaama, economic realities caused him to turn toward more 'mainstream' policy planning. He sought economic counseling from the World Bank. Believing that to transform

Tanzania from system of itinerant labor to a stable and developed workforce, the government has to institute a program of population relocation, which promised both economic growth and social welfare (McHenry, 1979).

The Bank and Nyerere, conceived of the program of Villagization, which led Bank technicians to relocate 85% of Tanzania's population, millions of people, to new 'working' villages. This was significant, as before 1973 only 10% of the population lived in villages, the rest of the population were largely nomadic (Hyden, 1980 and McHenry, 1979). One of the most successful and beloved aspects of villagization was the government's emphasis on social welfare, especially its health and education interventions. These were available in each village (McHenry, 1979). One of the least popular aspects of villagization was the dissolution of cooperatives, and the reinstatement of 'for export' cash-cropping. Ultimately, the population balked. Some walked out of villages, while others simply refused to grow cash crops.

Ultimately, the cash-crop system had to be curtailed. The cooperative arrangement was restored, which was the trigger for a larger program of decentralization (Massoi & Norman, 2009). Ultimately, villagization brought social welfare planning to Tanzanian communities, but did not create the kinds of economic opportunities envisioned. Women, who were marginalized in the cash-crop effort, suffered perhaps more acutely, than their men.

RETURNING TO AN INDIGENOUS PROGRESSIVISM: A FOCUS ON WOMEN

In 1988, four years after villagization, only 3% of all women in the Tanzania were employed in wage labor. By contrast, women made up twenty-four percent of central government employees (Rwebangira, 1996, p. 9). In agriculture, the discrepancies were even more acute. As late as 1993, a World Bank study suggested that women constituted 75% of the Tanzanian population tasked with subsistence farming, and that they produced 90% of the food requirements for the country (World Bank, 1993, p. 9). Thus, Tanzania's aspirational gaps were rooted in gender differences. These were fundamentally rooted in both income inequality, as well as the fact that women were still the nation's majority unpaid farm workers.

As it became clear that despite cash-cropping, fishing, or work in the tourist industries Zanzibari men, the nation's traditional and primary bread-winners, were unable to support their families, would have to find ways to earn money. At the time promoting women's empowerment seemed remote. What emerged was modest program, which promoted women's micro-credit programs, would gain the support of a variety of Tanzanian government agencies, and would become world-renowned.

As mentioned, the Seaweed Cluster project began as a 'small' effort, to address women's unemployment, creating a source of income, which neither relied on the tourist industry, no did it compete with men's employment. The two researchers, who founded the project, were a professor at the University of Dar-es-Salaam (USDM), Dr. Kesh Mshignei and later, his student, Dr. Flower Msuya, who began two seaweed pilot projects in 1984.

The university established one project on Pemba island and called it Fundo and created another on Unguja island called Fumba. The University's research toward commercial, environmentally responsible production of seaweed, began two years later. Four years after that, in 1990, two private firms, the Zanea Seaweed Company and the Zanzibar Agro-Seaweed Company (Zanzcol), (Msuya, 1999; Ngowi, 2005). Government Officials, including those from Zanzibar's Fisheries Department, as well as related agricultural and development ministries, supported the environmentally friendly seaweed farming industry By the late 1990s, the United Nations University declared seaweed farming one of the industries which should be awarded a "zero emission" designation. The seaweed farmers, organized into the Seaweed Cluster, and their mentors at Zanzibar's Institute of Marine Sciences as the University of Dar-es-Salaam (USDM-IMS) began receiving grants, from a variety of multilateral and international organizations, including various agencies of the United Nations (UNEP, 2000; Mshignei, 1996)

Seaweed farming was unique because, in addition to being an environmentally responsible trade, or eco-trade, it is an economic activity where, where the majority of its participants are women (Msuya, 1999). It is an exceptional example of social and economic intervention as Zanzibari women can net in two months what it takes Zanzibar's fisher-men six months or more to earn (Soekawa, 2008).

In 2000, Zanzibar's department of environment made women's access to coastal waters - for seaweed farming-on par with men's access to bays for commercial fishing (UNEP, 2000). Further, at the urging of an increasing population of seaweed farmers, the government has even considered regulating the tourist industry—the island's largest source of national income. Long before global climate change began influencing public consciousness Zanzibari women have taken on the challenge of protecting the environment. In 1992, a group of concerned citizens, of whom a number were seaweed 'farmers' living in historic Stone Town formed the Zanzibar Women's Corporation (ZAWCO), to help villages "solve environmental and economic problems" (ZAWCO, 2010). ZAWCO has enjoyed over 20 years of success, creating clean water and latrine systems, a private effort which has helped the government solve the island's difficulty containing raw sewage. Since the 1990s, Zanzibari women have formed numerous local NGOs, working in topics as broad as agriculture to Zanzibari politics.

In September 2006, various environmental groups, including those representing the seaweed and fishing industries, urged Zanzibar's parliament to implement a ban on plastic bags. The Zanzibari government announced a ban on plastic bags, in response to rampant littering of fresh water which led to the pollution of the coastal sea waters (TerraDaily, 2006). Zanzibar's plastic bag ban, was one of the first in the world, and created positive spillover effects, namely the emergence of a growing demand for, and production of, local sisal bags. However just as the Zanzibar's algae farmers were celebrating this notable victory, another, more ominous trend was emerging. Seaweed crops were showing distress, decline, retreat, and even extinction. It became clear, that pollution was not the culprit, but was, rather a more formidable foe, namely, Global Climate Change (Msuya, 2011).

THE FOCUS ON GCC: LOCAL KNOWLEDGE, SST AND THERMAL EXPANSION

Zanzibari seaweed decline, should be examined within a broader context. Residents of the many Indian Ocean islands, argue that the more pressing impact of GCC, thermal expansion and rising sea levels is the gradual disappearance of lands scattered across the Pacific Ocean. The Maldives, a string of a thousand plus islands, in the Western Indian Ocean (WIO), is experiencing such a rise in sea levels that the president has to declare a permanent state of emergency (Astazia, 2012). Meanwhile, Soloman Islands researchers have begun to report an acceleration in a documented, and decades long rise, in coastal waters. Most of the Pacific/Indian Ocean islands are vulnerable to rising water level. This includes popular tourist attractions such as Seychelles, Palau, Tuvalu, as well as Micronesia. Yet, the threat to small populations, those islands with less than a million persons, has yet to capture global attention. By contrast, encroaching waters in Bangladesh, and the resulting threat to tens of millions, is too worrying to ignore. Neighbors, and western nations, have offered to help Bangladesh consider mitigation strategies. These include various, dam projects, touted as an inexpensive and promising, but which have yet to be stated, much less, completed (HT Editors, 2010). One of the ironies of GCC, evident in Bangladesh, is that some parts of the country have too much water, and others do not have enough. Food security, indeed, national security, is threatened by the abundance of (sea) water on the one hand, and the lack of (fresh) water on the other.

As the debate over whether to dam, or not, vacillates between ambivalence and passion, other considerations, including the effort to work with "hydroponic system(s) which need much less (fresh) water than conventional soil-based agriculture," are coming to the fore (Haq, A.H.M. Rezaul; M. Asaduzzaman, & T. K. Ghosal, 2002; Astazia, 2012). Meanwhile among smaller WIO islands, remains a harbinger of a remote, but far scarier, existential threat. Many of these countries have even smaller infrastructure budgets than Bangladesh. In the Maldives, and on Unguja, or Zanzibar Islands, governments find that they have to consider mitigation/adaptation strategies that, are by definition, 'small,' as well as relatively simple and inexpensive (Gentle and Marasenit 2012).

Around the world, even 'wealthy nation' coastal cities have been slow to implement the type of 'large' mitigation/planning strategies that might address the long-term realities of rising tides. The notion that water temperature, and just as vitally, water levels are rising, is now largely, settled research. In fact, the general trend of rising waters has been widely acknowledged by various authorities, including National Oceanic and Atmospheric Administration/NOAA (Lindsey, 2013). Discussion has now shifted to discovering how fast the water is rising, and how best to proceed. NOAA and other climate agencies/organizations have focused on the Western Indian Ocean (WIO) where climate change, rising water temperatures, and land loss is the most dramatic. WIO research suggests this trend appears to be the most obvious off the coasts of the smallest islands, as well as the lowest lying provinces.

Previous instances of flooding, and the threat of incremental, and creeping saturation, now threaten to overwhelm vulnerable infrastructure. In the last decade, researchers have discovered some anomalies in Indian Ocean water patterns. Namely, in some provinces of the in the WIO, such as Zanzibar, rising sea surface temperatures (or SST) have not immediately been accompanied by expected rises in thermal expansion/rising water levels. Rather in Zanzibar, rising SST is now accompanied, by a five decades long pattern of shrinking levels of water. So far this is an anomaly exclusive to Zanzibar, and Seychelles. This phenomenon is explained as a function of their location, in the southwestern corner of the Indian Ocean (Han et al., 2010).

PARADOXES: IMMUTABLE REALTIES OF RISING TEMPERATURE AND WATERS

Zanzibar's unique situation has been the focus on a number of climate change studies. This is because the island offers a variety of co-existing, and paradoxical, weather anomalies. Many of these reports also address other factors which, while not strictly weather related, act as amplifiers to the sustained impact of climate change. These amplifies include shifts in wind speed and vagaries in water salinity, both of which describe the 'climate' in Zanzibar.

Notwithstanding these other factors, the work of Professor Weijing Han, and her colleagues at the Department of Atmospheric and Oceanic Sciences, at the University of Colorado, suggests that holding other metrological variables constant, and focusing on recent global climate change, there is a world-wide rise in the surface water temperature. They estimate this rise is one degree Celsius over the last 50 years. This measure, known as Sea Surface Temperature or SST, has caused in some regions, an expansion of water, expressed as coastal level water rise. In the case of Zanzibar, where there seems to be an anomaly – rising SST and shrinking levels of coastal waters – a comprehensive system. Professor Weiqing Han et al. suggest that the Indian Ocean basin should be viewed as whole. Rising waters in some regions bordering the Indian Ocean have caused water displacement in some provinces, creating, subsequently, falling levels of water, in other areas.

Professor Han et al. (2010), suggests that while there have been shrinking levels of coastal waters near Zanzibar over the last fifty years, in the last ten years, these waters have risen. This dynamic is consistent with converse and precipitously rising waters among many of the coasts of Indian Ocean provinces. According to Han:

[There is] newly detected rising sea levels in parts of the Indian Ocean, including the coastlines of the Bay of Bengal, the Arabian Sea, Sri Lanka, Sumatra and Java... The key player in the process is the Indo-Pacific warm pool, an enormous, bathtub-shaped area of the tropical oceans stretching from the east coast of Africa west to the International Date Line in the Pacific. The warm pool has heated by about 1 degree Fahrenheit, or 0.5 degrees Celsius, in the past 50 years, primarily caused by human-generated increases of greenhouse gases. (Han et al., 2010)

Ultimately Zanzibar's decades long rising water temperatures, (Han et al., 2010, p.5) is a fact that has been killing seaweed production. It may ultimately kill other sea vegetation, food for various sea creatures, and humans, alike. As the surface temperatures and levels of water continue to rise, it looks as if air temperature in Zanzibar, will rise as well. Recently the metrological measures at Zanzibar's airport have yielded a rise in (air) temperature of 0.2-0.5 degree Celsius (Han et al. pp 8-9). Duncan Geere, of the University of Göthenburg, whose research is funded by the Swedish International Development Agency, studies sea grass. Geere has found that like Zanzibari seaweed, other indigenous ocean vegetation, growing off the coasts of Unguja, have become increasingly vulnerable to climate change. According to Geere:

Invasive species attracted by rising water temperatures may spread pathogens and compete for resources, causing further harm to stressed populations... indications [are] that maximum temperatures are projected to rise to between 1.5C and 2C above pre-industrial values by 2050 and greater than 4C by 2100. Global climate models, statistically downscaled using local meteorological data for Zanzibar ... report that maximum monthly temperature is projected to increase uniformly throughout the year between 1.5 and 2C by the 2050s and 2 to 4C by the 2090s. (Geere, 2014, p. 16)

Beyond Han and Geere's work, other studies, suggest that despite a comparative rise in water in the southwest Indian Ocean basin, relative to provinces in the north east, the water is getting hotter, and the water is rising - everywhere. According to this research, there has even been a relatively recent spike in coastal waters, which is especially noticeable during seasonal changes (Watkiss et al., 2011, p.10)

Comparing the shrinking levels of water around Zanzibar island (Unguja), with rising water further northeast, near Sri Lanka, may make it appear that overall, Zanzibar is 'safe.' However, the rising water levels even as close as a small islet north of Zanzibar, named Panza, located at the Southernmost tip of Pemba, and less than 70 nautical miles to the north of Zanzibar, supports the assumptions of the Han et al. model. Water is rising rapidly in the northeast because of structural realities, but it is also rising in the southwest because GCC is a permanent reality. Thus, while climate change explains a consistent rise in SST, and a general rise in water levels, structural explanations also play a role in the permanent rising, or temporary waning, of water within the WIO. According to a UNPO report, there is significant variation as to sea levels as nearby as the neighboring coasts of Tanzania, and its other coastal Islands. In the case of Panza, a small dot of an island off Unguja (Zanzibar) Island, the UNPO reports that the island is rapidly losing coastal land because of rising water. While Panza may be low-lying and Zanzibar, relatively higher, rising waters will ultimately impact both islands.

By 2014, this thermal expansion/water level rise, which was threatening to drown Panza, began to threaten Zanzibari hotels, especially those built closest to its shores. The variation in coastal water levels on the islands of Panza, Unguja (Zanzibar), and Pemba suggest that Han's model of variance and displacement has merit. It deserves greater study both in tandem with analysis of aquaculture, and beyond, toward an intensive examination of other climate changes. Despite the fact that Zanzibar is on the high side of the WIO basin, the trend is for waters to rise is so powerful, that despite its structural and coastal advantages, one day Unguja could become as vulnerable as Panza.

Research by the IPCC, and other scientific organizations, indicate rising water levels due to rising SST, and resulting thermal water expansion, are projected by to exceed one degree Celsius, reaching closer to 3 degrees, over the next one hundred years. This is especially true when taking variations in rainfall, into consideration (UNPO, pp. 3-7, 13). Such remote projections of incremental rise, while vital to the overall impact of climate change, are recent and useful, but are not as informative as the rather astounding practical observation documented by scientists engaged in field work. Dr. Flower Msuya, Director of the Seaweed Cluster, argues that SST off Zanzibar has been much more dramatic then climate models project. According to Msuya:

Temperatures in the shallow seaweed farms have increased from below 30 degrees centigrade in the 1990s to about 38 degrees centigrade, and slightly, above recorded recently. (Kaunda, 2013, p. 2)

This more dramatic measure of SST also confirms another assessment of larger trends in the Western Indian Ocean made by various climate researchers and IPCC collaborating authors. They argue that:

According to some predictions if extensive sea-level rise would take place, Tanzania's island's off the coast, such as Zanzibar and Mafia, might submerge under water by 2100. In the western Indian Ocean (presently) some of the small islands experience sea level rise, whereas some islands experience a decline in sea-level; however, all islands will face a sea-level rise in the end of the century regardless of current locale-specific trends. (Mustelin 2009, p. 14)

The assessment of records, some of which indicate an eight-degree centigrade rise in SST off of Zanzibar, and which document recent floods, stand in contrast to other studies which continue to maintain a conservative one degree rise, and take not of the shrinking of water, during the same period. All totaled these suggest significant variations in the Indian Ocean and that more study is needed. For Zanzibari sea farmers, and concerned phycologists, high SST and long-term shrinkage, or lower SST along with immediate water rise, are all significant factors. That is because either way, and further, any combination of these trends, all result in a negative impact on seaweed production.

Moving past the sustained, and important, if now rudimentary reporting of current, and impending, climate change, some GCC researchers now simply write about the need for immediate mitigation/adaptation strategies. This effort is reflected in many of the goals of the UN 2030 sustainable development goals. In fact, the field of mitigation/adaptation is growing at such a rate that there are now discrete subfields. Some writing concerns overall strategies for identification and mitigation (Adger and Barnett, 2009; Barnett, 2010). Others address the political, social and possibly psychological impediments to mitigation/adaptation (Barrang-Ford et al., 2010; Pielike et al., 2007). Still others compare developing world, versus developed country, climate change mitigation/adaptation (Lemmen et al., 2008, Gagnon-Lebrun & Agrawala, 2007; Moser & Luers, 2008). Perhaps the most important unifying thread in much of this work is a realization that professionals in fields of oceanographic research, and coastal sea management, have always worked at the forefront of mundane weather disasters. Now they are the closest observers to some of the most dramatic, and permanent, weather related changes, these practical experts may find they have to teach non-scientific lay-persons, the 'relevant technical and vocational skills,' vital to promoting adaptation and mitigation, (Zhu, 2018) and (United Nations Department of Economic and Social Affairs, 2018).

Professionals in coastal management, in particular, have routinely been called to develop and implement mitigation/adaptation strategies, without much discussion, debate, or fan-fare (Tribbia & Moser, 2008). To an even greater extent than that of advanced economies, developing world climate change mitigation/adaptation, long feared to be complicated, may be more less expensive to promote, more difficult to learn (Parry et al., 2009; Stern, 2003)

SEEKING COLLABORATIVE SOLUTIONS TO CLIMATE CHANGE IN ZANZIBAR

Thermal expansion leading to loss of island shore line off Panza, the Maldives, Kirabati, Tuvalu, the Marshall Islands, as well as islets off of Fiji and Palau, have broad social implication. In an intriguing, if not disturbing political consequence, when a people lose their land, they can lose their sovereignty, and can in fact become 'stateless' peoples. So, the fear among island nation leaders concerns not only losing a physical home, but a psychological, social, or spiritual identity. For those islands not threatened with extinction, 'encroaching water' translates into a number of conditions that can lead to physical, as well as economic, resource and food insecurity. In Zanzibar creeping seawater is seeping into fresh groundwater and water wells. Zanzibari women, and children, are walking farther and farther to collect freshwater, and sometimes from sources that are polluted (Kabendra, 2013). The UNDP's Africa Adaptation Programme (AAP) has constructed pipelines from village to village, connected to a series of central water tanks to try to ameliorate the situation. The AAP programme also is working in both the islands off and within mainland Tanzania, as well as twenty other African countries, to promote 'climate smart policies.'

Various adaptation policies such as the freshwater pipelines implemented by AAP in Tanzanian villages in 2012, will be more in demand as climate change affects a greater number of vulnerable locales. Recently there has been a surge of literature that suggests that such interventions will be expensive, but that not implementing mitigation/adaptation projects, could become truly prohibitive. According to a European Commission report, damage from yearly floods throughout Europe, in the years approaching 2050, and the decades beyond, could without intervention, more than 45 billion annually. By contrast, mitigation/adaptation projects, would cost a fraction of that, or 3 billion dollars, as annual calculated cost (Giffords, 2014).

Advanced economy infrastructure demands interventions, which cost billions, whereas interventions in the developing world may cost - mere millions. Further, there are other reports, which suggest that addressing climate change may not be as expensive as initially estimated. Ultimately, researchers examining the developed and the developing world increasingly agree is that in order to ensure that mitigation/adaptation strategies remain relevant, climate change and cost models must be continually refined and updated (Giffords, 2014).

As global warming and mitigation/adaptation cost models continue to be improved, there will be room for greater collaboration among climate change researchers. In Zanzibar collaborative efforts have recently been initiated by donor countries including the United Kingdom, Finland and the US, most notably. The UK 'Economics of Climate Change in Zanzibar' has assiduously noted the technical ramifications of climate change on the island, including, sea level rise, temperature and acidification, as well as the cost of mitigating these. Other reports have focused on how to assist personnel in implementing effective mitigation strategies.

In a report on *Practical Measures to Tackle Climate Change: Coastal Buffer Zones and Shoreline Change in Zanzibar, Tanzania*, the Finnish government issued recommendations which not only addressed climate change, but also suggested how to convince the local population to promote mitigation/adaptation strategies. These included supporting capacity building for a range of GCC, developed by government agencies, as well as higher education institutions that have direct contact with the population (Mustelin et al., 2009, pp. 83-87).

The UN 2030 SDG platform, with its universal statement and sequential ordering of goals, echoes the practical measures published in a report by the Finnish government. Both, in their own way, address the impact of GCC on Tanzania weather and geography. The SDG 2030 platform, published in 2015, emphasizes economic advancement and community inclusion as a way to ensure that climate change adaption takes place. Similarly, the Finnish report, published in 2009, recommends that mitigation/adaptation strategies "strengthen livelihood opportunities options and recognize that poverty is a factor," and that mitigation/adaptation efforts should "support investment and initiative in alternative and renewable energy," (Mustelin, 2009, pp. 85-86). The Finnish report notes some of the impediments to community participation in climate change mitigation/adaptation, noting the difficulty Zanzibari seaweed farmers have sustaining their crop in space they must share with the tourist industry.

The authors of the Finnish report interviewed numerous seaweed farmers concerning their observation of weather, including changes in tide level, extreme variations in water and atmospheric temperature, as well as other factors which have an impact on farming. Despite these penetrating questions, the report does not suggest what response the seaweed farmers gave as to changes in coastal water. Ultimately the authors recommended that 'vulnerable' coastal places be rehabilitated through the planting of seedlings – at an estimated cost of less than 20,000 US dollars. It is not clear however, whether this recommendation was ever conveyed to the seaweed farmers, or whether it was implemented (Mustelin, 2009, p. 7).

Ultimately, the Finnish government issued the report five years before the 2014 seaweed crop blight, and six years before the UN's Rio + 20 conference issued its 2030 platform for sustainable development (BBC, 2014). Perhaps one of the most important contribution of the Finnish report is that it documents the commitment of the Tanzanian government to environmental protection, which stretches back to the early 1990s.

In 1992, the National Environmental Policy of Zanzibar, the government's formal proclamation on conservation, declared that coasts had to be shared by competing local groups involved in promoting tourism, fisheries and aquaculture. While this policy was well intentioned, the efforts to promote equal access to Zanzibar's natural resources, the very words of this proclamation, may have helped encouraged the kind of over-use typically associated with land-based agricultural (UNEP, 2000, p. vii).

In 2000, government reports began to take notice of coastal stress, and emphasized the need for collaborative efforts to reduce over use, and instead, to promote coastal protection. Unfortunately, at that time, little mention was made of global climate change. As a result, a decade of potential cooperation was lost – and most importantly a decade of aquaculture mitigation strategies was confined to rather the modest, if essential efforts, of groups such as the Zanzibari Seaweed Cluster. While GCC is a politically charged issue, it has clearly been present in collective consciousness of national leaders, and in 2010 the Tanzanian/Zanzibari government formally initiated its climate adaptation strategy.

Attention to climate change came at the moment when it was clear that rural to urban/coastal migration had increased. As the result of increased tourist, fishing and aquaculture opportunities, Zanzibar's rural populations decided to leave their villages, and moving to coastal cities (Levine, 2007). Research suggests that 'coastal urbanization,' wherever it takes place, leads to significant seaweed loss (Scherner et al., 2013) The cumulative effect of all this research on Zanzibar concerning, its vulnerability to global warming, means that locating culturally relevant mitigation strategies, need to consider a wide range of contingencies, actors and situations, in order to protect the island and its residents. For GCC researchers, Zanzibar may be a microcosm of the paradoxes and vagaries of climate change, especially the impact weather has on vegetation. It therefore deserves greater study, and support.

As *Practical Measures in Zanzibar*, suggests "...in Tanzania due to the traditional division of labor, women's tasks are dependent on the environment..." (Mustelin, 2009, p. 14). Further the report also suggests that owing to the fact that seaweed farming has been designated as merely "women's work," there has not been too much resistance to the fact that women are earning a good wage. Due to the fact that Cluster participants have "established linkages to better food safety and community standard of living," it has broad community support, and has received national and international attention, (Mustelin, 2009, p. 20). Seaweed farming is garnering even more attention now that its sponsors are working on making the island's premier site for climate change adaptation, and mitigation.

CURRENT RESEARCH REGARDING SEAWEED, GCC AND ADAPTATION

Researchers who specialize in genetic theory like to emphasize that "adaptation is *not* natural selection"(Orr, 2005, p. 118) Rather, it is the movement of a population, species, or in this case, flora, that can best be described by micro-mutations that may, or may not, ensure the species will survive over the long term (Orr, 2005, p. 119). The successful frequency of small adaptation, if successful, may lead to an evolution which then may compel natural selection – and in the long run survival. One of the most compelling theories, one which seems to describe the context for seaweed survival in Zanzibar, is the 'shifting balance theory.' The shifting balance theory 'maintains that the interaction between natural selection, genetic drift, and migration is more important than the action of any single force (Orr, 2005, p. 123).

It is the combination of meteorological, oceanographic, geographic and man-made conditions, that seem to threaten commercial seaweed production both in a theoretical sense, (Breeman, 1988) and in the practical, (Msuya 2011a, 2011b, and 2014). Given, the confluence of pernicious factors encroaching on aquaculture, perhaps the only 'hope' for sea farmers is if "adaptation leapfrogs many moderately beneficial mutations, arriving a at a strongly beneficial one," (Orr, 2005, p. 125) In the absence of the flora and fauna spontaneously adapting, instrumental actions must be created. Mitigation practices must be enhanced, or, conversely, other national resources must be promoted which better 'fit' emerging climate conditions. However, those that better fit the climate, are not always the most preferred, or lucrative, vegetation. Thus, when examining the interaction between traits and conditions that lead to adaptation, growth and ultimately population survival, GCC researchers are now examining the convergence of conditions, best described, as 'environmental tolerance.' Often the goal of this research is to predict 'phenotypic plasticity,' of the preferred species or vegetation (Chevin, Lande & Mace, 2010, p. 4).

Here examination of environmental stressors such as 'temperature' and 'salinity' are considered to determine under what conditions stability, and fitness of traits, will allow certain species to withstand, and even become tolerant, to emerging environmental conditions. Researchers call this process phenotypic plasticity. These are the conditions where 'tolerance curves emerge and evolve' (Chevin, Lande & Mace, 2010, p. 5). Ultimately, species reaction, and adaptation, is influenced by genetic constraints. While some traits reach (plastic phenotypic) saturation almost immediately, when new and extreme conditions emerge, other traits may thrive despite 'disruption to *homeostatis*.' Here plasticity, also known as the adaptive response, is 'amplified' (Chevin, Lande & Mace, 2010, p. 5) Those species which either saturate easily, or are given to amplified plasticity, may through the stubborn persistence, or abrupt decline, become heralds of this new trend of prolonged climate change.

Regarding global climate change, the paradox of warming waters, and ever-increasing amounts of CO_2 being absorbed into said waters. What is notable is that 'carbon dioxide dissolves more readily in cold, dense seawater' (Sarmiento, 1993; Block, 2014). Further, it is well established, that 'a warmer ocean absorbs and holds less CO_2 from the atmosphere simply because the solubility of CO_2 in seawater declines as temperatures rise (Sarmiento, 1993). In other words, warmer seas store carbon dioxide, which in turn increases the likelihood that SST will increase. This notion is consistent with research undertaken by oceanographic teams dedicated to tracking a century's worth of CO_2 absorption by the ocean. According to Columbia University oceanographer Samar Khatiwala:

Our method takes as input the relatively well-known atmospheric CO_2 concentration history. Given this history, we calculate the ocean absorption of industrial CO_2 consistent with this history… There are several factors that may be responsible for what's going on… Increasing acidity is only one of them. Faster emission growth rate is another, perhaps more important, cause, as could be changes in ocean temperature and circulation. (Block, 2014)

While colder waters become more acidic, warmer waters, such as those in the Indian Ocean, may be measurably less acidic than polar waters. However, as their SST continue to increase, they become less able to absorb as much CO_2, than cold water. Meanwhile even in warmer waters, acidity, like thermal expansion/water level rise, is an ever-increasing factor. The change in water acidity and temperature has affected not only seaweed, but also corals, seagrasses and other vegetation. Many of these feed ocean creatures, sometimes humans, and some are vital to the integrity of shorelines, especially when there are natural disasters. According to Francis Stuart Chapin et al., an analysis of dynamic, equilibrium and models, all totaled, show that rise in CO_2 means a loss in practically most varieties of vegetation. According to Chapin:

The loss in vegetation due to migration or actual loss, transient loss, in one area, may lead to more sustained loss in other areas where biodiversity is concerned. (Chapin, 2001, p. 378)

Research addressing *Environmental Stress, Adaptation and Evolution* in trees, Volker Biklsma suggests:

Information on patterns of genetic variation can be put to several uses. First it is important for conservation genetics...Second, this variation has implications for considering the consequences of climate change. (Bijlsma and Loeschcke, 1997, p. 58)

According to David T. Price and Michael J. Apps, whose research extends to ancient Boreal forests of Canada, if global climate change leads to an adaptation, where new species or plant varieties emerge, it will occur over the very long term. In the short term, there is a loss of vegetation. This is occurring with little promise of replacement that is until and perhaps, a new climate emerges- centuries later during a different planetary epoch. Price and Apps (1996) argue:

Equilibrium projections may take centuries or even millennia to manifest, whereas in practical socio-economic terms, the imminent, short-term, transient responses to a warming climate are likely to be of much greater importance.

A similar assertion may be made where seaweed is concerned. Seaweed is not only a species with a broad level genetic variation, it is very sensitive to fluctuations in environmental equilibrium. Phychology studies published in the early 1990s suggested that at an increase of 4 degrees Celsius would cause some varieties of seaweed beginning to migrate to cooler climates (Hurd et al., 2014, p. 322; Breeman, 1990). By 1997 the IPCC had compiled an extensive reference of scientific publications on climate change and ocean water rise (Watson et al., 2007) Studies published twenty years later suggest the current increase of merely one degree Celsius, may be responsible for migration, or loss, of over seventy varieties of seaweed (Hurd et al., 2014, p. 322).

Thomas Wernberg et al suggest that an ever- increasing number of seaweed 'groves' are retreating due to warming ocean water and thermal water expansion. As of 2011 Wernberg explains that:

Temperate species have experienced median shifts of 0.5° to 1.9° latitude poleward; that given future warming, up to ~25% of species might retract toward extinction and finally; [these] impacts are consistent with observed warming in both the Indian and Pacific Oceans. (Werberg et al., 2011, p. 1828)

Some varieties of seaweeds are more resilient than others. Ironically, some of the most sought-after variety are also the most vulnerable. Often, classifying each variety of seaweed is difficult and there is some ambivalence about which varieties should be excluded (Hurd et al., 2014). However, there is a general consensus, that are thousands of varieties of seaweed, and that number may, in fact, number be as high as 10,000 types, or more. However, all these seaweed varieties can be grouped into three general groups: red, brown and green.

Seaweed is both physically malleable and genetically pliable, or plastic. While these facts make it susceptible to weather, decline and extinction, these characteristics also make seaweed extract a valuable compound in many food products, medicine, and increasingly, in Bio-fuel. As such, applied science researchers are investing great effort contemplating the strengths and vulnerabilities of various seaweed varieties, research examining the properties of seaweed and its reactions to stressors, including the effects of varying degrees of salinity on the plant (Scherner et al., 2012), as well as seaweed's enormous capacity of absorption of toxins (Perkins et al., 2007).

As with other laboratory efforts, the problems of replicating the exact conditions of climate change, especially in the laboratory, are very difficult (Price & Apps, 1996, pp. 187-189). One such factor, which for various reasons, has not to be replicable in a lab, but which is especially relevant in Zanzibar, is the effect of increasing heat or retreat, decline and extinction. This is the invidious result of climate change, which promotes invasive species, blight or disease among existing and preferred seaweed (Rosenzweig et al., 2000). According to Professor Catriona Hurd:

[In order] to discuss or study it [seaweed], we must reduce it to smaller parts, to think about one variable at a time. And yet, each of these 'environmental factors' that we might consider – temperature, salinity, light and so forth (water motion, and nutrient value) is really a composite of many variables and they tend to interact.' (Hurd et al., 2014, p. 3)

While lab work may lead academic analysis, as water fluctuation in Zanzibar suggests, it cannot always account for all the unintended consequences, or unique, sometimes paradoxical, conditions of a given region. True of Zanzibar, it may be true of other islands especially in the WIO, especially those with plants as sensitive to environmental variability, as seaweed. Zanzibar seems to be leading the way in adaptive phycology, because its university system is able to cooperate with local and international NGOs, private firms, affiliated and independent scientists, as well as lay persons in the broader aquaculture community. All of these factors have helped the Zanzibar's Seaweed Cluster observe the impact of climate change on waterborne vegetation, and to begin to contemplate ways to adapt to these new realities.

This is especially true where water salinity, and Red seaweed, are concerned. In these instances, it is difficult to isolate this one factor and test its impact on seaweed growth, this is especially true in the absence of the other interacting water-based and meteorological factors (Abbot 1995 and Friedlander & Levy, 1995). Various studies generally testify to the fact that "[Red] seaweeds exhibit negative correlation with salinity…" (Jansi and Ramadhas, 2009, p. 472). Other studies suggest that even when high salinity does not appear to inhibit growth, this may be because high salinity can be mitigated if sea water is 'cold', an interaction which allows for photosynthesis and reproduction, despite high levels of salinity (Dawes et al., 1999, pp. 423-424). Given Dr. Msuya's observations that cooler water helps prevent the fouling [epiphytes] of Red seaweed, and given also, that cooler waters appears to mitigate salinity, it seems that the strategy of moving seaweed to cold water may be a necessary adaptation/mitigation. This may be especially true in settings where water conditions and weather could otherwise consistently inhibit growth.

On a philosophical level, Zanzibar's Seaweed Cluster, led by Dr. Flower Msuya, may be the realization of all the progressive impulses, both foreign and domestic, that have surged through the political life of the island, since the pre-independence era into the new millennium. On a more modest, and practical level, the Seaweed Cluster is a testament to the effort of government, public scholars and village communities, to advance community economic empowerment, and while promoting women's financial independence, and promoting environmental protection.

CURRENT RESEARCH: TEACHING [COLD WATER] MITIGATION/ADAPTATION

The United Nations published a compilation of all its reports in a web-based compendium entitled *Women, Gender Equality and Climate Change.* The synopsis of these various reports suggests that climate change will negatively impact women first, will affect them most often, and may hurt them the most comprehensively. The report further suggest that women must have access to technologies and training usually reserved for men. The UN report asserts:

In order to be effective, adaptation and mitigation technologies need to reach those who are most in need –the poor and vulnerable. This means that targeted efforts must ensure, firstly that it is understood that the situation of women may differ from that of men, secondly that technologies are designed in such a way as to be relevant to their [the women's] circumstances; and thirdly to ensure that they are given full access to knowledge, information and technologies related to adaptation. (UN WomenWatch, 2008, p. 8)

According to Dr. Flower Msuya, leading research scientist at Zanzibar's Institute of Marine Sciences, Zanzibaris have been gathering and exporting Red seaweed since the 1930s, and by the 1950s, Zanzibar was exporting several tons of the dried seaweed to Europe (Mshigeni, 1973, 1976). The systematic cultivation of Red Seaweed expanded throughout the 1970s in the Western Indian Ocean grew due to American and European demand. Dedicated research, which began in the late 1970s and 1980s, at the University of Dar-es-Salaam, ultimately led the value-added initiative. The effort to find new ways to manufacture red seaweed, making their products much more lucrative (Msuya, 2014). However, the Seaweed Cluster initiative is still a fairly young effort, which began with commercial cultivation in the late 1980s (Mshigeni 1976, 1983, 1985). According to Dr. Msuya:

Currently, two Red species – Eucheuma denticulatum, known commercially as Spinosum, and Kappaphycus alvarezii, known in Tanzania as Cottoniii – are farmed. The industry has become increasingly important by bringing foreign revenue into Zanzibar's economy and raising farmers' and communities' living standards, although its contribution to the mainland economy is still minimal. In Zanzibar, however, seaweed farming is second only to the tourism industry in terms of foreign exchange earnings. (Msyua, 2012)

As of the last decade, the export revenues created from exporting seaweed have reached past one Billion Tanzanian shillings – ten million in US currency (Msyua, 2012). Generally, there has been a preference in foreign markets for Cottonii because it is processed into Carrageenan, an emulsifier/stabilizer in many forms of food, medicine and cosmetics, and is generally considered 'safe' (Msuya, 2011, p. 46; Lerman, 2014). Finding a way to promote cold water mitigation for Zanzibari seaweed might mean protecting the world supply of Carrageenan, especially as water get hotter.

According to Dr. Msuya the cold-water seaweed mitigation/adaption strategy has been in effect since 2006, and has been pursued, in earnest, since 2010. Dr. Msuya contends that:

UDSM (University of Dar-es-Salaam) was involved in pioneering seaweed farming in southern Tanzania in 1995/96 through consultancy with the then Finnish funded project Rural Integrated Project Support (RIPS) Programme.... The UDSM has taken a key role in investigating the causes of die-offs and what can be done to alleviate the problem. Researchers from the Institute of Marine Sciences(ISM) of the UDSM working with the USAID-funded project, Smallholder Empowerment and Economic Growth through Agribusiness and Association Developments (SEEGAAD), studied the causes of die-offs. (Msuya, 2009, p. 6)

In much of her writing, Dr. Msuya describes how throughout the decade of the 2000s, the Seaweed Cluster worked with universities and governments of New

Found land, Canada, and Rhode Island, U.S. to collaborate and expand research. This group effort sought to discover why seaweed, both those varieties harvested, and those plants existing in the wild, were dying off (Msuya, 2009). The University of Rhode Island, and partners at the Costal Resource Center in Hawaii and Western Indian Ocean Marine Science Association (WIOMSA) based in Zanzibar, Tanzania all received SUCCESS designation. SUCCESS, the Sustainable Coastal Communities and Ecosystems (SUCCESS), Program is an USAID's Bureau of Economic Growth, Agriculture, and Trade/National Resource Management grant.

Thus in 2014, when the BBC reported that Zanzibari seaweed was dying due to warming waters and blight, it was not a great revelation to most of the oceanographic world. It was also certainly not 'news' to the Zanzibari Seaweed Cluster, which had been relying on real-world experimentation to remedy the decline of the seaweed. Though media reports have may have neglected to make the distinction, the type of seaweed dying off the coasts of Zanzibar (and Pemba), was the lucrative *Kappaphycus alvarezii,* or Cottonii, the type of seaweed, with the least tolerance for climate related changes in water temperature.

Back in 2010 Dr. Msuya, and colleagues at the Institute of Marine Sciences, in cooperation with a Tanzanian Seaweed company named Birr, began an effort to move seaweed to cooler, deeper water. The goal was to both Cottonii and Spinosum, the two export varieties of seaweed. The Seaweed Cluster moved it efforts north of Zanzibar, off the Tanzanian island of Pemba. This research drew on some preliminary efforts made as early as 2006. Researchers were looking at three variables: the nature and velocity of wind, weather, epiphytes (the general presence of a competing or invasive plant species) and fouling (pollution leading to invasion by seaweed/green alga *Cladophora*), (Msuya, 2013, p. 9). Today, deep, cool water seaweed farming is not only a real-world experiment, with all the problematic conditions that typically ensue, it is a real world commercial success.

The USDM and Dr. Msuya have assiduously documented all the problems connected with seaweed farming and has found that *Kappaphycus alvarezii,* or Cottonii is both intolerant to heat, and is unable to thrive in waters that are 'too calm' or still. Further, *K. alvarezii* is more succeptible to Cladophora fouling than other variety, *Eucheuma denticulatum*, or Spinosum, which happens to be the less lucrative variety (Msuya, 2013, p. 14). Dr. Msuya believes both an increase in pollution, and the rise in temperature, favor the blooms of fouling organisms, such as the green algae Cladophora (Msuya, 2011). Dr. Msuya has indicated that:

Working in five villages, the authors reported several causes of the die-off including high surface water temperatures, rapid changes in salinity, temperature etc. in shallow intertidal areas where Kappaphycus is farmed. (Msuya et al., 2014)

Further, as mentioned earlier, Sea Surface temperatures of 37–38 °C have been recorded in recent years in WIO (Msuya & Porter 2009; Msuya & Kyewalyanga 2010). Dr. Msuya reports that 'optimum water temperatures recommended worldwide for farming *Kappaphycus* are between 21 and 31 °C (Msuya & Porter, 2009). She adds that in recent years, water temperature in the Western Indian Ocean has been 37-38 degrees Celsius (Msuya, 2014).

Generally, the problems with Seaweed production can be divided into three categories. The first is in the area of planting, the next in harvesting, and the latter in production. In the production phase, the less lucrative Seaweed, *Spinosum* is heartier. Dr. Msuya is trying to boost returns on this product by creating value added soap and cosmetic products. Zanzibar has been the world's third largest exporter of *Spinosum*. However, recently the exports of Asian countries, which have 'cheaper production and transport costs' are making it difficult to compete (ATP 2016). Ultimately, if temperatures continue to rise, and both crops begin to decline, production value may become largely irrelevant.

In the first area, planting research seeks to apply research which has as its goal the application of new innovations. These intend to boost quantity, while ensuring quality, and maintaining sustainability/environmental conservation. Here an examination of research during the last four decades anticipating and projecting hypothetical remedies, and examining how climate change would affect growth of plants in general, and seaweed in particular, is necessary. The other part of this review requires an exploration of current research which seeks a practical remedy, or adaptation to global climate change and warming waters. In the latter instance research goes beyond projections to applied efforts, both in the field, and in the lab.

The work of the Seaweed Cluster has been widely acknowledged and is funded, by a variety of multilateral and governmental organizations, and yet the 'best practices' of the Cluster have only recently been observed or studied outside the sometimes-cloistered world of WIO real world aquaculture experts. However, if long-term GCC research begins to examine Seaweed Cluster practices through a theoretical lens, then experiments may yield results for replication and instruction. This is especially true where other similarly situated agricultural and aquaculture efforts are in progress, throughout the Western Indian Ocean, and beyond. Recently, scholar-practitioners of the Social and Environmental Research Institute have published an article which examines a case study of climate change adaptation in South Carolina. According to lead researcher Tom Webler:

We report on progress to design and evaluate a process framework intended as a systematic approach to access local knowledge by providing opportunities for facilitated, deliberative learning-based activities with local decision makers on the topic of climate change vulnerability and adaptation. (Webler et al., 2014, p. 2)

Webler suggests that the methodology is based on a deliberative process "which the US National Research Council has recommended in contexts of risk-related decision making, characterized by high levels of uncertainty and disputed values" (NRC, 1996, 2010). The methodology of studying a particular environmental effort, or organization, is dubbed the Vulnerability, Consequences, and Adaptation Planning Scenarios (VCAPS) process. Professor Webler adds:

VCAPS is a systematic approach to integrate local knowledge with scientific understanding by providing opportunities for facilitated, deliberative learning-based methods to address adaptation and climate change. (Webler, 2014, p. 37)

Webler contends further that efforts at adaptation/mitigation will go further when they:

Solicit local knowledge are likely to perform better when they integrate insights from communications, group learning, decision making, and collaborative planning to structure effective deliberative learning processes and foster a shared understanding among a very wide scope of local community members, including all interested and affected stakeholders, as well as relevant government and tribal officials. (Webler, 2014, p. 39)

The Seaweed Cluster decades long program to involve the community in improvement in mirrors VCAP's mitigation/adaptation program. Both these and the UN 2030 sustainable goals seek a 'deliberative process of cooperation' that the UN's SDG 2030 also advances (Webler, 2014, p. 1). Today the Seaweed Cluster, and Zanzibar's larger community, are poised to expand seaweed production from an economic to a social experiment. Reflecting some of the same ideas articulated in the 2030 SDG platform, the Cluster has extended an opportunity to teach women to sail, involves men in seaweed production and to utilize unemployed youngsters to dive, while older family members remain on boats, organizing and sorting the harvest (Msuya, 2014).

Funding opportunities for the Seaweed Cluster are more and better than they were twenty years ago, when the cost of a boat was impediment to sailing into deep waters, where only men used to sail (Msuya, 1993). From planting in the shallows, to sailing in the deep, the Seaweed Cluster has availed itself of this deliberative process of cooperation, all the while facing increasing odds, a situation all farmers experience.

SEAWEED AND THE POLITICS OF LOCAL ADAPTATIONS TO GCC

In 2014, two years after Rio+ 20, when the language of the UN's SDG 2030 platform was being developed, Zanzibar brought a delegation to the third international conference on small island developing states, SIDS. There the conference goal was to "build partnerships and get recognized as a small island nation" (UN News 2014). At that meeting, the Zanzibari delegation formally announced that climate change was having a negative impact on the its economic development.

The language of SIDs three conference documents deal with climate change. Meanwhile, more than half of SDGE 2030 goals are committed to addressing climate change, while six out of the 2030 seventeen mandates, promote community participation and inclusion. Historically, promoting community inclusion in state sponsored development, has historically been fraught. In an effort to promote community inclusion across the UN system, and more specifically regarding the 2030 sustainable development goals, the UN has established a High Level Political Forum to help implement the SDG platform. As it stands, the HLPF is convened by those interested in climate policy, but largely at the level of 'heads of state' (UNGA A Res/ 67/290, p. 3). Similarly, meetings such as the small island developing states also feature membership, which is largely at the level of state, and its executives.

Again, one of oft-discussed 'paradoxes' of development, and increasingly, of climate change adaptation, is that is that it is important to promote the participation, and commitment, of a sometimes, invisible or intangible community. Yet, in the past, the official definitions of a problem, especially those by large if well-meaning multilateral organizations, may have over-looked input by those most impacted by problems. These oversights are most acute when dealing with implementing GCC adaptation/mitigation strategies (Ayers, 2010, pp. 42, 118).

Fortunately, the Maritime Institute, the official sponsor of Zanzibar's Seaweed Cluster, has rather regular access of government ministries, especially those concerned with agriculture and the environment. On the other hand, the Seaweed Cluster believes more government involvement and protection would enhance community participation, as well as the Cluster's ability to promote its mitigation efforts. In 2016, Dr. Msuya suggested that seaweed production could be improved if the Zanzibar State Trade Corporation (ZSTC), which promotes its original cash crop, cloves, now promotes, its most important cash crop, namely Cottonii seaweed.

Still, the Seaweed Cluster enjoys the consistent support of the government as it strives to find its own adaptation and mitigation strategies. By moving the seaweed into deeper water and by expanding the deep-water floating lines, the Seaweed Cluster has anticipated what adaptive migrations the plant, *Kappaphycus alvarezii*/Cottoniii, may pursue over time. The effort to transplant seaweed by physically moving the crop toward adaptation rather than death, has certainly improved the possibility of species survival (Msuya 2011, 2013 and 2014). The Cluster has innovated, and in doing so, has provided a model for what climate change adaptation can look like, in 'the real world.' Beyond merely surviving, the scientists leading the Cluster, namely Dr. Msuya, have striven to help farmers make seaweed production a value-added activity, where exporting goods, such as cosmetics and foods, is consistently profitable.

Despite reversals, and declines in the production of Seaweed, the Seaweed Cluster, remains attractive to donor agencies, because of is green policies and its commitment to gender advancement, and most vitally economic development. Further, Dr. Msuya and her colleagues at the Seaweed Cluster, tend to write about mitigation/adaptation in the same context mentioning climate change - careful, perhaps to avoid emphasizing global warming aspect to western donor agencies. An ironic, if diplomatic, tactic as these agencies are now discussing all three: GCC, mitigation, and adaptation, and doing so, rather bluntly. This is especially true of the UN's most important instrument, its 2030 Sustainable Development Goals. Eleven of the seventeen make a pledge to address sustainability, or climate change or environmental protection. Some goals reflect all three (SDG 2015).

Aside from the strides creating real world experiments where non-scientists are participants, where funding agencies are transnational, there is another important aspect of the deep cold-water seaweed farming experiment. It is social. After a decade and a half where the project was 90% women, wading in shallow coastal waters, the deep cold-water terrain demands boats. Men who find that they no longer can or want to fish are joining the formerly largely female Seaweed Cluster.

There are questions that will have to be asked. It turns out however, these are not whether Zanzibari women will learn to like swimming in cold water, but whether they will learn to sail, be encouraged to buy their own boats, feel comfortable working in a trade with men, where work in Zanzibar has been generally segregated and gendered. Dr. Msuya reports that despite this, and potential conflicts with fishermen. Some of these problems have been ameliorated if not resolved (Msuya, 2006; Msuya et al., 2007). A brave new world which is getting warmer demands that Zanzibari women may have to be very brave, too. Dr. Msuya continues to innovate, encouraging women to test deeper waters. In the last few years, the Seaweed Cluster has adopted Brazilian technology. Specifically, Dr. Msuya has adapted Brazilian fishing nets to tether seaweed that now requires harvesting in deeper, cooler and rougher waters, (Ott, 2018). So far, the government of Tanzania is not publicizing any grand construction efforts to protect coastlines, and mitigation efforts, like the Seaweed Cluster's, mitigation/adaptation interventions, generally remain 'small.'

To date, Zanzibaris living on the coast are benefitting from an anomaly of geography and tectonic placement where water is rising rather slowly. If, as Dr. Msuya suggests water temperature has not risen by one degree, but rather by as many as eight degrees Celsius - in the last 20 years - then the fifty-year water shrinkage detected by Han et al. (2010) may begin to appear to reverse. At the point, the importance of whether seaweed thrives may decline in national consciousness, as matters of creeping water, flooding and land loss begin to occupy the Tanzanian political agenda. What remains important about seaweed farming in Zanzibar, and the Seaweed Cluster's government, NGO, scholars and activist supporters, is that while they may be discussing global climate change, they are using a scientific and deliberative process to mitigate, and when possible, they are adapting to it.

Zanzibar's Seaweed Cluster is one of many examples in the developing world, where geography, economic status and political history will determine whether nations rely more on individual (human) adaptation or collective (structural) mitigation. Further, the case of the Seaweed Cluster suggests that promoting education and awareness or climate change, begins 'at home,' and will vary from place to place. Achieving both knowledge of, and response to, a variety of climate change events will most likely succeed where communities of academics, as well as public and private actors, focus on intervention first, and then move to publicizing their successes and failures.

Where experts contribute their knowledge to climate change projects in the developing world, these may initially take the form of small-budget, individual adaptation effort. Whereas wealthier nations may be able to increase infrastructure-enhancing mitigation efforts, even large nations can benefit from small adaptations. In both cases, participants in these endeavors will need to find a forum to share successes and failures.

Though school- based instruction will ultimately be needed to educate beyond the immediate site of climate change, local communities, the sites most impacted by these altered states may be best equipped to raise local awareness - even as they devise a range of response protocols. In the case of Zanzibar, a small island, with a modest population, successes and failures at adaptation/mitigation will be hard to disguise. In the past, Zanzibari women enjoyed success when they led a national campaign to ban plastic bags, an effort covered by the local press and by the international media. A similar strategy may be needed as the rise in sea surface temperature has a sustained influence beyond seaweed harvests, and begins to impact sea water salinity, and coastal flooding. Where projects seek adaptation and mitigation, they will succeed and thrive, as a concerned public considers the connection between their own climate change realities and those in the far corners of the world. The populations of small islands in the Indian Ocean may have to eschew cultural reticence, or academic distance, in favor of a vigorous drive which recalls Zanzibar's plastic bag campaign.

This is especially true given the UN's effort to advance sustainable development goals (2030) through broad public participation. Recent research suggests that seaweed, of all varieties, may prove a universally beneficial crop. Studies by the University of the South Pacific, examining Seaweed production as a CO_2 mitigating agent, suggest:

That if 9% of the ocean were to be covered in seaweed farms, the farmed seaweed could produce 12 gigatonnes per year of biodigested methane which could be burned as a substitute for natural gas. The seaweed growth involved would capture 19 gigatonnes of CO_2. A further 34 gigatonnes per year of CO_2 could be taken from the atmosphere if the methane is burned to generate electricity and the CO_2 generated captured and stored. (Flannery 2018)

This, and other research suggests that what is needed is a greater number of farmers cultivating commercial seaweed. In 2008, the University California San Diego (UCSD) joined with various institutions of the Scripps organization and established a Center for Algae Biotechnology (SD-CAB). In 2013, the center expanded incorporating research collaborations and commercial partnerships across the state of California and is now the California Center for Algae Biotechnology or Cal-CAB (USCD, 2018).

Part of an initiative dubbed 'Food and Fuel for the 21st Century,' Cal-CAB is partnering with another UCSD initiative, called Educating and Developing workers for the Green Economy (EDGE) which, in turn, is promoting a program it calls Women In Algae. The Cal-CAB/ Edge/WIA program seeks to 'connect aspiring [female] algae scientists, engineers and professionals in the algae community… and promote an inclusive and diverse culture [by] starting in the early days of the growing algae industry (USCD, 2018b).

Promoting diversity, inclusivity and sustainability, and doing so by promoting women in seaweed, is a growing global phenomenon. In Indonesia, women and men engaged in growing seaweed, provide 37% of global production. The Indonesian government encourages seaweed farming because it is sustainable and lucrative. It is an 'important alternate to 'blast fishing' where fishermen use explosives or poison to stun or kill fish, decimating corals and other habitats in the process (Waters 2018). Meanwhile, the Indonesian Navy has even heralded seaweed farming as an alternative livelihood to piracy, training ex-pirates how to farm with the intent of keeping them from returning to a life of crime (Waters, 2018).

Throughout Asia, in Europe, as well as in Canada, and North America women seaweed farmers are harvesting this crop in a variety of independent initiatives which remain largely small efforts. In South Korea, where women have been gathering seaweed since the 17th century there has been in a steady decline. Rather than harvesting it in the shallows, the Hanenyeo (sea women), dive in the cold, deep East China Sea. The work is grueling and potentially hazardous, and, owing to tradition, it is largely done by women, who pass on the trade on to their daughters. Whereas in the 1970s, there were upwards of 14,000 Haenyeo, today there are less the 5,000. Yet, these remarkable women, most over 50 years old, with some as old as 90, can make up to 25,000 dollars a year. However, they have to subsidize this work, with second jobs such as farming or fishing. Still, in a country, where the elderly has to live frugally, diving for seaweed offers financial independence and the company of a productive community (Mundy, 2015).

The efforts of San Diego's Cal-CAB, to promote 'aspiring female' farmers and sciences will be in greater demand, and finding sustainable alternatives to food and fuel become truly urgent. In 2017, drawing inspiration from 'the hardscrabble female aquafarming operation in Zanzibar,' Tessa Emmer, Avery Resor and Catherine O'Hare, founded Salt Point Seaweed (Sammon, 2018). The trio harvest wild seaweed in the waters off Mendocino County, and sell it to local restaurants, seafood CSAs, and at retail (Sammon, 2018). More than inspiration, Ms. Resnor traveled to Zanzibar, to learn the skill of harvesting seaweed (Sammon, 2018). This transfer of knowledge, from developing world to developed, will be enhanced as the demand for both raw seaweed, and its by-products, increases. Parity for women in developing world communities, might be achieved, as their governments strive to find new markets for their products.

Part of promoting seaweed, will require that producers and governments, as well as NGOs, enumerate the value of seaweed as a healthy, carbon neutral fuel and a food._Unfortunately, recent reports regarding a climate driven surge in wild algae blooms, may have damaged the cause. In 2017-2018, throughout the Eastern Caribbean, and along the coasts of America's southeastern states, the swell of wild sargassum, has threatened fisheries, killed protected wildlife, and ultimately fouled beaches, especially after rotting (Allen, 2018). According to Hazel Oxenford of the University of the West Indies, the surge is a response to anthropogenic behavior:

We're talking about mass increase in nutrient levels from deforestation in the Amazon, from urbanization in the Congo... Plus climate change, particularly an increase in surface water temperatures. Because the two things that plants or this floating seaweed will respond to is an increase in nutrients and an increase in water temperature. (Allen, 2018)

Controlling wild algae blooms will demand the persistent intervention of coastal managers, as well as prolonged attention of government agencies. Meanwhile, if researchers at the University of the South Pacific have their way, expanding commercial seaweed, could actually help address the excesses of anthropogenic behavior, and possibly global warming. Ultimately, any effort which seeks to expand purposeful and controlled seaweed production, especially in its capacity as a global CO_2 mitigating agent, is a positive development. Like all such efforts, however, this will require a vast public, government and private commitment. This is just the kind of effort the 2030 United Nations SDG platform envisions for the coming decades.

REFERENCES

Abbott, I. A. (1995). A decade of species of Gracilaria (sensu latu). In Taxonomy of Economic Seaweeds (Vol. 5). California Seagrant Publication.

Adger, W. N., & Barnett, J. (2009). Four reasons for concern about adaptation to climate change. *Environment & Planning A*, *41*(12), 2800–2805. doi:10.1068/a42244

AFP. (2016). Seaweed Farmers in Hot Water As Zanzibar Struggles. *The Daily Mail*. Retrieved from http://www.dailymail.co.uk/wires/afp/article-3567780/Seaweed-farmers-hot-water-Zanzibar-struggles.html

Allen, G. (2018). Masses Of Seaweed Threaten Fisheries And Foul Beaches. *NPR*. Retrieved from https://www.npr.org/2018/07/21/630575085/masses-of-seaweed-threaten-fisheries-and-foul-beaches

Amadiume, I. (1997). *Reinventing Africa: Matriarchy, Religion and Culture*. London: Zed Books.

Barnett, J. (2010). Adapting to climate change: three key challenges for research and policy: an editorial essay. *Climatic Change*, *1*(3), 314–317.

BBC. (2014). Seaweed- Zanzibar's Gift from the Ocean. Retrieved from http://www.bbc.com/news/world-africa-26770151

Bennett, N. R. (1978). *A History of the Arab State of Zanzibar*. Methuen Publishing.

Berger, I., & White, E. F. (1999). *Restoring Women to History*. Indiana Univ. Press.

Berrang-Ford, L., Ford, J. D., & Paterson, J. (2010). Are we adapting to climate change? *Global Environmental Change*, *21*(2), 25–33.

Bijlsma, R., & Loeschcke, V. (Eds.). (1997). *Environmental Stress, Adaptation and Evolution*. Birkhauser Verlag. doi:10.1007/978-3-0348-8882-0

Breeman, A. M. (1990). (Section). In J. J. Beukema & J. J. Brouns (Eds.), *Expected Effects of Climatic Change on Marine Coastal Ecosystems*. Kluwer Academic Press.

Chapin, F. S., Sala, O. E., & Huber-Sannwald, E. (Eds.). (2001). Global Biodiversity in a Changing Environment: Scenarios for the 21st Century Ecological Studies. Ecological Studies, 152.

Chevin, L.-M., Lande, R., & Mace, G. M. (2010). Adaptation, Plasticity, and Extinction in a Changing Environment: Towards a Predictive Theory. *PLoS Biology*, *8*(4). doi:10.1371/journal.pbio.1000357 PMID:20463950

CLGF. (2008). Commonwealth Local Government Forum. Local Government System in Tanzania. Retrieved from http://www.clgf.org.uk/userfiles/CLGF/File/2008_Country_Files/TANZANIA.pdf

CRC/Department of Environment, Institute of Marine Sciences, Sub-commission for Forestry, Sub-commission for Fisheries, Integrated Planning Unit. (1996). Toward Integrated Management and Sustainable Development of Zanzibar's Coast. Narragansett: Coastal Resource Center. Retrieved from http://www.crc.uri.edu/download/ZAN_0056.PDF

Dawes, C. J., Orduña-Rojas, J., & Robledo, D. (1998). Response of tropical red seaweed *Gracilaria cornea* to temperature, salinity and irradiance. *Journal of Applied Phycology*, *10*(5), 419–425. doi:10.1023/A:1008021613399

Degenhardt, S. (n.d.). *A multiple-case study of exploring social-ecological resilience of seaweed farming in Unguja*. Norwegian University of Life Sciences Faculty of Social Sciences, Department of International Environment and Development Studies. Retrieved from https://brage.bibsys.no/xmlui/bitstream/handle/11250/2381701/Stephanie%20DegenhardtMSC_BRAGE.pdf?sequence=1

FAO. (2018). Zanzibar mariculture leaps forward, FAO in Tanzania. Retrieved from http://www.fao.org/tanzania/news/detail-events/en/c/1125890/

Flannery, T. (2017). How farming giant seaweed can feed fish and fix the climate. The Conversation. Retrieved from https://theconversation.com/how-farming-giant-seaweed-can-feed-fish-and-fix-the-climate-81761

Friedlander, M., & Levy, I. (1995). Cultivation of *Gracilaria* in outdoor tanks and ponds. *Journal of Applied Phycology*, *7*(3), 315–3. doi:10.1007/BF00004005

Gagnon-Lebrun, F., & Agrawala, S. (2007). Implementing adaptation in developed countries: An analysis of progress and trends. *Climate Policy*, *7*(5), 392–408. doi:10.1080/14693062.2007.9685664

Gallucci, M. (2013). *6 of the World's Most Extensive Climate Adaptation Plans*. Retrieved from https://insideclimatenews.org/news/ 20130620/6-worlds-most-extensive-climate-adaptation-plans

Gardner, H. (2014, March 14). Facing Rising Seas, Bangladesh Confronts the Consequences of Climate Change. The New York Times.

Geere, D. (2014). Adaption to climate-related changes in seagrass ecosystems in Chwaka Bay (Zanzibar). Retrieved from http://www.bioenv.gu.se/digitalAssets/1488/1488564_duncan-geere.pdf

Geiger, S. (2002). *Women in African Colonial Histories*. Indiana University.

Gentle, P., & Maraseni, T. N. (2012). Climate Change, Poverty and Livelihoods Adaptation Practices by Rural Mountain Communities in Nepal. *Environmental Science & Policy*, *21*, 24–34. doi:10.1016/j.envsci.2012.03.007

Germanwatch. (2004). *Sea Level Rise in Bangladesh and the Netherlands: One Pheonomenon, Many Consequences*. Berlin: Germanwatch.

Giffords, G. (2014). Flooding could cost Europe $32 billion a year by 2050—four times as much as it does now. Retrieved from http:/flooding-could-cost-europe-32-billion-a-year-by-2050-four-times-as-much-as-it-does-now/+&cd=1&hl=en&ct=clnk&gl=us

Han., Meehl, G. A., Rajagopalan, B., Fasullo, J. T., Hu, A., Lin, J. … Yeager, S. (2010). Patterns of Indian Ocean sea-level change in a warming climate. Nature Geoscience, 3(8), 546–550.

Haq, A. H. M. R., Ashaduzzaman, M., & Ghosal, T. (2002). *Soil-less Agriculture in Bangladesh. A Grameen Trust*. Bangladesh Publication.

Hashim, N. (2009). *Language and Collective Mobilization: The Story of Zanzibar*. Lexington Press.

Hawkins, J., Roberts, C., Van'T, T., De Meyer, K., Tratalos, J., & Aldam, C. (1999). Effects of recreational Scuba diving on Caribbean coral and Fish Communities. *Conservation Biology*, 889–997.

HT Editors/Himalayan Times. (2010). *Bangladesh dams to reclaim 600 square kms of land*. Retrieved from http://www.thehimalayantimes.com/rssReference.php?headline=Bangladesh+dams+to+reclaim+600+ square+kms+of+land&NewsID=256638

Hughes, K. (2003). Terror Warnings slow Zanzibar's Progress. *The Guardian*. Retrieved from http://www.guardian.co.uk/theobserver/2003/sep/21/worldview

Hurd, C. L., Harrison, P. J., Bischof, K., & Lobban, C. S. (2014). *Seaweed ecology and physiology*. London: Cambridge University Press. doi:10.1017/CBO9781139192637

Hyden, G. (1980). *Beyond Ujamaa in Tanzania*. Berkeley, CA: University of California Press.

Ibhawoh, B., & Dibua, J. I. (2003). Deconstructing Ujamaa: The Legacy of Julius Nyerere in the Quest for Social and Economic Development in Africa. *American Journal of Political Science*, *8*(1).

IPCC. (1996). Climate change impacts on forests. In *Climate Change 1995: Impacts, Adaptations and Mitigation of Climate Change: Scientific-Technical Analyses. Contribution of Working Group II to the Second Assessment Report of the Intergovernmental Panel on Climate Change*. Cambridge University Press.

IPCC. (1997). Revised 1996 IPCC Guidelines for National Greenhouse Gas Inventories. Intergovernmental Panel on Climate Change. Bracknell, United Kingdom: Meteorological Office.

Jansi, M., & Ramadhas, V. (n.d.). Effect of salinity & dissolved nutrient on the occurrence of some seaweeds in Manakkudy Estuary. *Indian Journal of Marine Sciences*, *38*(4).

Kabendra, E. (2014). Zanzibar's Encroaching Ocean Means Less Water. Retrieved from http://www.ipsnews.net/2013/06/zanzibars-encroaching-ocean-means-less-water

Kaunda, D. (2013). Warming seas frustrate Zanzibar's seaweed farmers. Thomson Reuters Foundation. http://www.trust.org/item/?map=warming-seas-frustrate-zanzibars-seaweed-farmers

Keshodkar, A. (2013). *Tourism and Social Change in Post-Socialist Zanzibar: Struggles for Identity, Movement, and Civilization*. Lexington Books.

Killingray, D. (1986). The Maintenance of Law and Order in British Colonial Africa. *African Affairs*, *85*(340), 411–413. doi:10.1093/oxfordjournals.afraf.a097799

Kiss, P., & Hunt, A. (2012). *Scoping study: reviewing the coverage of economic impacts in the CCRA*. Paul Watkiss Associates.

Lemmen, D., Warren, F., & Lacroix, J. (Eds.). (2008). *From impacts to adaptation: Canada in a changing climate 2007*. Ottawa: Government of Canada. doi:10.4095/226455

Levine, A. (2007). Staying Afloat: State Agencies, Local Communities and International Involvement in Marine Protected Area Management in Zanzibar, Tanzania. *Conservation & Society*, *5*(4), 566–567.

Lindsey, R. (2014). Global warming or just weather. *Climate.gov*. Retrieved from https://www.climate.gov/news-features/features/global-warming-or-just-weather

Lofchie, M. (1978). Agrarian Crisis and Economic Liberalisation in Tanzania. *The Journal of Modern African Studies*, *16*(3), 451–475. doi:10.1017/S0022278X00002524

Madeweya, K., Hiroyasu, O., & Mistsuo, M. (1994). *Proceedings of the International Workshop of Ecotourism and the Environmental Conservation in Zanzibar*. Department of Commercial Crops, Tanzania.

Mapolu, H. (1990). Tanzania: Imperialism, the state and the peasantry. In *African agriculture: The critical choices*. Zed Books. Retrieved from http://archive.unu.edu/unupress/unupbooks /uu28ae/uu28ae0h.htm#8

Margareth, K. (2011). *Institute of Marine Sciences – Zanzibar*. Retrieved From: https://twas.org/sites/default/files/twas_sig_ims.pdf

Massoi, L., & Norman, A. S. (2009). *Decentralisation by Devolution in Tanzania: Reflections on community involvement in the planning process in Kizota Ward in Dodoma*. Mzumbe University.

McHenry, D. (1979). *Tanzania's Ujamaa Villages: The Implementation of a Rural Development Strategy*. Berkley, CA: Institute of International Studies.

McMahon, E., & Decker, C. (2009). Wives or workers? Negotiating the social contract between female teachers and the colonial state in Zanzibar. *Journal of Women's History*, (21), 2.

MDG. (2015). Millennium Development Goals. UNDP. Retrieved from http://www.undp.org/content/undp/en/home/sdgoverview/mdg_goals.html

Moser, S. C., & Luers, A. L. (2008). Managing climate risks in California: The need to engage resource managers for successful adaptation to change. *Climatic Change*, 87.

Mshigeni, K. E. (1973). Exploitation of seaweeds in Tanzania: The Genus *Eucheuma* and algae. *Tanzania Notes and Records*, *72*, 19–36.

Mshigeni, K. E. (1976). Seaweed farming: A possibility for Tanzania's coastal Ujaama Villages. *Tanzania Notes and Records*, *79-80*, 99–105.

Mshigeni, K. E. (1983). Algae resources, exploitation and use in East Africa. In F. E. Round & D. J. Chapman (Eds.), *Progress in Phycological research* (Vol. 2, pp. 387–419). Amsterdam: Elsevier Science.

Mshignei, K. (1996). Zero Emissions Projects in Tanzania. In *Proceedings of the 2nd Second Annual UNU.*

Msuya, F. (2008). Seaweed Farmers use seaweeds directly instead of selling raw product. Retrieved from http://www.wiomariculture.org/forum/?p=114

Msuya, F. (2011b). *Experimental Farming Of The Seaweed Kappaphycus In Floating Rafts In Pemba Island, Zanzibar, Tanzania - BIRR-MACEMP Project Phase II - Final Report.* Institute of Marine Sciences, University of Dar es Salaam.

Msuya, F. E. (1996). Seaweed farming in Lindi and Mtwara regions, Phase-two, Implementation and Expansion. Consultancy report for Rural Integrated Project Support (RIPS) Programme, Mtwara, Tanzania. IMS 1997/01.

Msuya, F. E. (2006). The Impact of Seaweed Farming on the Social and Economic Structure of Seaweed Farming Communities in Zanzibar, Tanzania. In A. T. Critchley, M. Ohno, & D. B. Largo (Eds.), *World seaweed resources: An authoritative reference system.* Amsterdam: ETI Bio-Informatics.

Msuya, F. E. (2011a). Environmental changes and their impact on seaweed farming. *World Aquaculture, 42*(4), 34–37, 71. Retrieved from http://www.academia.edu/2043063/Environmental_changes_and_ their_impact_on_seaweed_farming_in_Tanzania on 5/28/2014

Msuya, F. E., Buriyo, A., Omar, I., Pascal, B., Narrain, K., Ravina, J. J. M., ... Wakibia, J. G. (2014). 'Cultivation and Utilisation of red seaweeds in the Western Indian Ocean (WIO)Region. *Journal of Applied Phycology, 26*(2), 699–705. Retrieved from https://www.ft.com/content/e1ec5434-50f8-11e5-b029-b9d50a74fd14 doi:10.100710811-013-0086-4

Msuya, F.E., Ngoile, M.A.K., & Shunula, J.P. (1996). *The impact of seaweed farming on the macrophytes and macrobenthos of the East Coast of Unguja Island, Zanzibar, Tanzania.* Report submitted to the Canadian International Development Agency (CIDA), Institute of Marine Sciences, University of Dar es Salaam, Zanzibar, Tanzania, IMS 1997/05.

Msuya, F. E., & Porter, M. (2009). Impacts of environmental changes on the farmed seaweed and seaweed farmers in Songosongo Island, Tanzania. Report submitted under a Collaborative Project on Sustaining Coastal Fishing Communities, Memorial University of Newfoundland, University of Dar-es-Salaam.

Mustelin, J. (2009). Practical measures to tackle climate change: coastal forest buffer zones and shoreline change in Zanzibar, Tanzania. *UNESCO*. Retrieved from http://www.unesco.org/csi/climate-frontlines/PracticalMeasuresZanzibar_Ebook.pdf

Ngowi, R. (2005). Seaweed Farming helps Women in Tanzania. Retrieved from http://myhero.com/myhero/heroprint.asp?hero=Seaweed_Farming_AP

NOAA. (2013). *State of the Climate: Sea Surface Temperature – 2013 National Oceanic and Atmospheric Administration*. Retrieved from https://www.climate.gov/news-features/understanding-climate/2013-state-climate-sea-surface-temperature

NRC. (1996). *Understanding Risk: Informing Decisions in A Democratic Society*. Washington, DC: National Academy Press.

Orr, H. A. (2005). The Genetic theory of Adaptation: A Brief History. *Nature Reviews. Genetics*, *6*(2), 119–127. doi:10.1038/nrg1523 PMID:15716908

Ott, H. J. (2018). Climate change eroding women's status in Zanzibar. *CBS News*. Retrieved from https://www.cbsnews.com/news/climate-change-eroding-womens-status-in-zanzibar/

Parpart, J., & Staudt, K. (1988). *Women and the State in Africa*. Lynne Reiner Publishers.

Parry, M., Lowe, J., & Hanson, C. (2009). Overshoot, adapt and recover. *Nature*, *458*(7242), 1102–1103. doi:10.1038/4581102a PMID:19407775

Pielke, R. A., Prins, G., Rayner, S., & Sarewitz, D. (2007). Climate change 2007: Lifting the taboo on adaptation. *Nature*, *445*(7128), 597–598. doi:10.1038/445597a PMID:17287795

Price, D. T., & Apps, M. J. (1996). Boreal forest responses to climate-change scenarios along an ecoclimatic transect in central Canada. *Climatic Change*, *34*(2), 179–190.

Rosenzweig, C., Iglesias, A., Yang, X. B., Epstein, P., & Chivian, E. M. (2000). *Climate Change and U.S. Agriculture: The Impacts of Warming and Extreme Weather Events on Productivity Plant Diseases, and Pests, Center for Health and the Global Environment*. Harvard.

Rwebangira, M. (1996). *The Legal Status of Women and Poverty in Tanzania*. Nordiska Afrikainstitutet.

Sammon, A. (2018). Meet the Women Growing the California Seaweed Economy. *Civil Eats*. Retrieved from https://civileats.com/2018/03/23/meet-the-women-behind-californias-first-open-water-seaweed-farm/

Sarmiento, J. L., Slater, R. D., Fasham, M. J. R., Ducklow, H. W., Toggweiler, J. R., & Evans, G. T. (1993). A seasonal three-dimensional ecosystem model of nitrogen cycling in the North Atlantic Euphotic Zone. *Global Biogeochemical Cycles*, *7*(2), 417–450. doi:10.1029/93GB00375

Scherner, F., Horta, P. A., de Oliveira, E. C., Simonassi, J. C., Hall-Spencer, J. M., Chow, F., ... Pereira, S. M. B. (2013). Coastal urbanization leads to remarkable seaweed species loss and community shifts along the SW Atlantic. *Marine Pollution Bulletin*, *76*(1-2), 106–115. doi:10.1016/j.marpolbul.2013.09.019 PMID:24090881

Sheriff, A., & Ferguson, E. (1991). *Zanzibar Under Colonial Rule*. London: Currey.

Smith, S. (1986). *The Development of Capitalism in Africa*. London: Methuen.

Soekawa, E. (2008). Women Rake it in. Marie Claire Magazine. Retrieved from http://www.accessmylibrary.com/coms2/summary_0286-36094964_ITM

Stern, N. (2006). *Stern review: the economics of climate change*. Cambridge, UK: Cambridge University Press.

TerraDaily. (2006). Zanzibar Plastic Bag Ban takes Effect as Environment Woes Mount. Retrieved from http://www.terradaily.com/reports/Zanzibar_Plastic_Bag_Ban_Takes_Effect_As_Environment_Woes_Mount_999.html

The Zanzibar Women Corporation. (2010). The International Collaborative for Science Education and the Environment.

UN Fact Sheet. (2009). Women, Gender Equality and Climate Change. Retrieved from www.un.org/womenwatch/feature/climate_change/downloads/Women_and_Climate_Change_Factsheet.pdf

United Nations Department of Economic and Social Affairs. (2015). Sustainable Development Goals. Retrieved From https://sustainabledevelopment.un.org/?menu=1300

United Nations Department of Economic and Social Affairs. (2018). Division for Sustainable Development Goals: Transforming our world: The 2030 Agenda for Sustainable Development. Retrieved from https://sustainabledevelopment.un.org/about

United Nations Department of Economic and Social Affairs. (2018). Sustainable Development Goal Four. Retrieved from https://sustainabledevelopment.un.org/sdg4

United Nations Environmental Program/ United Nations Environmental Program/ Food and Agriculture Organization of the U.N./Priority Action Programs/CDA. (2000). *Progress in Integrated Coastal Management (ICAM) for Sustainable Development of Zanzibar's Coast: Unguja Island Coastal Profile and Management Strategy*. East African Regional Seas Technical Reports, Split, Croatia.

United Nations. (n.d.). UNGAR 67/290 High Level Political Forum - General Assembly Resolution 67/290. Retrieved from http://www.un.org/ga/search/view_doc.asp?symbol =A/RES/67/290&Lang=E

USCD. (2018). *About (Cal-CAB), California Center for Algae Biotechnology, UC San Diego*. Retrieved from https://algae.ucsd.edu/about/index.html

USCD. (2018b). *Women in Algae, California Center for Algae Biotechnology, UC San Diego*. Retrieved from https://algae.ucsd.edu/events/womeninalgae.html

Wallevik, H., & Jiddawi, N. S. (1999). Impacts of tourisms on woman activities at the South Coast of East of Unguja. In *Proceedings of the 20th Anniversary Conference on Advances in Marine Sciences in Tanzania* (pp. 535-550).

Walton, M. (2008). Algae: The Ultimate in Renewable Energy. *CNN*. Retrieved from http://www.cnn.com/2008/T ECH/ science/04/01/algae.oil/index.html

Waters, T. (2018) Seaweed Farming: A Gateway to Conservation and Empowerment, Cool Green Science (Smarter by Nature), *The Nature Conservancy*. Retrieved from https://blog.nature.org/science/2018/02/12/seaweed-farming-a-gateway-to-conservation-and-empowerment/

Webler, T., Tuler, S., Dow, K., Whitehead, J., & Kettle, N. (2014). Design and evaluation of a local analytic-deliberative process for climate adaptation planning. *Local Environment: The International Journal of Justice and Sustainability*. doi:10.1080/13549839.2014.930425

World Congress on Zero Emissions, United Nations University Zero Emissions Initiative. (n.d.). Retrieved from http://www.unu.edu/zef/publications _e/96_congress_proc.pdf

Xue, Y., Hu, Z., Kumar, A., Banzon, V., Smith, T. M., & Rayner, N. A. (2014). Sea surface temperatures. *Bulletin of the American Meteorological Society*, S51–S54.

Chapter 11
Resilience and Sustainability Development:
Lessons From Climate Change Adaptation Research

Lynn A. Wilson
SeaTrust Institute, USA & Walden University, USA

ABSTRACT

Informed action by the leaders of the future is critical for creating resilient communities. Preparing these future leaders through formal and informal education, research, and environmental/climate change programs that interweave local knowledge with the most current global science positions them to becomes the catalysts that propel community leaders to engage a wider range of possible futures. This chapter integrates findings from a SeaTrust Institute research project with the sustainable development goals in an analysis supporting dynamic and reconfigurable combinations of agents that promote the attributes of elasticity, future orientation, and motivation to address the high stakes choices for resilience to climate and environmental/social change. Author objectives in this chapter are to illustrate the optimum roles of youth in the process and what preparations and conditions are needed to instill and support youth in their ability to flip a process at the point of catastrophe to restore equilibrium and promote resilience.

DOI: 10.4018/978-1-5225-7727-0.ch011

INTRODUCTION

Walk with the dreamers, the believers, the courageous, the cheerful, the planners, the doers, the successful people with their heads in the clouds and their feet on the ground. Let their spirit ignite a fire within you to leave this world better than when you found it... —— Wilferd Peterson

Informed action by new leaders of the future, the youth of today, is critical for closing the loop with policy and research to create resilient communities. Preparing these future leaders through formal and informal education, research and climate change programs that interweave local knowledge with the most current global science positions them to becomes the catalysts that propel community leaders to make plans for and react to a wider range of possible futures.

While scientists express high confidence that climate change is and will continue to escalate at an increasing rate (IPCC, 2014), the myriad ways in which this will play out in specific regions and communities remain uncertain. People tend to discount events that are seen as unusual, uncertain, or perceived to be only relevant in the distant future. Well documented cases in economics and debates about the gravity or even existence of climate change (Sherwood 2007; Heal 2009) attest to the tendency to set aside difficult, complex issues; people face difficulty in responding to incremental change processes (Glantz 1999) until crisis is imminent or in progress.

One way that climate change is being addressed is through the Sustainable Development Goals, (SDGs). The SDGs are 17 globally agreed-upon goals with 169 targets that encapsulate the major social issues in the world today, and to provide a roadmap for integrating interwoven responses to those issues through addressing sustainability and resilience challenges. The goals include poverty, hunger, health, education, gender, water and sanitation, energy, work, infrastructure, inequality, cities, consumption and production, climate change, oceans, terrestrial ecosystems, peace, and partnerships. The SDGs replaced the Millennium Development Goals (MDGs), learning from both the successes and failures of the more general MDGs in moving towards a more sustainable, fair, and equitable world.

SDG Goal# 4 is Quality Education, defined as a goal designed to "Ensure inclusive and equitable quality education and promote lifelong learning opportunities for all" (UN General Assembly, 2015). While progress has been reported for access to education, particularly at the primary school level, for both boys and girls, it still leaves millions of children in over 40 countries without pre-primary education. And, access does not necessarily mean quality of education or even completion of primary school leaving over 100 million youth worldwide without basic literacy skills; more than 60 percent of those are women which puts Target 1 of Goal 4 (to ensure that, by 2030, all girls and boys complete free, equitable, and quality primary and secondary education) in jeopardy. Indicators for SDG Goal #4 targets for learning outcomes (target 1), early childhood education (target 2), and effective learning environments rely on incomplete and aggregate data, making it difficult to analyze and identify the children at greatest risk of being left behind. This highlights some of the challenges with a global evaluation process that depends upon local data and local implementation under voluntary reporting and in the absence of context-sensitive global oversight. Recognizing this foible, the UN has called for gathering more information from countries about how non-state actors can contribute to all the SDGs, including quality education.

Nimble organizations that can adapt quickly to both local and global situations and that combine an educational and research focus are in a position to engage community youth, leaders and professionals in working together to build resilience through country consultation, community-based projects and capacity building, conducting youth education, facilitating meaningful youth-led projects, and providing technical support for climate change adaptation plans and tangible actions within the local cultural context. Partnerships between these research and education NGOs and community-based organizations with support from university and professional experts allows for the combination of optimum flexibility (NGO), community knowledge and trust (CBO), and access to a wide variety of expert knowledge (professional experts) needed to address the complex issues encountered in climate change adaptation and resilience development.

This chapter frames findings from the SeaTrust Institutes pilot project of the Halcyon Adaptation Scenario(SM) within the general systems model of dynamic and reconfigurable combinations of the right kind of disconnected diverse functioning units that can self-organize, often in ways otherwise never seen except in times of crisis. It illustrates that, as communities develop adaptive capacity, it is vital to prepare youth, who have the attributes of elasticity, future orientation and motivation to become agents that can flip the system at the edge of climate change crisis back to a state of equilibrium, to integrate climate change knowledge with social, economic and political realities, and to collaborate with experts and leaders on community solutions. The research was conducted while the SDGs were being framed. Because the lead researcher was part of the SDG development process which began during the 2012 United Nations Conference on Sustainable Development known as RIO+20, the research was conducted with the evolving process in mind. This reflective look on that research aims to extract lessons related to education and youth as the SDG process has evolved from planning into implementation.

SeaTrust Institute researchers used the Halcyon Adaptation Scenario(SM) in a 2013 pilot project in the Chiawa region of the Lower Zambezi River in Zambia. The Halcyon Adaptation Scenario(SM) provides a roadmap for collaborative climate change adaptation development at the community level as well as the conceptual framework for action-oriented, science driven youth education that prepares them to active agents in creating and implementing community climate change resilience. Key features of the approach include:

1. Integration of local data into the most relevant and recent globally vetted science
2. Trust development and safety of participants
3. Using human health as a measurable gauge of successful adaptation strategies

Analysis of field research conducted by SeaTrust Institute in 2013 supports scholarly observations that, as climate change knowledge and leadership capabilities increase, those who are subject to high stakes choices can manifest strategies of reorganization, trust and diversity as new behaviors within the framework of intentionality and complementarity. Author objectives in this chapter are to illustrate the optimum roles of youth in the process and what preparations and conditions are needed to instill and support youth in being able to flip a process at the point of climate change and associated social crises.

BACKGROUND

Climate change affects the interwoven physical and social conditions of any particular location such that, as crisis points are approached, already stressed systems are propelled towards or over their resilience threshold more rapidly than traditional processes are equipped to handle. A map published in October 2013 in the journal Nature (Mora 2013) addresses the question of when global climate change will begin to significantly alter local climates in a measurable, definable way, beginning as early as 2020 in West Papua, Indonesia.

While some causes of catastrophic events and associated social and environmental interdependencies are global such as ENSO (El Nino Southern Oscillation) events, global economic dependencies, warming and increasingly acidic oceans, and melting ice sheets and glaciers, specific effects and appropriate adaptation responses are local. At the community level, creating the context for real, actionable leadership that utilizes the most current global science within local social and cultural norms within informal educational parameters means the community and its leaders must understand local vulnerabilities, resources and implications of choices made (Hollander, 2014).

Findings published in the 2014 IPCC AR5 Working Group II, and in publications from PROVIA and UNEP express strong agreement that social vulnerability and climate change adaptation (CCA) are most appropriately addressed from a local perspective (PROVIA 2013; IPCC 2014; United Nations Environment Assembly of the United Nations Environment Programme 2014). This approach informed the 2015 global focus as the Millennium Development Goals transformed into the Sustainable Development Goals and a new iteration of the disaster risk Hyogo Framework for Action and related processes emerge. Progress was made during 2016 by local and global civil society partnerships working in purposeful collaboration with local educational institutions and other community- based organizations who know the details about their own physical, geographical and social attributes and constraints and are intimately familiar with local and national government structures and their role in deciding the course of action for their own climate adaptation strategies.

The 2017 High Level Political Forum (HLPF) on Sustainable Development was held in New York City featuring the Higher Education Sustainability Initiative (HESI), a partnership between United Nations Department of Economic and Social Affairs, UNESCO, and other multilateral agencies and initiatives to provide higher education institutions with an interface between higher education, science, and policy. The HLPF highlighted the 2030 Agenda for Sustainable Development (UN General Assembly, 2015), showing how the SDGs inform educational planning and implementation of sustainability strategies, research, teaching, pedagogy, and campus practices. HESI strives to place higher education institutions as key drivers for achieving the SDGs.

Those with the highest stakes for the future are local youth who require the skills, knowledge and leadership development to work with adult leaders to build, refine and implement locally driven and effective climate change and disaster risk adaptation and resilience programs. Core ideas of the scientific learning required to empower these youth include disciplinary significance and scholarly relevance; direct relevance to people's lives; include key tools for investigating the complex problems of climate change and problem-solving (National Research Council 2011). These same skills, knowledge and motivation must be reflected in the larger community if climate change resilience and adaptive behavior are to occur at the local level. When youth become agents of educating the larger community, particularly in combination with community projects and programs that support resilience and climate change adaptive choices and learning at a tangible, relevant local level, the system can iteratively adapt to reflect increased community resilience as conditions change as thresholds are being approached that threaten community stability through a deterioration of life sustaining ecosystem services. Dynamic initiatives such as Global Power Shift (Ebel 2014) which in June 2013 engaged nearly 500 youth from over 130 nationalities in developing climate leadership, the more than 3 billion people worldwide under the age of 24 are becoming the best positioned and most educated generation in history to be in service of sustainability

The purpose of this study was to analyze a field test of the *Halcyon Adaptation Scenario* SM as both a process and a tool for assisting communities in developing climate change adaptation plans and resilience strategies and actions, providing a key role for community youth. Results of this study have been used to make modifications in the system to improve resilience outcomes at the community level and to use the process to inform educational curricula for pre-university youth in SeaTrust Institute's AWARE (*Action Within a Resilient Environment*) program for pre-university youth and for internship training.

Findings from this study inform the community of NGOs and educators who are involved in capacity building, climate change adaptation and disaster risk reduction and resilience as well as the global policy community involved in these endeavors. By examining SeaTrust Institute's pilot of the Halcyon Adaptation Scenario process within a conceptual framework provided by other adaptation research, guiding principles which are critical for climate adaptation work point to changes that researchers and practitioners alike may need to make in the way in which climate change adaptation planning, policy and programs are being conducted. In particular, because youth have the highest stake in securing a preferable future, the findings are directly relevant to young people who are learning, engaging and bringing new perspectives to climate change adaptation and community resilience development.

Research Questions

In interpreting the pilot, this study explores the following questions:

- What are the key considerations if communities in developing countries to meaningfully engage in climate change adaptation development for sustainability and resilience?
- How are practitioners and researchers adapting previous strategies to incorporate new learning, knowledge and practices that are truly useful to addressing the urgencies of people under siege from climate change?
- How can youth be integrated into climate change adaptation development and implementation to catalyze a community's CCA plans and actions?
- What are the wise practices for integrating interdisciplinary thinking into the process to reflect the learning and knowledge gained through the process of developing a self-directed, expert guided resilience development and climate change adaptation scenario and strategy?

Methodology

This study examines the processes and outcomes of the pilot research and action planning system and tool, the Halcyon Adaptation Scenario Process, within an interpretive review to test the first three and a half stages of the 6-stage community based participatory research process. By investigating the transformative Halcyon process through an interpretive paradigm, this case study is informed by phenomenology and hermeneutics that respects and includes different values and assumes multiple socially constructed realities. Data was collected primarily through participant observation, key informant interviews and the infrequently available documents.

Researchers worked with 24 native Zambian men over a 5-month period who engaged with the process under investigation that engaged traditional western mixed methods research with participatory rural appraisal and transformative participatory action research. The Halcyon process itself operates within a transformative paradigm combined with key elements of the indigenous research paradigm in order to be informed by the relational nature of indigenous knowledge systems while simultaneously building credibility and confidence in the participants to support them in changing elements of their society through developing dialectical understanding aimed at critical praxis (Chilisa 2012).

During the pilot, researchers guided participants by analyzing participant input using mixed qualitative and quantitative methods in what has been described by scholars as researcher as provocateur (Mertens 2010). Iterative outputs that reflected the participants' decisions and consequences of those decisions produced credible and useful interpretation that helped participants define the most important climate change adaptation questions for their community, decide upon the local and international resources they trusted to inform their learning and decision making, and prioritize a series of interrelated health, social, political, and environmental factors within both observed and scientifically informed climate change information. New information from global science, individual and group observations were used to iteratively reframe questions that reflected a growing understanding and integration of western science and traditional knowledge. Women and other villagers were visited by the researchers in their villages, using local translators when necessary. Key informants were regional leaders including village and chief headmen, the doctor, nurses, the Catholic priest and local regulatory officials who contributed valuable information through interviews, onsite visits, and less formal interactions.

Analysis of the process through interpreting participant comments, outcomes and concerns within the framework provided through key informant interviews yields information on what matters most to the community. Results suggest ways to make the process more valuable and durable to the pilot population and to similar communities.

ANALYSIS OF THE HALCYON ADAPTATION SCENARIO^{SM} PILOT PROJECT IN CHIAWA REGION, LOWER ZAMBEZI, ZAMBIA

NGO Profile

SeaTrust Institute, a United States - based research and educational nongovernmental organization, has a guiding philosophy of *Local to Global and Back Again*^{SM}. The organization uses a bottom-up approach through the Halcyon Adaptation Scenario ^{SM} process that informs the local community relationship development and training processes as well as the educational products that are used in building local capacity. In 2013, interns and staff spent 5 months in the Chiawa region of the Lower Zambezi River valley in Zambia with community members to test the Halcyon Adaptation Scenario process efficacy in engaging local participants in ascertaining vulnerabilities and alternative strategies for responding climate change adaptation.

Participant Profile and Selection

The participants were twenty-four native Zambians, nearly all of whom self-identified as belonging to the local Goba people, formed the core group of local participants. These were men, employed by a local lodge but who lived in local villages throughout the region from Chiawa to Chirandu, the border crossing into Zimbabwe. Participants were selected based upon regular accessibility by the research team, English language proficiency, the ability to be present at regular sessions throughout the pilot period and the ability to facilitate transportation and serve as translators for the researchers for work in surrounding villages. Effort was made to include individuals representing different villages and with different areas of expertise. All participants were working at a lodge along the Zambezi River within the Chiawa region.

Women, other villagers, and key regional leaders including the Chief Headman, a local village headman, the doctor, a nurse, the Catholic Priest and a natural resources technician were also part of the study. These were also Goba people. The Goba who live in villages instead of isolated homesteads do so for mainly for access to roads, water, and protection from wildlife (Lancaster 1981). The local governance structure is headed by the Chieftainess, who owns all of the land upon which the villagers live and farm. She makes critical decisions for the citizens of the villages

Table 1. Participants in Halcyon Adaptation Scenario(SM) Pilot in Zambia

Nationality	Town	Expertise	Job Title	Age	Family at Home	Languages
Zambian	Chirundu	Accounting	Assistant Manager	33	2 children: 1 son & 1 daughter; younger brother	Goba; Njanja; English; Tonga
Zambian	Chiawa Village	Field Farmer of Maize	Guide and Boat Driver	32	3 children: 2 girls & 1 boy	Goba; Njanja; English
Zambian	Chiawa Village	Field Farmer of Sorghum	Self-Catering Attendant	30	1 child: girl	Goba; Njanja; English
Zambian	Chiawa Village	n/a	Kitchen Helper	31	2 children: both boys	Goba; Njanja; English
Zambian	Pontoon Village	Field Farmer of Maize	Builder/ Gardener	36	3 children: 2 boys & 1 girl, father also	Goba; Njanja; English
Zambian	Chiawa Village	Field Farmer of Maize	Room Attendant	27	2 children: 1 boy & 1 girl	Goba; Njanja; English
Zambian	Chiawa Village	Knowledgeable on Flora and Fauna	Workshop/ Guide	42	4 children: 2 boys & 1 girl	Goba; Njanja; English
Zambian	Chiawa Village	n/a	Campsite/ Room Attendant	31	3 children: 2 girls & 1 boy	Goba; Njanja; English
Zambian	Lusaka	n/a	Room Attendant	38	3 children: 2 girls & 1 boy	Goba; Njanja; English
Zambian	Lusaka	Gardening/ Field Farmer of Maize	Head Gardener	31	4 children: 1 girl & 3 boys	Goba; Njanja; English; Tonga; Lozi
Zambian	Chiawa Village	Knowledgeable on Local Flora & Fauna	Guide/ Boatman	35	3 children: 2 boys & 1 girl	Goba; Njanja; English
Zambian	Chiawa Village	n/a	Gardener	22	Single	Goba; Njanja; English
Zambian	Chiawa Village	Field Farmer of Maize	Guide/ Boatman	36	2 children: 1 boy & 1 girl	Goba; Njanja; English
Zambian	Chiawa Village	Field Farmer of Maize	Kitchen Helper	32	3 children: 2 girls & 1 boy	Goba; Njanja; English
Zambian	Chiawa	Knowledgeable on flora and fauna, field farmer of sorghum, carpentry	Guide	38	3 children: all girls; brother and law and his son	Goba; Njanja; English
Zambian	Chiawa Village	Field Farmer of Maize and Sorghum	Storeman/ Barman	53	5 children: 4 girls & 1 boy	Goba; Njanja; English
Zambian	Chiawa Village	n/a	Mechanic	40	5 children: 3 girls & 2 boys	Goba; Njanja; English

continued on following page

Table 1. Continued

Nationality	Town	Expertise	Job Title	Age	Family at Home	Languages
Zambian	Chiawa	n/a	Waiter	42	6 children: 4 girls & 2 boys	Goba; Njanja; English
Zambian	Chiawa Village	Field Farmer of Maize	Laundryman	36	4 children: 1 girl & 3 boys	Goba; Njanja; English
Zambian	Chiawa Village	n/a	Mechanic	40	5 children: 4 girls & 1 boy	Goba; Njanja; English
Zambian	Chiawa Village	Field Farmer of Maize	Head Room Attendant	29	2 children: 1 girl & 1 boy	Goba; Njanja; English
Zambian	Chiawa Village	Field Farmer of Maize	Gardener	32	4 children: 1 girl & 3 boys	Goba; Njanja; English
Zambian	Chiawa Village	Field Farming of Maize	Cook	33	1 child: girl	Goba; Njanja; English
Zambian	Chiawa Village	n/a	Cook	39	4 children: 2 boys & 2 girls	Goba; Njanja; English

including the locations of bore holes that supply fresh water and, with input from the Headman of each village, makes the decisions about most community matters. The Chief Headman is the conduit of information from the village level headmen to the Chieftainess.

The Halcyon Adaptation Scenario Process(SM)

Considering health and adaptation as tools for social learning and for evaluating strategies within the 6-step Halcyon Scenario uncovers locally useful information for local governments as well as communities, because it accounts for barriers and opportunities for adaptation. Researchers guiding the process collect, make available and correlate global and local data that pertains to the region and scale of interest, and use health as a gauge for what constitutes effective action.

Participants prioritize important, sometimes mutually exclusive choices to establish thresholds beyond which the system basically flips or is no longer sustainable in its current form and construct an iterative narrative of climate change adaptation scenarios based upon their most pressing needs, trusted information, cultural relevance and practical likelihood of successful implementation.

The first stage is facilitated framing of the focal question or questions that are most relevant to the group. Participants' choice of questions determines what they interrogate with data from global scientific sources (provided by experts) intermixed

Figure 1. Halcyon Adaptation Scenario(SM) process
(Wilson 2013)

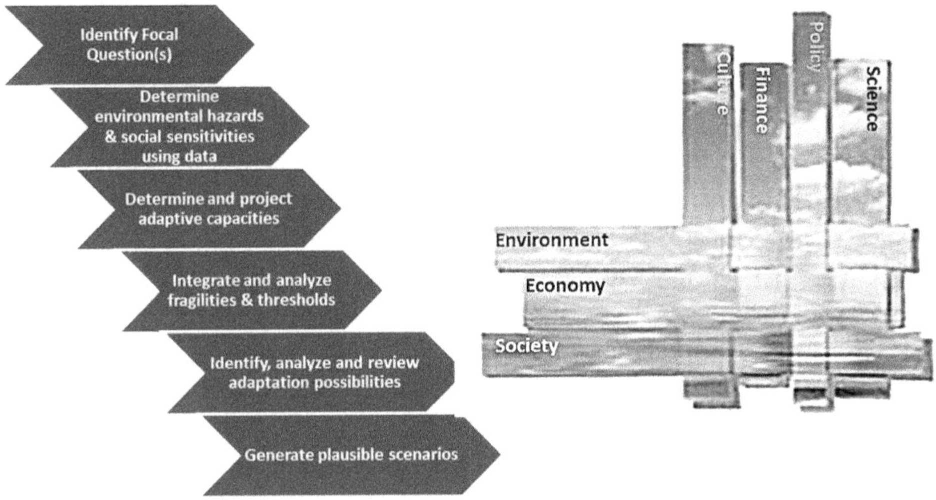

with trusted local information, including anecdotal data during stage two. The process of engaging the public in scientific reasoning by asking/answering questions and evaluating the evidence coupled with inviting the public to understand science as a way of knowing has been shown by the Sciencenter's Nanoscale Informal Science Education network as two of the "strands of learning" that help forge linkages between science, society, daily life and the environment (Hollander 2014). In this stage, participants are creating their own vulnerability profile against which to design a climate change adaptation and resilience development strategy.

During the third stage, participants prioritize important, sometimes mutually exclusive choices to establish thresholds beyond which the system flips or is no longer sustainable in its current form and develop a scenario to manage the system so that is does not breech the threshold. Through stages three and four, Halcyon focuses on the changes rather than the static situations, specifically the changes among the interconnected elements of climate, environment, social, economic and political factors, as evidenced in human health effects. Participants make choices at critical junctions that are quantitatively and qualitatively analyzed by researchers during stage 5 and returned to the participants for course correction and redirection of the scenario before producing a new future direction based on evidence and choices vetted by the group during stage 6. At all point in the process, participants may revisit choices of questions, data sources, choices and directions as new information emerges or as physical, social or political changes create new situational realities.

Problem Statement

In the Lower Zambezi region of Zambia, lack of usable water at the right times of year, compounded by a land tenure system that gives absolute power of the land and its resources to the local chiefs means that the choices to stay, as a village headmen interviewed through a local interpreter adamantly asserts, are potentially fraught with serious repercussions that can be measured by starvation, dangerous encounters with wild animals competing for scare food resources and diseases from using contaminated river water when bore holes were fouled or too far from villages (CITE: interview with Chiawa Village Headman, 2013). For these people, political and economic issues influence what they count as useful information and their priorities about what to even consider.

Droughts, floods and increased temperature are the major physical climate change issues in the region. In this seasonally arid area, impassable roads and floods in the wet season are punctuated by a new interim dry season that means crops like sorghum that take 6 months to mature are no longer viable. People in Chiawa acknowledged to the researchers that more and increasingly extreme climate events are affecting their food supply; high concern for their continued ability to grow maize (which according to UN data is likely to be highly marginalized by 2050) led to conversations about developing aquaculture or micro businesses so they could buy maize from the Chieftainess who owns the land and who will irrigate portions of her holdings to grow food to sell to the people.

However, in such a remote and distressed area, to whom do they sell these products? Tourism, though important, is insufficient to create a thriving economy. For them, migration is not an option. One of the pilot participants said "we'll just do with less," which was echoed by the core pilot group who said "We will never leave – no matter how bad it gets; this is our Motherland." These people have been displaced from their ancestral land in recent times by the government to establish a national park so they are even more adamant about not relocating and certainly not to an urban slum.

Malaria is the most significant health problem in the region. According to the World Health Organization, "Most deaths occur among children living in Africa where a child dies every minute from malaria (World Health Organization 2014). A new strain of drought induced malaria, dysentery, diarrheal and other diseases from contaminated water, skin and eye diseases were reported by the only regional doctor as the other most common health problems, all exacerbated by climate change (Wilson 2013). A local nurse also cited respiratory infections, increasing heat, less rain, malnutrition from reduced maize yields, and increased injury from encounters with wildlife, especially elephants competing for food (Wilson 2013) while participants also reported cholera and tuberculosis as active health issues. The

Table 2. Major health, social, cultural and political factors

Primary Health Issues:	Social/ Cultural/ Economic & Political Factors
• Malaria • Cholera • Skin Rashes • Tuberculosis • Upper Respiratory Tract Infections • Diarrhea • Dysentery • HIV	• Lack of government support • No voice in issues • Little knowledge on when the Kariba Dam is opening • High Unemployment rate • Clean Water • Expensive health care • Expensive schooling

health profile also includes a significant portion of the population testing positive for AIDS, which, while not climate change related, increases the vulnerability of the population to additional stressors.

Finding ways to infuse and incorporate new knowledge to help the villages in the Chiawa region to develop a more hopeful future by preparing for, and to the extent possible exploiting climate variability is necessary if those communities are to develop resilience to climate change. Factors that influence the ability of individuals and groups to learn about climate and environmental issues (McDaniels 2004; Berkhout 2006) and further exacerbated by the land tenure, social and economic conditions of the region.

FINDINGS

The Halcyon Adaptation Scenario Pilot as a Transformational Approach to Creating Climate Change Resilience

What Are the Key Considerations if Communities in Developing Countries to Meaningfully Engage in Climate Change Adaptation Development?

Resilience is defined in this study as maintaining core principles and integrity – of all kinds - under dramatically changed circumstances. Successful climate change resilience development requires that its practitioners adhere to core principles and activities. The first two of the nine listed elements in Figure 2 provide the conceptual framework within which the other elements emerge in local situations.

Key considerations for meaningful community engagement in climate change adaptation development begin with the principle of *intentionality*, not only as reflected by the researchers' internal intentions but as communicated in word choices and definitions. In adaptation development, resilience is generally associated

Figure 2. Elements needed for climate change resilience development

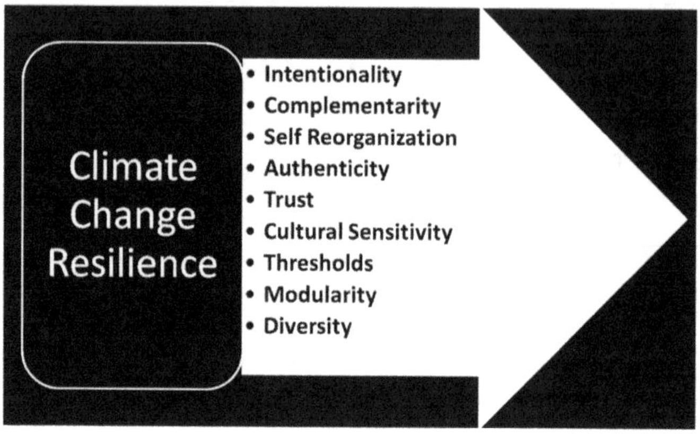

with adaptation in the same way sustainability is associated with mitigation. The overarching assumption is that those engaged in global adaptation strategy and capacity-building have an intention to create resilient communities. In the pilot, the researchers' intention was to learn what it would take to create a resilient community in the Chiawa region of the Lower Zambezi River, a place that is already experiencing extreme effects from climate change with extreme droughts and floods, exacerbated by a new extra drought season.

The pilot participants exemplified global citizens who are becoming consciously aware of the connections between climate pattern changes, extreme events and the health of their communities and citizens. The awareness is coming from what they are seeing for themselves – the effects of their own fragilities under increasing stress. Introducing relevant research and data through a socially and culturally sensitive approach showed intentionality on the part of the researchers. The data was a mix of community-accepted global science and local anecdotal data.

Halcyon Adaptation Scenario Zambia Pilot: Self-Defined Issues and Selected Data Sources

Participant Questions for Scenario Development:

- How can we get the government to be more involved in our lives here in Chiawa?
- What is the best way to solve our water issue of sanitized water?

Issue 1: Droughts (Climate Issue)

Events: 1991 and 2011 there were severe droughts that resulted in starvation

Observations: Droughts are severe especially here because people live in the valley; it is a very dry area. The drought also causes dangerous animals to hang around the river making it harder for women and children to access the water.

Applicable Data:

- "Top 10 Disasters" from the International Disaster Database-*Accepted: "we remember some of these disasters"*
- "Environmental Performance Index- *Accepted: "The plot shows that Zambia is improving and we agree"*
- "Zambia Lowres Report" from the UNDP- *Accepted: "The climate and other information seems appropriate"*
- "Zambia PPCR Strategic Program"- *Accepted after a review of the summary was made*
- "MDG Report 2012- *Accepted "we like the goals of this paper"*
- "Zambia GAIN Index"- *Rejected because it was not properly understood*
- Local Source: The wind is read which tells them when it will rain if it is not the rainy season. *Accepted by all of the members*

Issue 2: Floods (Climate Issue)

Events: 1988, 1995, 2005: all floods caused by the Kariba dam that completely inundated our crops

Observations: The government told us that the gates were going to open but only after we had planted our crops, so we lost all of it.

Applicable Data:

- "Top 10 Disasters" from the International Disaster Database-*Accepted: "we remember some of these disasters"*
- "Environmental Performance Index- *Accepted: "The plot shows that Zambia is improving and we agree"*
- "Zambia Lowres Report" from the UNDP- *Accepted: "The climate and other information seems appropriate"*
- "Zambezi River Basin Flood Forecasting and Early Warning System Strategy"- *Accepted because it discusses flooding specific to the Zambezi*
- "Zambia NAPA Final Report"- *Accepted, they agreed very much with table 1.1 which contained a summary of vulnerability factors*

- "Zambia PPCR Strategic Program"- *Accepted after a review of the summary was made*
- "MDG Report 2012"- *Accepted "we like the goals of this paper"*
- "Zambia GAIN Index"- *Rejected because it was not properly understood*
- "Zambezi Zambia Report"- *Accepted because it focuses on the flooding of the Zambezi which is very relevant*
- Local Source: The wind is read which tells them when it will rain if it is not the rainy season. *Accepted by all of the members*

Issue 3: Temperature Increases (Climate Issue)

Events: The hot season has been getting hotter every year, and some believe that the cool season has been getting hotter as well. Others disagree and think the cool season is getting cooler.

Observations: The temperatures being hotter causes more crops to fail and it also affects people working outside.

Applicable Data:

- Lower Zambezi National Park Temperature and Precipitation Data by Weather2Travel- *Accepted, they agreed with the data from this chart*

Issue 4: Malaria (Health Issue)

Events: Malaria is most common in the rainy season but is becoming more common throughout the year

Observations: Some say that malaria is getting worse, the doctor and nurse says that there is a new strand of malaria in the area that has unique symptoms and is usually only common in dry parts of Africa

Applicable Data:

- "Zambia National Health Strategic Plan 2011-2015"- *Accepted because it seems appropriate and was written by the Zambia Ministry of Health*
- "Yearly Disease Aggregations"- *Accepted because this information is local*
- "Mortality Rates Zambia" WHO data- *Accepted*
- "Fertility Rates Zambia" WHO data- *Accepted*
- "Life Expectancy Zambia" WHO data- *Accepted*
- "Local Health Issues"- *Accepted the diseases that are prevalent*

Issue 5: Diarrhea and Dysentery (Health Issue)

Events: Diarrheal diseases are most common in the rainy season; dysentery is common all year long.

Observations: The doctors, nurses and group say that diarrhea is becoming more of a problem.

- "Zambia National Health Strategic Plan 2011-2015"- *Accepted because it seems appropriate and was written by the Zambia Ministry of Health*
- "Yearly Disease Aggregations"- *Accepted because this information is local*
- "Mortality Rates Zambia" WHO data- *Accepted*
- "Fertility Rates Zambia" WHO data- *Accepted*
- "Life Expectancy Zambia" WHO data- *Accepted*
- "Local Health Issues"- *Accepted the diseases that are prevalent*

Issue 6: Upper Respiratory Infections (Health Issue)

Events: Upper respiratory infections have become more common especially in the winter/ dry season.

Observations: URTI's have become more common as it gets drier and there is more dust in the area. This is a very dusty region.

- "Zambia National Health Strategic Plan 2011-2015"- *Accepted because it seems appropriate and was written by the Zambia Ministry of Health*
- "Yearly Disease Aggregations"- *Accepted because this information is local*
- "Mortality Rates Zambia" WHO data- *Accepted*
- "Fertility Rates Zambia" WHO data- *Accepted*
- "Life Expectancy Zambia" WHO data- *Accepted*
- "Local Health Issues"- *Accepted the diseases that are prevalent*

Issue 7: Skin Rashes (Health Issue)

Events: Skin rashes including heat rashes, eczema, chicken pox, ringworm and head rashes.

Observations: Chicken pox and heat rashes especially have become more common.

- "Zambia National Health Strategic Plan 2011-2015"- *Accepted because it seems appropriate and was written by the Zambia Ministry of Health*
- "Yearly Disease Aggregations"- *Accepted because this information is local*
- "Mortality Rates Zambia" WHO data- *Accepted*

- "Fertility Rates Zambia" WHO data- *Accepted*
- "Life Expectancy Zambia" WHO data- *Accepted*
- "Local Health Issues"- *Accepted the diseases that are prevalent*

Issue 8: Cholera (Health Issue)

Observations: Mentioned

- "Zambia National Health Strategic Plan 2011-2015"- *Accepted because it seems appropriate and was written by the Zambia Ministry of Health*
- "Yearly Disease Aggregations"- *Accepted because this information is local*
- "Mortality Rates Zambia" WHO data- *Accepted*
- "Fertility Rates Zambia" WHO data- *Accepted*
- "Life Expectancy Zambia" WHO data- *Accepted*
- "Local Health Issues"- *Accepted the diseases that are prevalent*

Issue 9: Measles (Health Issue)

Observations: Mentioned

- "Zambia National Health Strategic Plan 2011-2015"- *Accepted because it seems appropriate and was written by the Zambia Ministry of Health*
- "Yearly Disease Aggregations"- *Accepted because this information is local*
- "Mortality Rates Zambia" WHO data- *Accepted*
- "Fertility Rates Zambia" WHO data- *Accepted*
- "Life Expectancy Zambia" WHO data- *Accepted*
- "Local Health Issues"- *Accepted the diseases that are prevalent*

Issue 10: Tuberculosis (Health Issue)

Observations: Mentioned

- "Zambia National Health Strategic Plan 2011-2015"- *Accepted because it seems appropriate and was written by the Zambia Ministry of Health*
- "Yearly Disease Aggregations"- *Accepted because this information is local*
- "Mortality Rates Zambia" WHO data- *Accepted*
- "Fertility Rates Zambia" WHO data- *Accepted*
- "Life Expectancy Zambia" WHO data- *Accepted*
- "Local Health Issues"- *Accepted the diseases that are prevalent*

Issue 11: Injuries (Health Issue)

Observations: Mentioned

- "Zambia National Health Strategic Plan 2011-2015"- *Accepted because it seems appropriate and was written by the Zambia Ministry of Health*
- "Yearly Disease Aggregations"- *Accepted because this information is local*
- "Mortality Rates Zambia" WHO data- *Accepted*
- "Fertility Rates Zambia" WHO data- *Accepted*
- "Life Expectancy Zambia" WHO data- *Accepted*
- "Local Health Issues"- *Accepted the diseases that are prevalent*

Issue 12: HIV/ AIDS (Health Issue)

Observations: Mentioned

- "Zambia National Health Strategic Plan 2011-2015"- *Accepted because it seems appropriate and was written by the Zambia Ministry of Health*
- "Yearly Disease Aggregations"- *Accepted because this information is local*
- "Mortality Rates Zambia" WHO data- *Accepted*
- "Fertility Rates Zambia" WHO data- *Accepted*
- "Life Expectancy Zambia" WHO data- *Accepted*
- "Local Health Issues"- *Accepted the diseases that are prevalent*

Issue 13: Clean Water (Social/ Cultural Issue)

Events: Most of them do not have access to pumps for sanitized water. It is a big problem during the rainy season when the Zambezi is very murky and contains more bacteria that gives them illnesses.

Observations: Diarrhea and also cleanliness are problems due to the lack of clean water

Applicable Data:

- "Improved Sanitation Facilities" by The World Bank- *Accepted because they believe that about half of Zambia does not have access to water sanitation facilities*

Issue 14: High Unemployment Rate (Social/ Cultural Issue)

Events: Many know people who have gone to school and then been unable to find jobs after to pay off student loans

Observations: The president has made a few more jobs but he has not done what he has promised.

Applicable Data:

- "Employment" From World Bank- *Accepted because they think the figures look appropriate*
- "Climate Change, Economic Growth and Poverty in Zambia"- *Rejected because it is not understood*
- "Zambia" Economic data via the United Nations- *Accepted*
- UN Comtrade "Zambia Exports"- *Rejected because it was not understood*
- "African Statistics 2012: Zambia"- *Accepted because the data seemed accurate*
- "Population: Zambia" WHO data- *Accepted*
- "HDI Zambia": *Accepted*

Issue 15: Expensive Schooling (Economic Issue)

Events: We have to pay every month for our children to go to school and we have to pay any hospital fees we have which can be very expensive. Secondary school is most expensive and many do not finish it because of the costs.

Applicable Data:

- "Literacy Rate" by the World Bank- *Accepted, seems legitimate*

Issue 16: Expensive Health Care (Economic Issue)

Events: We have to pay every month for our children to go to school and we have to pay any hospital fees we have which can be very expensive. Secondary school is most expensive and many do not finish it because of the costs.

Issue 17: Lack of Government Support (Political Issue)

Events: Little help with malaria, floods and droughts

Observations: For example the government sends someone down to check for eye problems once a year, but if you miss them the one-day that they come then they cannot help you

Applicable Data:

- "Zambia NAPA Final Report" *Accepted: Agreed very much with Table 1.1 which contained a summary of the vulnerability factors*
- "Zambia PPCR Strategic Program": *Accepted: Once the "summary" was talked about and discussed*
- "Zambezi Zambia Report": *Accepted: it talks about flooding in the Zambezi- very appropriate for here*

Issue 18: No Voice on Issues (Political Issues)

Events: Inability for Chiawa locals to get in contact with government officials

Observations: The members have had no contact with the Chieftainess or any officials through the Zambian government. They have no way to share their thoughts or opinions.

Applicable Data:

- MDG Report 2012- *Accepted because they like the goals of the paper*

However, it requires more than just bringing information about fragilities and options and walking away. Research and projects conducted in the developing world are often criticized for doing just that: collecting data and leaving the communities without new tools or useful ways to address the now-illuminated issues. Responsible expert guided CCA processes must locate keys that unlock their system of incorporating new information and combined knowledge systems into sustained and iterative action.

The other key principle is *complementarity*, which is defined as combining elements in such a way as to enhance or emphasize the qualities of the other. In physics complementarity is demonstrated in the combination of colors it takes to create white light. White light is an appropriate metaphor for the research and educational practitioners' role as an adjustable lens through which to focus the global science within the landscape of local knowledge and conditions while keeping in view the intention of developing a resilient community.

To be complementary, a climate change adaptation process must continuously adapt itself to the situational changes, physically, culturally and socially. It is here that human health functions as the nexus for examining physical and social changes and, as it was discovered in the pilot, as a means for examining changes made through the participants' learning processes. An example of health as a gauge for the interconnected changes is shown through a significant addition to the malaria issues brought about by the new interim drought. In an interview with Dr. Charles, researchers learned about the "new" malaria that has very recently appeared in conjunction with the new drought season – *vivax* malaria. Endemic to other parts of the world, it is morphing quickly in the lower Zambezi, becoming resistant to traditional malaria drugs and spreading as drought spreads.

The standard "solutions" were not working in this region, even for malaria treatment. They aren't working to increase food security, or manage spreading diseases. Even the medical literature hadn't caught up with the actual spread of the *vivax*. This event is an example that affects the second question in this analysis: *How are practitioners and researchers adapting previous strategies to incorporate new learning, knowledge and practices that are truly useful to addressing the urgencies of people under siege from climate change?*

All sources point to the need for more local data that illuminate the dynamic changes. Water quality and availability figure heavily in adaptation decisions. Chemical and microbial pollution overtly affect health but poor or nonexistent baseline information makes drawing the connections between changing climates and socioeconomic / political contexts and decisions difficult to document. Medical care amounts to triage, with a single doctor who has been trained only as a nurse responsibility for the health of the entire population. Certainly the emergence of new strains of disease like *vivax* malaria need to be documented and accounted for in adjusting dynamic adaptation strategies. Bore holes access deep groundwater sources in arid regions but quality and testing is irregular and unreliable so the problem of microbial pollution in small scale systems like ground wells compounds negative effects from other alternatives like using the heavily polluted Zambezi river water. Political issues affecting bore hole drilling location choices by the Chieftainess/ landowner mean that water in unevenly available to villages.

Halcyon Adaptation Scenario proved to be appropriate for collecting, maintaining, and making dynamic the global and local information available not only locals but to other regions experiencing similar issues. The design is appropriate for the community's needs, could be largely administered by a lay person, and was accepted as respectful of community norms. During the first phase of the pilot, participants agreed on the following questions for developing their climate change adaptation scenario:

How can we get the government to be more involved in our lives here in Chiawa?

What is the best way to solve our water issue of sanitized water?

The pilot participants were engaged several times per week, added data and anecdotal evidence, discussed difficult choices and wanted to continue the process after the conclusion of the pilot. They began asking questions about how to change the situation and address their vulnerabilities which raised questions from those who benefitted from the status quo.

Tensions arising between participants and foreign property owners affecting the possible continuation of employment resulted in decreased participation during the last 5 weeks. The pilot was ended prematurely resulting in stages 5 and 6 being omitted and insufficient prioritization completed in stage 4 for quantitative analysis. After prioritizing issues and options, and looking at them through the health issues in the community to assess their relative impact, the participants made the following exit comments about what is important them in completing the process of building a community that is resilient to climate change.

POLITICAL

Lack of Government Support and Having No Voice on Issues

We would like it if the Chieftainess could come to Chiawa more and visit and listen to our problems, but we rarely see her. She should be the one supporting us and being our voice for the big government. We would also like the government to take more interest in this area and see what we really need like more boreholes.

ECONOMIC

Expensive Health Care and Schooling

Getting teachers trained better so that we can send our kids to public school instead of private school would be good. We need better training colleges for our teachers. Also if the teachers at the public schools were paid more then maybe they would care more about the wellbeing of our kids.

HEALTH

Inadequate Care Facilities

What we really need in Chiawa Village Clinic is a vehicle that can take the really sick people to a different hospital. We can only do this with money though and right now the government does not give us enough. Help from a non-profit would be could and maybe we could charge people for each ride. We also need more medicines like inhalers. If the Chieftainess came here more then she would see that we need help in these areas.

If Changes Were Made Including Water Harvesting, Different Crops, Better Relationship With Higher Officials, Would That Create a More Resilient Community to Climate Change?

We can see how all these changes and adaptations would make our lives better for the future. We worry about having healthy children because so many kids these days are malnourished. We worry when they get sick from water in the rainy season. If we could adapt to not have these worries anymore then we would be much happier. We also want our kids to get better schooling than us so that they can be more successful. It would make us most happy to see our children get good college educations and good paying jobs. "If we can just continue to provide for our families then we will be happy but it must be here in Chiawa because we will never leave.

Because the last two stages contain largely analytical functions performed by the researchers, with some feedback from participants about changes they would like to make, the unanticipated ending did not affect the learning from the pilot about the efficacy of the Halcyon process in engaging participants and testing the process within the community. However, the event attests to the degree to which flexibility and the opportunity to reflect upon experiences are central to learning and behavior changes (Fazey 2005). It was clear from

The Missing Element

Although largely successful as gauged by the activity and output of the pilot participants, analysis reveals that for the learning and new knowledge through the Halcyon Adaptation Scenario(SM) process to actually affect the community and help behaviors to turn at the threshold of a crisis, something more is required. If imposed from above, resilience strategies most often fail. Why? Myriad reasons exist but largely it is because top down imposition is simply not reflective of the way in which systems work to be self-sustaining. In the system under investigation, resilience attitudes and actions must be embedded authentically into the relationships that mediate people's everyday lives; health is a measurable proxy for those relationships and an equalizer of social stratification. This was an assumption by the researchers who sought to make this a bottom-up process which was largely successful. However, for the system to continue without the direct intervention of the researchers, even in the absence of the external threats that were encountered by the pilot team, the community must sense, interdict and intervene at critical thresholds and to do this it has to be a dynamic and reconfigurable combination of the right kind of disconnected diverse functioning units that can self-organize, often in ways otherwise never seen except in times of crisis.

Although learning is essential, creating resilience and appropriate climate change adaptation strategies requires more than training or exposure to issues. An agent within the community must be engaged in continuously learning, data collection and new knowledge dissemination within the community. Success depends upon deeply embedded social factors and the ways in which those factors interact with change processes. That agent must be trusted, have a vested interest in the future of the community, and be an integral part of daily life.

As shown above and the elements of climate change resilience, the agent needs to have the combination of reorganization and complementarity, something defined in biology as a "sleeping functional group." An example of a group that combines the two is the batfish, which in 2006 saved an algae-infested patch of coral near Orpheus Island on Australia's Great Barrier Reef (Bellwood 2006). Parrot fish usually keep the algae mowed in that area, but only in healthy coral ecosystems. Orpheus Island was a severely declining coral ecosystem and parrot fish were not eating the algae. Near the point of ecosystem collapse, something else began eating the algae that was choking the reef. A single species of batfish, *Platax pinnatus*, which is comparatively rare on the Great Barrier Reef, was thinning the algae from below, although the fish was not previously known to eat anything except invertebrates. The technical term for this behavior under crisis is "sleeping functional group," a group exhibiting a function that only emerges when required and one which can self-organize to change behavior to avoid a crisis. It is possible that within this concept may reside a clue for successful community climate change adaptation.

In human communities, sleeping functional groups require other elements in the list of climate change adaptation requirements in order to be cohesive and self-organizing. The group must be based in trust and respect and yet not be so tightly woven that rigidity sets in. The unusual behaviors of sleeping functional groups only happen as thresholds are being reached, but natural and human systems are not static. New and changing stressors such as political and economic changes constantly redefine thresholds in human systems. Resilience is not a final state, but a delicate, constantly renegotiated balanced reaction to changing conditions. Only with sufficient diversity can these latent countercyclical agents appear that can exhibit unusual behavior to disrupt calamity to keep the system from flipping into disasters as thresholds are approached--- after which those agents return to disconnected or even opposing positions.

One of the most powerful counter-cyclical agents we can mobilize in communities under stress from climate change is the group that manifests most extreme health issues from climate change – and that by morphing the cited victims into the very agents of change and resilience development, the system may be able to equilibrate (Wilson 2013).

Youth are the most frequently cited victims of climate change both in the present moment and as targets for policy makers creating climate agreements that address generational equity. *How can youth be integrated into climate change adaptation development and implementation to catalyze a community's CCA plans and actions?* This is the third question under investigation in this study. Youth are the agents that, like the batfish, can flip the system when educated in climate science and local social conditions and given the opportunity to deliver that knowledge to their communities. In the Halcyon Adaptation Scenario(SM) pilot, youth training youth was the missing element. An intern with the research team who was significantly younger than the participants assisted with the research by adult members; no local youth were involved except passively during village visits with translators present. The addition, peer youth-to-youth training, first from the NGO to the community and subsequently ongoing within the community, would significantly strengthen the effectiveness and the longevity of the adaptation strategy development process. Youth training their own communities would vet the process, knowledge and new learning within the community social system.

RECOMMENDATIONS

The fourth question in this analysis guides the recommendations from this interpretation of the field pilot. *What are the wise practices for integrating interdisciplinary thinking into the process to reflect the learning and knowledge gained through the process of developing a self-directed, expert guided climate change adaptation scenario and strategy?*

In order to construct a scenario for contending with climate change and building community resilience to future changes, local people must know what they are facing from climate changes and associated health impacts and what options are available. Learning systems at the community level must engage all levels of society as active participants. Although as noted by studies referenced in this chapter, people discount events that are seen as rare, uncertain, or only relevant in the distant future, the pilot participants were well aware of climate change and its effects on their health, current situation and the prospects for future generations. Accurate and believable information is essential in catalyzing decisions that will create a resilient future.

It is also critical that researchers and community members jointly acknowledge what is not viable in their current political, economic and cultural state. Dr. Charles

Chiyama, the only doctor serving the region, showed the researchers the "epidemic kit:" one small packing box with a few saline bags on the floor of an almost empty pharmacy at the "hospital" serving over 900 people - the Chiawa Water and Beds Project -outfitted with three, malaria bed nets. It is clear that even with assistance villagers are going to have to work within local governance, economic and cultural contexts and constraints.

Climate change adaptation processes must be durable and self- regenerating as conditions change. During the Halcyon analysis phase (stages 5 and 6, the expert process), researchers examine the consequences of the interdependencies among physical, biological, social, political and economic systems, and offer this information back into the community for their further consideration. Although measurement and evaluations are being added to projects of all scales in an attempt to locate and remediate causes of sub-optimum outcomes, expertise and knowledge offered to assist local people in developing and implementing climate change adaptation plans and engaging in behaviors that lead to community resilience to environmental change is often not as effective as intended. If climate adaptation, including disaster risk reduction activities, are not embedded into local development plans and policies the likelihood that the activities will have long term effects is almost assuredly miniscule; if the multilateral process does not support these local efforts either by inertia or by the inability to change to more flexible and cross sectoral approaches, the same less-than-optimum outcomes result. These is apparent in results from aid of all kinds, and even with educational efforts if a local leadership group and structure is not engaged with "owning" the tools and methods of change that are offered by outside entities. Even when received graciously by communities in need, even the best global tools for creating climate change adaptation and resilience quickly come to resemble rusted cars in the yard or rusted boats in the harbor – given with good intentions but wasted for lack of the tools to maintain their continued use.

Funding must be appropriate and consistent. A bottom-up approach is widely considered to be appropriate for reducing social vulnerability through approaching climate change adaptation as a process rather than an endpoint that is compatible with community and cultural standards. When participants interweave local

Table 3. Example of participant considerations of viable alternatives

Climate Changes	Health Impacts	Viable Options	Non-Viable Options
Increasing drought	Malaria (traditional) Emerging dryland vivax malaria (new) Respiratory diseases	clay and sand water filtration systems; building water tanks; indigenous knowledge about coping mechanisms to droughts and floods	private land ownership, and access to basic resources such as clean water, medicines

knowledge with expert scientific and policy knowledge and sources and document shared learning, it becomes useful in informing both local governance and global knowledge. Difficulties come with projects that are funded for a given duration (a common and sometimes necessary constraint of global development funding) but that fail, for any number of often reasons, to imbed the new learning or behavior changes into the lives of the local people.

Health is an appropriate gauge of progress and a compass for course corrections and should be led by local citizens and health professionals. Using health in this way requires stepped-up and appropriate data surveillance and global communications, increasingly available and even crowdsourcing in efforts like that of Kenyan-based Ushahidi during the earthquakes in Haiti (Zolli 2012), or a project with one of SeaTrust Institute's partners using tablets for information-sharing about health. Both of these approaches demonstrate how local people are frequently in the best position to create parameters and populate data that can be used to construct the baseline and follow the patterns of climate variability.

Perhaps the most important recommendation that emerged from the examination of this pilot is the need to train and engage youth in every facet of climate change adaptation. Youth behavior changes at the thresholds; innovation that can flip the system emerges as they are empowered with knowledge – they are the most appropriate individuals to deliver the message within a community for optimum cultural relevance, adoption and engagement over the long term. Attaching the message to youth and having that message delivered by youth to the community helps sensitize decision makers to opportunities to tack on climate adaptation to other governance initiatives through finance, technology transition and inclusion of local knowledge in ways that position governmental officials in their own best interest with groups they deem important. This closes the loop in policy and research with iterative implementation: education is capacity building. Linking education to policy decisions is truly, as SeaTrust Institute's motto says, "local to global and back again(SM)."

FUTURE RESEARCH NEEDS

Further research is needed to identify the best ways to engage youth in creating their own preferable future by placing health at the heart of resilience development and climate change adaptation. The SDGs provide a current vehicle; these goals will morph into their next iteration when they expire in 2030, and it is important that society is equipped to continue to engage youth through the formal and informal educational processes in the evolving challenges and successes. Rio+20 did not elaborate specific goals but stated that the SDGs should be limited in number, aspirational and easy to communicate. The goals were designed address, in a balanced way, all three dimensions of sustainable development (environmental, economic and social or better known to some as "people, planet and profit") and be coherent with and integrated into the UN development agenda from 2015-2030. Under work overseen by the UNFCCC and its subsidiary bodies, these goals were developed and became part of the 2030 Agenda, adopted by the United Nations in September 2015 and memorialized within the Paris Agreement at the COP21 meetings in Paris under the aegis of the UNFCCC. Research to identify what is working within these processes at local and global levels, and what modifications can best prepare for the future, is needed.

One approach to this research might involve horizon scanning as a technique based on retrieving, analyzing and archiving emerging information from multiple disciplines that are relevant to an issue under investigation (Owen 2014), for example, climate change. Integrating science, environment, technology and society helps inform decision makers and is compatible with other tools and methods such as risk analysis and scenario building. Horizon scanning is not limited to traditional academic research but may also be approached through activities including local and global education and community programs, partnerships that train health professionals as first responders, and participatory and community-driven climate change resilience development.

At the same time, applying a theory of change approach allows for back-filling activities from a desired future state or set of future state options and is compatible with logframes, a standard sustainable development project management device (Brown, 2016). Theory of Change is a comprehensive description and illustration of how and why a desired change is expected to happen in a particular context. It is focused in particular on mapping out or "filling in" what has been described as the "missing middle" between what a program or change initiative does (its activities or interventions) and how these lead to desired goals being achieved. Applying ToC thinking to the current implementation phase of the SDGs would help scholars and country-level practitioners better assess their planning and progress towards achieving what they have agreed to in their Nationally Determined Contributions in support of achieving the SDGs and the Paris Agreement. ToC also identifies interventions for achieving the pre-conditions for long term goals. Leaders working to achieve their NDCs and SDGs face difficult choices regarding fixed (or set and unchanging standards) vs adaptive evaluation processes that change as conditions develop.

As adaptation planning progresses into implementation, it is imperative to link climate change adaptation with social and development needs in developing countries and everywhere on the planet. Many countries have done so, and many, like France, Vanuatu, Kenya, and the UK have taken them into aspects of their formal National Adaptation Plans. Developing health gauges for these plans would help to guide measurement and evaluation by a universal human metric provides a structure for vulnerability analysis that combines systemic, behavioral, and institutional factors. This enables insights into the sources of resilience and vulnerability that identifies targeted points of entry at different scales where interventions can build resilience.

In conjunction with shared-learning and iterative planning, the role of health in creating an adaptive mechanism for translating insights into practical courses of action over time bears further research. Educational activities used as an instigator of learning and dialogue providing a lens through the use of systems like Halcyon's gauge for community and system resilience - human health opens up even more questions about risk perception and tolerances; social learning, and how to remain flexible as conditions change.

Further investigations into the ways to engage, motivate and retain youth in activities that promote their community leadership in climate change adaptation and resilience are needed. For scientists, reef managers and for the adaptation community, the batfish has thrown up an interesting challenge: how to identify other 'sleeping functional groups' that may prove lifesavers for crisis situations faced by climate change, but about whose habits and abilities at crisis thresholds we as yet know little or nothing. For the framework to be useful, however, considerable skill is required in adapting its use to different contexts; models and measurement approaches cannot be applied blindly. Substantial translation and interpretation is required for use with practitioners or policy actors.

Finally, work is required to clarify the characteristics that contribute to resilience that may be hazard-, culture-, or otherwise specific. Relating these characteristics to elements such as diversity, reorganization, flexibility, modularity and the ability to learn that are fundamental to both human and systems resilience requires additional research.

CONCLUSION

To each there comes in their lifetime a special moment when they are figuratively tapped on the shoulder and offered the chance to do a very special thing, unique to them and fitted to their talents. What a tragedy if that moment finds them unprepared or unqualified for that which could have been their finest hour.
—— *Winston Churchill*

In this interpretive review of the Halcyon Adaptation Scenario as a transformative system to guide climate change adaptation and community resilience development, the author of this chapter has uncovered a handful of guiding principles that point to changes that both scholars and practitioners may need to make in the approach to climate change adaptation. CCA operates under dynamic and changing conditions that defy traditional risk management processes and requires a deeper understanding of social learning as society transitions from awareness to action. Education is often cited as "the" answer, not only for climate change but for the other sustainability initiatives as well. Not only does this place unachievable expectations upon the sector, the notion is antithetical to integrating SDG#4, education, into all other sustainability activities and interweaving it with the other 16 goals, including climate change, places education in a place to be the most effective in building for resilience and sustainability.

Using health as a gauge allows for a query of social learning in a measurable way to test the efficacy of CCA approaches. Health works well in this effort because it levels the conversation and includes all social, generational, gender and cultural groups. Youth, the most vulnerable group to climate change now and into the future, can engage directly with health effects to measure success and translate that knowledge to all levels of policy making to effect positive changes for their own futures.

Local approaches are widely considered to be the most appropriate method for reducing social vulnerability and involve approaching CCA as a process rather than an endpoint. When communities interweave local knowledge with expert scientific and policy knowledge and sources and document shared learning, the result is useful in informing both local governance and global knowledge. When youth learn and then also teach their communities, the results become embodied within the culture and resilience becomes an organic, iterative process that dynamically responds to changing conditions. Guided peer to peer youth coordination and training is particularly effective as a starting point.

In May 2017, the UNFCCC Secretariat held a mandated event during the intersessional SB46 meetings a for country delegates about the role of non-party stakeholders in helping countries achieve the SDGs because it had become apparent that the SDGs can only be achieved by engaging non-governmental stakeholders. Education professionals are key stakeholders: they inspire new learners to reform societies and governments. It is up to educators to "walk the talk" through forwarding sustainable practices on campuses and at schools, integrating the SDGs throughout curricula and research agendas, and putting themselves forward as experts and seasoned collaborators to draw upon for policy change processes. Between 6-17 November 2017, thousands of government delegates and leaders from all sectors of society gathered in Bonn (Germany) for the 2017 UN Climate Change Conference, COP24. Under the Presidency of Fiji, the UNFCCC secretariat with the support of Germany hosted this annual meeting with one clear objective: making progress for a successful, inclusive and ambitious implementation of the Paris Agreement.

This includes negotiations on the rule book—or operating manual —for transparent climate action under the Paris Agreement, as well as showcasing cooperative climate action, including on vulnerability and resilience which continued at the May 2018 UNFCCC intersessional meetings. This work is a preamble to the first benchmark SDG reporting period beginning in 2020. When the education SDG #4 comes up for HLPF review in July 2019, the rule book will be in place and educators will be implementing their initiatives for sustainability under those global guidelines.

The concept of youth as the sleeping function unit, the "batfish" for CCA, led SeaTrust Institute to link field research with integration of climate change science and local decision policy processes with educational systems through formal and informal internships for university students and community-based projects for pre-university youth, and to expand it to adults including community elders, policy and business professionals through the AWARE family of educational programs. As AWARE educational programs have expanded into adult components for community, regional and global capacity building for adaptation and resilience development, its youth origins are apparent in the intergenerational focus of adult AWARE programs that support government and private sector as well as older adults within communities. The launch of a new version of AWARE that is increasingly experiential, geographically and culturally context dependent, and engages with the changes in science and society in conjunction with the global focus on SDG #4 in 2019 is purposefully designed with the flexibility and robustness to support resilience development in this complex and changing environment.

Adaptation professionals must remain nimble and willing to adapt as conditions change and new knowledge emerges. Using models is appropriate; relying on them in lieu of integrating knowledge into human systems hasn't worked to date and is not likely to. Adaptation strategies must themselves adapt to sometimes radical physical and social changes, which are constant and rapid as crisis points are approached. That means that for adaptation as a process, today, and each day, is day number one.

The Crossing Boundaries project at St. Lawrence University defines a Siberian Yupik word for the type of changes to leading internationally and facilitating locally: *Aksik*: to steer the boat quickly in a new direction to avoid immediate danger (St. Lawrence University 2014). In this context it means to create a new type of relationship with local people as actual partners rather than recipients of aid with "strings" attached. This partnership of community with global expertise, technologies and leadership must be established openly, with a clear recognition of the joint benefits and goals. This is not an easy task; initially, it requires a readjustment of political "business as usual" practices that shift the balance from proscriptive to inclusive and even to locally-led as the responsibility for action shifts to local areas. In addition, these relationships include the private sector more directly, and in ways which may vary from local to multinational. These conceptual changes profoundly affect the structures that make a new system of durable resilience development possible.

REFERENCES

Bellwood, D. R., Hughes, T. P., & Hoey, A. S. (2006). Sleeping functional group drives coral-reef recovery. *Current Biology, 16*(24), 2434–2439. doi:10.1016/j.cub.2006.10.030 PMID:17174918

Berkhout, F., Hertin, J., & Gann, D. M. (2006). Learning to adapt: Organisation adaptation to climate change impacts. *Climatic Change, 78*(1), 135–156. doi:10.100710584-006-9089-3

Brown, A. M. (2016, March 18). *What is this thing called 'Theory of Change?'* [Blog post]. Retrieved from https://usaidlearninglab.org/lab-notes/what-thing-called-theory-change

Chilisa, B. (2012). *Indigenous research methods*. Thousand Oaks, CA: Sage.

Ebel, A., & Rinke, T. (2014). *Listening to the voices of young and future generations. State of the world 2014: Governing for sustainability. Worldwatch Institute.* Washington, DC: Island Press/Center for Resource Economics.

Fazey, I., Fazey, J. A., & Fazey, D. M. A. (2005). Learning more effectively from experience. *Ecology and Society, 10*(2), 4. doi:10.5751/ES-01384-100204

Hollander, R. D., Benya, F. F., & Fletcher, C. H. (Eds.). (2014). *The climate change educational partnership: Climate change, engineered systems, and society: A report of three workshops*. Washington, DC: National Academy of Engineers.

Lancaster, C. S. (1981). *The Goba of the Zambezi*. Norman, OK: University of Oklahoma Press.

McDaniels, T. L., & Gregory, R. (2004). Learning as an objective within a structured risk management decision process. *Environmental Science & Technology, 38*(7), 1921–1926. doi:10.1021/es0264246 PMID:15112789

Mertens, D. M. (2010). *Research and evaluation in education and psychology: Integrating with quantitative, qualitative and mixed methods*. Thousand Oaks, CA: Sage.

Mora & Frazier. (2013). The projected timing of climate departure from recent variability. *Nature, 502*, 183-187.

National Research Council. (2011). *A framework for K-12 science education: Practices, cross cutting concepts, and core ideas. Committee on a Conceptual Framework for New K-12 Science Education Standards. Board on Science Education, Division of Behavioral and Social Sciences Education.* Washington, DC: National Academies Press.

Owen, R. (2014). Emerging issues in ocean and human health: Managing uncertainty and new knowledge. In R. Bowen, M. Depledge, C. Carlarne, & L. Fleming (Eds.), *Oceans and human health: Implications for society and well-being*. Hoboken, NJ: John Wiley & Sons, Ltd.

PROVIA. (2013). *PROVIA guidance on assessing vulnerability, impacts and adaptation to climate change. Consultation document.* Nairobi, Kenya: United Nations Environment Programme. Retrieved from http://www.unep.org/provia

St. Lawrence University. (2014). *Aksik: Stories about adaptation and subsistence: Native voices from the frontiers of climate change.* Retrieved from http://aksik.org/

Summary for policymakers. (2014). In Field, C. B., Barros, V. R., Dokken, D. J., Mach, K. J., Mastrandrea, M. D., Bilir, T. E., ... White, L. L. (Eds.), *Climate Change 2014: Impacts,Adaptation, and Vulnerability. Part A: Global and Sectoral Aspects. Contribution of Working Group II to the Fifth Assessment Report of the Intergovernmental Panel on Climate Change. IPCC* (pp. 1–32). Cambridge, UK: Cambridge University Press.

UN General Assembly. (2015, October 21). *Transforming our world: The 2030 Agenda for Sustainable Development.* A/RES/70/1. Retrieved from http://www.refworld.org/docid/57b6e3e44.html

United Nations Department of Economic and Social Affairs. (n.d.). *United Nations Conference on Sustainable Development, Rio+20.* Retrieved from https://sustainabledevelopment.un.org/rio20.html

United Nations Environment Assembly of the United Nations Environment Programme. (2014). *Programme of research on climate change vulnerability, impacts and adaptation.* Nairobi: Author.

Wilson, L. A. (2013). Every day is day #1: Resilience, climate change and health: How must our own work adapt? Presented at the *United Nations Framework Convention on Climate Change COP19*, Warsaw, Poland.

Wilson, L. A. (2013). *Interview with Dr. Charles Chiyama, Chiawa region, Zambia, July 13, 2013*. Lower Zambezi, Zambia: Chiawa Water and Beds Project.

World Health Organization. (2014). *Malaria Fact Sheet*. Retrieved from http://www.who.int/mediacentre/factsheets/fs094/en/

Zolli, A., & Healy, A. M. (2012). *Resilience: Why things bounce back*. New York: Free Press.

Compilation of References

Abbott, I. A. (1995). A decade of species of Gracilaria (sensu latu). In Taxonomy of Economic Seaweeds (Vol. 5). California Seagrant Publication.

Adams, R. M., Prelip, M. L., Glik, D. C., Donatello, I., & Eisenman, D. P. (2018). Facilitating partnerships with community-and faith-based organizations for disaster preparedness and response: Results of a national survey of public health departments. *Disaster Medicine and Public Health Preparedness*, *12*(1), 57–66. doi:10.1017/dmp.2017.36 PMID:28735594

Adger, W. N., & Barnett, J. (2009). Four reasons for concern about adaptation to climate change. *Environment & Planning A*, *41*(12), 2800–2805. doi:10.1068/a42244

AFP. (2016). Seaweed Farmers in Hot Water As Zanzibar Struggles. *The Daily Mail*. Retrieved from http://www.dailymail.co.uk/wires/afp/article-3567780/Seaweed-farmers-hot-water-Zanzibar-struggles.html

Agenda 21: programme of action for sustainable development; Rio Declaration on Environment and Development; Statement of Forest Principles: The final text of agreements negotiated by governments at the United Nations Conference on Environment and Development (UNCED). (1992). Rio New York, NY: United Nations Dept. of Public Information.

Aitsi-Selmi, A., Murray, V., Wannous, C., Dickinson, C., Johnston, D., Kawasaki, A., ... Yeung, T. (2016). Reflections on a Science and Technology Agenda for 21st Century Disaster Risk Reduction Based on the Scientific Content of the 2016 UNISDR Science and Technology Conference on the Implementation of the Sendai Framework for Disaster Risk Reduction 2015–2030. *International Journal of Disaster Risk Science*, *7*(1), 1–29. doi:10.100713753-016-0081-x

Aldrich, D. P., & Meyer, M. A. (2015). Social capital and community resilience. *The American Behavioral Scientist*, *59*(2), 254–269. doi:10.1177/0002764214550299

Allen, G. (2018). Masses Of Seaweed Threaten Fisheries And Foul Beaches. *NPR*. Retrieved from https://www.npr.org/2018/07/21/630575085/masses-of-seaweed-threaten-fisheries-and-foul-beaches

Amadiume, I. (1997). *Reinventing Africa: Matriarchy, Religion and Culture*. London: Zed Books.

American Society of Civil Engineers. (2017). 2017 Infrastructure Report Card.

Anagbogu, M. A., Ada, C., & Nwokolo, A. (2014). *School children's perceptions of climate change and possible remediation*. Retrieved from www.academicresearch

Andersen, A. D., & Andersen, P. D. (2012). Innovation-system foresight: Explicating and systemizing the innovation-system foundations of foresight and exploring its implications. *Technical University of Denmark*. Retrieved from http://orbit.dtu.dk/files/10590515/Innovation_system_foresight.pdf

Anderson, A., & Strecker, M. (2012). Sustainable development: A case for education. *Environment*, *54*(6), 3–16. doi:10.1080/00139157.2012.730010

Andrade, A. (2018). *Technical Expert Meeting on Adaptation 2018: Session #1 Adaptation planning for vulnerable ecosystems* [Webcast]. Retrieved from http://tep-a.org/technical-expert-meetings-on-adaptation/agenda/

Ardoin, N. M., & Heimlich, H. E. (2013). Views from the field: Conservation Educators' and practitioners' perceptions of education as a strategy for achieving conservation outcomes. *The Journal of Environmental Education*, *44*(2), 97–115. doi:10.1080/00958964.2012.700963

Auerbach, D. (2016). *The Theory of Everything and Then Some*. Slate.

Banda, F. (2013). *Climate Change in Africa: A Guidebook for Journalists. United Nations Educational Scientific and Cultural Organization*. UNESCO.

Barnett, J. (2010). Adapting to climate change: three key challenges for research and policy: an editorial essay. *Climatic Change*, *1*(3), 314–317.

Bartosh, O. (2003). *Environmental Education*. Retrieved from http://www.seer.org

Bassey, A., Ejue, J., & Bassey, I. (2005). Attitudinal patterns of counsellors and teachers towards counselling curriculum for schools. *Journal of World Council for Curriculum and Instruction Nigeria Chapter Forum*, *5*(2), 48–44.

BBC. (2014). Seaweed- Zanzibar's Gift from the Ocean. Retrieved from http://www.bbc.com/news/world-africa-26770151

Bell, W. (1997). *Foundations of Futures Studies*. New Brunswick, N.J. and London UK: Transaction Publishers.

Bell, W. (1997). *Foundations of futures studies*. New Brunswick, NJ: Transaction Publishers.

Bellwood, D. R., Hughes, T. P., & Hoey, A. S. (2006). Sleeping functional group drives coral-reef recovery. *Current Biology*, *16*(24), 2434–2439. doi:10.1016/j.cub.2006.10.030 PMID:17174918

Bengston, D. N., & Fan, D. P. (1999). An innovative method for evaluating strategic goals in a public agency: Conservation leadership. *Evaluation Review*, *23*(1), 77–10. doi:10.1177/0193841X9902300104

Bennett, N. R. (1978). *A History of the Arab State of Zanzibar*. Methuen Publishing.

Berger, I., & White, E. F. (1999). *Restoring Women to History*. Indiana Univ. Press.

Berkhout, F., Hertin, J., & Gann, D. M. (2006). Learning to adapt: Organisation adaptation to climate change impacts. *Climatic Change*, *78*(1), 135–156. doi:10.100710584-006-9089-3

Berrang-Ford, L., Ford, J. D., & Paterson, J. (2010). Are we adapting to climate change? *Global Environmental Change*, *21*(2), 25–33.

Berry, W. (1991, February). Out of your car, off your horse: Twenty-seven propositions about global thinking and the sustainability of cities. *The Atlantic*. Retrieved from https://www.theatlantic.com/magazine/archive/1991/02/out-your-car-your-horse/309159/

Besada, H., & Sewakambo, N. (2009). *Climate Change in Africa: Adaptation, Mitigation and Governance Challenges. The Centre for International Governance Innovation*. CIGI.

Bijlsma, R., & Loeschcke, V. (Eds.). (1997). *Environmental Stress, Adaptation and Evolution*. Birkhauser Verlag. doi:10.1007/978-3-0348-8882-0

Biodiversity and Climate Change. (2007). *International Day for Biological Diversity*. Convention on Biological Diversity.

Bizzotto, M. (2018). *Resilient Cities Report 2018*. Bonn: ICLEI World Secretariat.

Black, H. (2000). Environmental and public health: Pulling together the pieces. *Environmental Health Perspectives*, *108*(11), A513–A515. doi:10.1289/ehp.108-a512

Bourn, D., & Brown, K. (2011). Young People and International Development: Engagement and Learning. Development Education Research Centre (Research Paper 2). Institute of Education, University of London.

Bowen, R. E., Depledge, M. H., Carlarne, P., & Fleming, L. E. (2014). *Oceans and human health; implications for society and well-being*. Chichester, England: John Wiley & Sons.

Breeman, A. M. (1990). (Section). In J. J. Beukema & J. J. Brouns (Eds.), *Expected Effects of Climatic Change on Marine Coastal Ecosystems*. Kluwer Academic Press.

British Council. (2004). *Africa Talks Climate.The public understanding of Climate change in ten countries*. World Service Trust.

Brockman, J. (1995). *The third culture: Beyond the scientific revolution*. New York, NY: Simon & Schuster.

Bromley, E., McCreary, M., Wells, K. B., Eisenman, D. P., Williams, M., Chandra, A., ... Kim, B. (2017). How do communities use a participatory public health approach to build resilience? The Los Angeles County Community Disaster Resilience Project. *International Journal of Environmental Research and Public Health*, *14*(10), 1267. doi:10.3390/ijerph14101267 PMID:29065491

Brown, A. M. (2016, March 18). What is this thing called 'Theory of Change?' [Blog post]. Retrieved from https://usaidlearninglab.org/lab-notes/what-thing-called-theory-change

Brown, D., & Kulig, J. (1996). 1997). The concept of resiliency: Theoretical lessons from community research. *Health and Canadian Society (Winnipeg, Man.)*, *4*(1), 29–52.

Brown, L. M., Haun, J. N., & Peterson, L. (2014). A proposed disaster literacy model. *Disaster Medicine and Public Health Preparedness*, *8*(03), 267–275. doi:10.1017/dmp.2014.43 PMID:24992944

Brown, O., Hammill, A., & McLeman, R. (2007). Climate change as the 'new' security threat: Implications for Africa International Affairs. *Roy Inst Int Aff*, *83*(6), 1141–1154. doi:10.1111/j.1468-2346.2007.00678.x

Bruntland, G. H. (1987). *Our common future: Report of the 1987 World Commission on Environment and Development*. Oxford, UK: Oxford University Press.

Buddemeier, Kleypas, & Aronson. (2004). *Coral Reefs and Global Climate Change: Potential Contributions of Climate Change to Stresses on Coral Reef Ecosystems*. Pew Centre.

Buikstra, E., Ross, H., King, C., Baker, P., Hegney, D., McLachlan, K., & Rogers-Clark, C. (2010). The components of resilience: Perceptions of an Australian rural community. *Journal of Community Psychology*, *38*(8), 975–991. doi:10.1002/jcop.20409

Burbidge, R. (2016). Climate-proofing the airport of the future. *Journal of Airport Management*, *11*(2), 114–128.

Burke, S. E. (2015). Enemy Number One for the Electric Grid: Mother Nature. *SAIS Review*, *35*(1).

Burns, T. W., Connor, D. J., & Stocklmayer, S. M. (2003). Science communication: A contemporary definition. *Public Understanding of Science (Bristol, England)*, *12*(2), 183–202. doi:10.1177/09636625030122004

Cape Town, on verge of running out of water, braces for "chaos" (2018, January 19). *CBS News*. Retrieved from https://www.cbsnews.com/news/cape-town-drought-water-shortage-south-africa/

Cassar, G. (2018). "The Davos 2018 environment agenda - what you need to know and how to follow online." World Economic Forum. Retrieved from https://www.weforum.org/agenda/2018/01/the-davos-2018-environment-agenda-online-what-you-need-to-know

Caton, W. (1982). *Overshoot: The Ecological Basis of Revolutionary Change*. University of Illinois.

Ceballos, G., Elrich, P. R., & Dirzo, R. (2017). Biological annihilation via the ongoing sixth mass extinction signaled by vertebrate population losses and declines. In *Proceedings of the National Academy of Sciences*. doi:10.1073/pnas.1704949114

Centre for Research on the Epidemiology of Disasters, & United Nations Office for Disaster Risk Reduction (2015). *The human cost of weather related disasters, 1995-2015*. Retrieved from www.unisdr.org/files/46796_cop21weatherdisastersreport2015.pdf

Chandra, A., Williams, M., Plough, A., Stayton, A., Wells, K. B., Horta, M., & Tang, J. (2013). Getting actionable about community resilience: The Los Angeles County community disaster resilience project. *American Journal of Public Health Association*, *103*(7), 1181–1189. doi:10.2105/AJPH.2013.301270 PMID:23678906

Chan, E. Y., Yue, J., Lee, P., & Wang, S. S. (2016). Socio-demographic predictors for urban community disaster health risk perception and household based preparedness in a Chinese urban city. *PLoS Currents*. doi:10.1371/currents.dis.287fb7fee6f9f4521af441a236c2d519 PMID:28856059

Chanza, N., & de Wit, A. (2015). Enhancing climate governance through indigenous knowledge. *South African Journal of Science*, *112*(1/2), 1–7.

Chapin, F. S., Sala, O. E., & Huber-Sannwald, E. (Eds.). (2001). Global Biodiversity in a Changing Environment: Scenarios for the 21st Century Ecological Studies. Ecological Studies, 152.

Chapman, P. (2014). Environmental Education and Sustainability. *Independent School*, *73*(4), 20.

Chavas, D. R. (2009). *Seasonal climate prediction dissemination to rural farmers in sub-Saharan Africa: a "bottom-up" perspective and the emergence of the mobile phone*. Cambridge, MA: Massachusetts Institute of Technology.

Chawla, L., & Cushing, D. F. (2007). Education for strategic environmental behavior. *Environmental Education Research*, *13*(4), 437–452. doi:10.1080/13504620701581539

Checkoway, B., & Gutiérrez, L. (2006). Youth Participation and Community Change: An Introduction. *Journal of Community Practice*, *14*(1/2), 1–9. doi:10.1300/J125v14n01_01

Chege, N. (2016). The Implications of the implementation of UNCLOS: A review of the approach to resolving environmental issues in African nations. Manuscript submitted for publication.

Chen, J., Chen, T. Y., Vertinsky, I., Yumagulova, L., & Park, C. (2013). Public-private partnerships for the development of disaster resilient communities. *Journal of Contingencies and Crisis Management*, *21*(3), 130–143. doi:10.1111/1468-5973.12021

Chevin, L.-M., Lande, R., & Mace, G. M. (2010). Adaptation, Plasticity, and Extinction in a Changing Environment: Towards a Predictive Theory. *PLoS Biology*, *8*(4). doi:10.1371/journal.pbio.1000357 PMID:20463950

Chicago Conservation Corps. (2014). Partner organizations. Retrieved from http://chicagoconservationcorps.org/blog/about-this-weblog/c3-partner-organizations/

Chicago Environmental Network. (2014). Contact CEN. Retrieved from http://www.chicagoenvironment.org/contact/index.cfm

Chilisa, B. (2012). *Indigenous research methods*. Thousand Oaks, CA: Sage.

Chinowsky, P. S., Price, J. C., & Neumann, J. E. (2013). Assessment of climate change adaptation costs for the U.S. road network. *Global Environmental Change*, *23*(4), 764–773. doi:10.1016/j.gloenvcha.2013.03.004

Chirisa, I., Matamanda, A., & Mutambwa, J. (2018) Africa's Dilemmas in Climate Change Communication: Universalistic Science Versus Indigenous Technical Knowledge. In Handbook of Climate Change Communication: Vol. 1. Climate Change Management. Springer.

Cinderby, S., Haq, G., Cambridge, H. & Lock, K. (2015). Building community resilience: can everyone enjoy a good life? *Local Environment, International journal of justice and sustainability, 21*(10), 1252-1270. doi:10.1080/13549839.2015.1100597

City of Chicago. (2014). City service. Retrieved from http://www.cityofchicago.org/city/en/depts/cdot/provdrs/conservation_outreachgreenprograms/svcs/volunteer_for_thechicagoconservationcorps.html

City of Chicago. (2014). Environment: Programs and initiatives. Retrieved from http://www.cityofchicago.org/city/en/progs/env.html

CLGF. (2008). Commonwealth Local Government Forum. Local Government System in Tanzania. Retrieved from http://www.clgf.org.uk/userfiles/CLGF/File/2008_Country_Files/TANZANIA.pdf

Cobia, D., & Henderson, D. (2003). Handbook of school counselling. Academic Press.

Colwell, R. R. (1996). Global climate and infectious disease: The cholera paradigm. *Science, 274*(5295), 2025–2031. doi:10.1126cience.274.5295.2025 PMID:8953025

Conservation International Volunteer Program. (2014). Destinations. Retrieved from http://www.conservationvip.org/destinations/

Convention on Biological Diversity Secretariat. (2014). *Global Biodiversity Outlook 4. A mid-term assessment of progress towards the implementation of the Strategic Plan for Biodiversity 2011-2020.* Montreal, Quebec, Canada: CBD Convention on Biological Diversity.

Convention on Biological Diversity. (2010). *COP 10 Decision X/2: X/2.Strategic Plan for Biodiversity 2011-2020.* Retrieved from https://www.cbd.int/decision/cop/?id=12268

Convention on Biological Diversity. (2016, December 10). *Decisions adopted by the Conference of the Parties to the Convention on Biological Diversity XIII/4. Biodiversity and climate change* CBD/COP/DEC/XIII/4. Retrieved from www.cbd.int/doc/decisions/cop-13/cop-13-dec-04-en.pdf

Convention on Biological Diversity. (n.d.). *The Aichi Biodiversity Targets.* Retrieved from: https://www.cbd.int/sp/targets/

Corcoran, P., Weakland, J., & Wals, A. E. J. (2017). *Envisioning futures for environmental and sustainability education.* Wageningen: Wageningen Academic Press. doi:10.3920/978-90-8686-846-9

Craft, L. (2017). *Perceived Threats to Food Security and Possible Responses Following an Agro-Terrorist Attack* [Doctoral dissertation]. Walden University.

Craft, L. (2018, March 13). Accelerated Weirding: Policy Administration in Hothouse Futures [Panel]. In *American Society for Public Administration Annual Conference.* Denver, CO.

CRC/Department of Environment, Institute of Marine Sciences, Sub-commission for Forestry, Sub-commission for Fisheries, Integrated Planning Unit. (1996). Toward Integrated Management and Sustainable Development of Zanzibar's Coast. Narragansett: Coastal Resource Center. Retrieved from http://www.crc.uri.edu/download/ZAN_0056.PDF

Creswell, J. W. (2013). *Research Design: Qualitative, Quantitative, and Mixed Methods Approaches*. Thousand Oaks, CA: Sage.

Crutchley, T. M., Rodgers, J. B., Whiteside, H. P. Jr, Vanier, M., & Terndrup, T. E. (2007). Agroterrorism: Where are we in the ongoing war on terrorism? *Journal of Food Protection*, *70*(3), 791–804. doi:10.4315/0362-028X-70.3.791 PMID:17388078

Csala, J. (2017). *Water and sustainable development* [Policy brief]. Port Townsend, WA: SeaTrust Institute.

Cutter, S. (1996). Vulnerability to environmental hazards. *Progress in Human Geography*, *20*(4), 529–539. doi:10.1177/030913259602000407

Cutter, S. L., Barnes, L., Berry, M., Burton, C. G., Evans, E., Tate, E. C., & Webb, J. (2008). *Community and regional resilience: Perspectives form hazards, disasters and emergency management. CARRI Research Report 1*. Oak Ridge: Community and Regional Resilience Institute. Retrieved from http://www.resilientus.org/wp-content/uploads/2013/03/FINAL_CUTTER_9-25-08_1223482309.pdf

Cutter, S., Ismail-Zadeh, A., Alcántara-Ayala, I., Altan, O., Baker, D. N., Briceño, S., ... Wu, G. (2015, June). Pool knowledge to stem losses from disasters. *Nature*, *522*(18), 277–279. doi:10.1038/522277a PMID:26085255

Das, B., Fernández Fernández, C., van der Gaag, N., McIntyre, P., & Rychlewski, M. (Eds.). (2016). *Water Integrity Global Outlook 2016*. Berlin, Germany: Water Integrity Network.

Dator, J. (1995). *What futures studies is, and is not*. Honolulu: Hawaii Research Center for Future Studies.

Dator, J. (2009). Alternative futures at the Manoa School. *Journal of Futures Studies.*, *14*(2), 1–18.

Dator, J. (2009a). Alternative Futures at the Manoa School. *Journal of Futures Studies*, *14*(2), 1–18.

Dator, J. (2009b). The Unholy Trinity, Plus One. *Journal of Futures Studies*, *13*(3), 33–48.

Dawes, C. J., Orduña-Rojas, J., & Robledo, D. (1998). Response of tropical red seaweed *Gracilaria cornea* to temperature, salinity and irradiance. *Journal of Applied Phycology*, *10*(5), 419–425. doi:10.1023/A:1008021613399

De Wit, M., & Jacek, S. (2006). *Changes in surface water supply across Africa with predicted climate change—AEON—Africa Earth Observatory Network. University of Cape Town, Science Express Report, Cape Town Federal Ministry for Economic Cooperation and Development. 2015. Disaster risk management: Approach and Contribution of German development Cooperation*. Berlin: Federal Ministry for Economic Cooperation and Development.

Degenhardt, S. (n.d.). *A multiple-case study of exploring social-ecological resilience of seaweed farming in Unguja*. Norwegian University of Life Sciences Faculty of Social Sciences, Department of International Environment and Development Studies. Retrieved from https://brage.bibsys.no/xmlui/bitstream/handle/11250/2381701/Stephanie%20DegenhardtMSC_BRAGE.pdf?sequence=1

Delors, J. (1996). *Learning: The treasure within – report to UNESCO of the international commission on education on education for the twenty-first century*. Paris: UNESCVO.

Denzin, N., & Lincoln, Y. (2011). *The SAGE Handbook of Qualitative Research* (4th ed.). London: SAGE.

Dewey, J. (1934). *Art as Experience*. New York: Minton, Malach, and Co.

Diamond, J. (2005). *Collapse*. New York, NY: Penguin Group.

Diez-Roux, A. V. (2000). Multilevel analysis in public health research. *Annual Review of Public Health*, *21*(1), 171–192. doi:10.1146/annurev.publhealth.21.1.171 PMID:10884951

Dlouha, J., Barton, A., Janouskova, S., & Dlouhy, J. (2013). Social learning indicators in sustainability-oriented regional learning networks. *Journal of Cleaner Production*, *49*, 64–73. doi:10.1016/j.jclepro.2012.07.023

Earthwatch Institute. (2014). Expeditions. Retrieved from http://earthwatch.org/expeditions

Ebel, A., & Rinke, T. (2014). *Listening to the voices of young and future generations. State of the world 2014: Governing for sustainability. Worldwatch Institute*. Washington, DC: Island Press/Center for Resource Economics.

ECOSOC. (2016). Climate change resilience for sustainable development. In *World Economic and Social Survey 2016: Climate change resilience – an opportunity for reducing inequalities* (E/2016/50). UN Economic and Social Council. Retrieved from: http://www.un.org/en/development/desa/policy/wess/

Edwards, K. (2012). Moving beyond green: Sustainable development toward healthy environments, social justice, and strong economies. *New Directions for Student Services*, *2012*(137), 19–28. doi:10.1002s.20011

Ehrlich, P., & Ehrlich, A. (2013). Can a collapse of global civilization be avoided? In *Proceedings of the Royal Society*.

Ellis, B., & Herbert, S. (2011). Complex adaptive systems (CAS): An overview of key elements, characteristics and application to management theory. *Informatics in Primary Care*, *19*(1), 33–37. PMID:22118334

Estrella, M., Renaud, F. G., Sudmeier-Rieux, K., & Nehren, U. (2016). Defining New Pathways for Ecosystem-Based Disaster Risk Reduction and Adaptation in the Post-2015 Sustainable Development Agenda. In F. G. Renaud, K. Sudmeier-Rieux, M. Estrella, & U. Nehren (Eds.), *Ecosystem-Based Disaster Risk Reduction and Adaptation in Practice. Advances in Natural and Technological Hazards Research* (Vol. 42, pp. 553–593). Switzerland: Springer. doi:10.1007/978-3-319-43633-3

European Environmental Agency. (n.d.). *Water stress*. Retrieved from: https://www.eea.europa.eu/themes/water/wise-help-centre/glossary-definitions/water-stress

FAO. (2018). Zanzibar mariculture leaps forward, FAO in Tanzania. Retrieved from http://www.fao.org/tanzania/news/detail-events/en/c/1125890/

Faulkner, L., Katrina, B., & Tara, Q. (2018). Analyzing community resilience as an emergent property of dynamic social-ecological systems. *Ecology and Society*, *23*(1), 24. doi:10.5751/ES-09784-230124

Fazey, I., Fazey, J. A., & Fazey, D. M. A. (2005). Learning more effectively from experience. *Ecology and Society*, *10*(2), 4. doi:10.5751/ES-01384-100204

Federal Emergency Management Agency. (2009). Mitigation. Retrieved from http://www.fema.gov/what-mitigation

Federal Emergency Management Agency. (2011). A whole community approach to emergency management: Principles, themes, and pathways for action. Retrieved from https://s3-us-gov-west-1.amazonaws.com/dam-production/uploads/20130726-1813-25045-0649/whole_community_dec2011_2_.pdf

Feffer, J. (2014). *Earth: Game Over?* Foreign Policy in Focus.

Flannery, T. (2017). How farming giant seaweed can feed fish and fix the climate. The Conversation. Retrieved from https://theconversation.com/how-farming-giant-seaweed-can-feed-fish-and-fix-the-climate-81761

Flint, C. G., & Lutoff, A. E. (2005). Natural Resource-Based Communities, Risk, And Disaster: An Intersection Of Theories. *Society & Natural Resources*, *18*(5), 399–412. doi:10.1080/08941920590924747

Folke, C. (2007). Social-ecological systems and adaptive governance of the commons. *Ecological Research*, *22*(1), 14–15. doi:10.100711284-006-0074-0

Fortier, J. D., Grady, S. M., Lee, S. A., & Marinac, P. A. (1998). *WisconsinÕs Model Academic Standards for Environmental Education*. Wisconsin Department of Public Instruction.

Friedlander, M., & Levy, I. (1995). Cultivation of *Gracilaria* in outdoor tanks and ponds. *Journal of Applied Phycology*, *7*(3), 315–3. doi:10.1007/BF00004005

Friedman, T. (2007). The People We Have Been Waiting For. *New York Times*.

Frisk, E. & Larson, K. (2011). Educating for Sustainability: Competencies & Practices for Transformative Action. *Journal of Sustainability Education, 2*.

Frumkin, H., & Patz, J. (2016). Climate change and human health. In Frumkin (Ed.) Environmental health: Global to local. Hoboken, NJ: Wiley.

Gagnon-Lebrun, F., & Agrawala, S. (2007). Implementing adaptation in developed countries: An analysis of progress and trends. *Climate Policy*, *7*(5), 392–408. doi:10.1080/14693062.2007.9685664

Gallucci, M. (2013). *6 of the World's Most Extensive Climate Adaptation Plans*. Retrieved from https://insideclimatenews.org/news/20130620/6-worlds-most-extensive-climate-adaptation-plans

Gardner, H. (2014, March 14). Facing Rising Seas, Bangladesh Confronts the Consequences of Climate Change. The New York Times.

Geere, D. (2014). Adaption to climate-related changes in seagrass ecosystems in Chwaka Bay (Zanzibar). Retrieved from http://www.bioenv.gu.se/digitalAssets/1488/1488564_duncan-geere.pdf

Geiger, S. (2002). *Women in African Colonial Histories*. Indiana University.

Gell-Mann, M. (1995). *Quark and the Jaguar*. New York, NY: W.H. Freeman.

Gentle, P., & Maraseni, T. N. (2012). Climate Change, Poverty and Livelihoods Adaptation Practices by Rural Mountain Communities in Nepal. *Environmental Science & Policy*, *21*, 24–34. doi:10.1016/j.envsci.2012.03.007

George, N., & Stark, A. (2016). Community resilience and crisis management: Policy lessons from the ground. *Policy and Politics*, *44*(4), 591–607. doi:10.1332/030557315X14351420409945

Germanwatch. (2004). *Sea Level Rise in Bangladesh and the Netherlands: One Pheonomenon, Many Consequences*. Berlin: Germanwatch.

Gidley, J. M. (2017). *The future: A very short introduction*. Oxford: Oxford University Press. doi:10.1093/actrade/9780198735281.001.0001

Gidley, J. M., Fien, J., Smith, J., Thomsen, D. C., & Smith, T. F. (2009). Participatory Futures Methods: Towards Adaptability and Resilience in Climate-Vulnerable Communities. *Environmental Policy and Governance*, *19*(6), 427–440. doi:10.1002/eet.524

Giffords, G. (2014). Flooding could cost Europe $32 billion a year by 2050—four times as much as it does now. Retrieved from http:/flooding-could-cost-europe-32-billion-a-year-by-2050-four-times-as-much-as-it-does-now/+&cd=1&hl=en&ct=clnk&gl=us

Glanville, J. L., Andersson, M. A., & Paxton, P. M. (2013). Do social connections create trust? An examination using new longitudinal data. *Social Forces*, *2*(545). doi:10.1093fot079

Global Humanitarian Forum Human Impact Report. (2009). *Climate change, the anatomy of a silent crisis*. Author.

Global Volunteer Network. (2014). Most popular programs. Retrieved from http://www.globalvolunteernetwork.org/programs/index.php?type=Environmental

GoAbroad.com. (2014). Volunteer abroad in environment. Retrieved from http://www.goabroad.com/volunteer-abroad/search/environment/volunteer-abroad-1

Goh, B., Pomsagun, A., Le Tissier, M., Dennison, W.C., Kremer, H.H., & Weichselgartner, J. (Eds.). (2008). Science Communication in Theory and Practice. LOICZ Reports and Studies No. 31. LOICZ, Geesthacht, Germany.

Goldberg, M. (2009). Social conscience. The ability to reflect on deeply-held opinions about social justice and sustainability. In A. Stibbe, (Ed.), The handbook of sustainability literacy: skills for a changing world (pp. 105-110). Devon: Green Books.

Granados, J. (n.d.). The challenges of higher education in the 21st century. *Global University Network for Innovation*. Retrieved from http://www.guninetwork.org/resources/he-articles/the-challenges-of-higher-education-in-the-21st-century

Grant, M. (2009). Internationalising higher education for sustainability: Youth Encounter on Sustainability. In P. Corcoran & P. Osano (Eds.), *Young People, Education, and Sustainable Development: Exploring Principles, Perspectives, and Praxis*. Wageningen Academic Publishers.

Graugaard, J.D. (2012). A tool for building community resilience? A case study of the Lewes Pound, Local Environment, *International journal of justice and sustainability,* 17(2), 243-260. doi:10.1080/13549839.2012.660908

Grayson, J. (2011). Charting a green course. *T.H.E. Journal, 38*(4), 26.

Guha-Sapir, D., Hoyois, Ph., & Below, R. (2014). *Annual Disaster Statistical Review 2013: The Numbers and Trends*. Brussels: CRED.

Gumbhir, V. K. (2014, May). From Students to Scholars: Undergraduate Research and the Importance of Regional Conferences. *The American Sociologist, 45*(2-3), 298–300. doi:10.100712108-014-9212-2

GVI. (2014). GVI youth programs. Retrieved from: http://www.gviusa.com/find-a-program/search-results/

Hák, T., Janousková, S., & Moldan, B. (2016). Sustainable Development Goals: A need for relevant indicators. *Ecological Indicators, 60*, 565–573. doi:10.1016/j.ecolind.2015.08.003

Hales, S., Kovats, S., Lloyd, S., & Campbell-Lendurm, D. (2014). *Quantitative risk assessment of the effects of climate change on selected causes of death, 2030s and 2050s*. Geneva: World Health Organization.

Han., Meehl, G. A., Rajagopalan, B., Fasullo, J. T., Hu, A., Lin, J. … Yeager, S. (2010). Patterns of Indian Ocean sea-level change in a warming climate. Nature Geoscience, 3(8), 546–550.

Haq, A. H. M. R., Ashaduzzaman, M., & Ghosal, T. (2002). *Soil-less Agriculture in Bangladesh. A Grameen Trust*. Bangladesh Publication.

Harness, H., & Drossman, H. (2011). The environmental education through filmmaking project. *Environmental Education Research*, *17*(6), 829–849. doi:10.1080/13504622.2011.618626

Harvey, B., Ensor, J., Carlile, L., Garside, B., Patterson, Z., & Naess, L. O. (2012). *Climate change communication and social learning–Review and strategy development for CCAFS*. CCAFS Working Paper No. 22. CGIAR Research Program on Climate Change, Agriculture and Food Security (CCAFS), Copenhagen, Denmark. Available online at www.ccafs.cgiar.org

Hashim, N. (2009). *Language and Collective Mobilization: The Story of Zanzibar*. Lexington Press.

Hawkins, J., Roberts, C., Van'T, T., De Meyer, K., Tratalos, J., & Aldam, C. (1999). Effects of recreational Scuba diving on Caribbean coral and Fish Communities. *Conservation Biology*, 889–997.

Hayden, M., Houwer, R., Frankfort, M., Rueter, J., Black, T., & Mortfield, P. (2011). Pedagogies of Empowerment in the Face of Climate Change Uncertainty. *Journal for Activism in Science and Technology Education*, *3*(1), 118–130.

Head, B. (2011). Why not ask them? Mapping and promoting youth participation. *Children and Youth Services Review*, *33*(4), 541–547. doi:10.1016/j.childyouth.2010.05.015

Heuer, Richards J. & Center for the Study of Intelligence (U.S.). (1999). *Psychology of intelligence analysis*. Center for the Study of Intelligence, Central Intelligence Agency. Retrieved from http://www.loc.gov/catdir/enhancements/fy0634/99489116-t.html

High School Students Go to the Reef. (2006). Online Submission. *Paper presented at the New Zealand Association for Environmental Education (NZAEE) Conference*, Auckland, New Zealand, Jan 22-25.

Hinchberger, B. (2017). *Disaster insurance against climate change attracts African countries*. Available from: https://www.un.org/africarenewal/magazine/may-july-2017/disaster-insurance-against-climate-change-attracts-african-countries

Hogge, K. (2018, February 2). Not cool: Climate change turning 99% of these sea turtles female [blog post]. *Ocean Conservancy*. Retrieved from https://oceanconservancy.org/blog/2018/01/25/not-cool-climate-change-turning-99-sea-turtles-female/

Holdeman, E. (2011). Black swan events. *Emergency Management*, *6*(3), 10.

Hollander, R. D., Benya, F. F., & Fletcher, C. H. (Eds.). (2014). *The climate change educational partnership: Climate change, engineered systems, and society: A report of three workshops*. Washington, DC: National Academy of Engineers.

Hopkinson, P., Hughes, P., & Layer, G. (2008). Sustainable Graduates: Linking Formal, Informal and Campus Curricula to Embed Education for Sustainable Development in the Student Learning Experience. *Environmental Education Research*, *14*(4), 435–454. doi:10.1080/13504620802283100

Hostettler, S. (2015). Technologies for Development. What Really Matters? In S. Hostettler, S., Hazboun, E. & Bolay, J.-C. (Eds.) Technologies for Development: What is Essential? Paris, Heidelberg, New York, Dordrecht, London: Springer. doi:10.1007/978-3-319-16247-8

Howard, G., Calow, R., Macdonald, A., & Bartram, J. (2016). Climate Change and Water and Sanitation: Likely Impacts and Emerging Trends for Action. *Annual Review of Environment and Resources*, *41*(1), 253–276. doi:10.1146/annurev-environ-110615-085856

HT Editors/Himalayan Times. (2010). *Bangladesh dams to reclaim 600 square kms of land*. Retrieved from http://www.thehimalayantimes.com/rssReference.php?headline=Bangladesh+dams+to+reclaim+600+square+kms+of+land&NewsID=256638

Hughes, K. (2003). Terror Warnings slow Zanzibar's Progress. *The Guardian*. Retrieved from http://www.guardian.co.uk/theobserver/2003/sep/21/worldview

Huq, S., Kovats, S., Reid, H., & Satterthwaite, D. (2007). Editorial: Reducing risks to cities from disasters and climate change. *Environment and Urbanisation, 19*(1), 3-15.

Hurd, C. L., Harrison, P. J., Bischof, K., & Lobban, C. S. (2014). *Seaweed ecology and physiology*. London: Cambridge University Press. doi:10.1017/CBO9781139192637

Hyden, G. (1980). *Beyond Ujamaa in Tanzania*. Berkeley, CA: University of California Press.

Hyvärinen, J., & Vos, M. (2015). Developing a conceptual framework for investigating communication supporting community resilience. *Societies (Basel, Switzerland)*, *5*(3), 583–597. doi:10.3390oc5030583

Ibhawoh, B., & Dibua, J. I. (2003). Deconstructing Ujamaa: The Legacy of Julius Nyerere in the Quest for Social and Economic Development in Africa. *American Journal of Political Science*, *8*(1).

Iltus, S. (2013). *Climate Change and Environmental Education*. UNICEF.

Impacts of a Warming Arctic: Arctic Climate Impact Assessment. (2004). Cambridge University Press.

Independent Commission on Multilateralism & International Peace Institute. (2016). *The 2030 Agenda for Sustainable Development and Addressing Climate Change*. Discussion Paper. Retrieved from https://www.icm2016.org/the-2030-agenda-for-sustainable-development-and-addressing-climate-change

Inside Climate News. (2018). *Six of the G7 commit to climate action. Trump wouldn't even join conversation*. Available from: https://insideclimatenews.org/news/10062018/g7-summit-climate-change-communique-trump-allies-estranged-germany-france-canada

Intergovernmental Panel on Climate Change. (2014). Summary for policymakers. In *Climate change 2014: Impacts, adaptation and vulnerability. Contribution of Working Group II to the Fifth Assessment Report of the Intergovernmental Panel on Climate Change*. Geneva, Switzerland: IPCC.

International Organization for Migration. (2014). Human mobility: Shaping vulnerability and resilience to disasters. Retrieved from https://www.iom.int/files/live/sites/iom/files/What-We-Do/docs/background-paper-for-HFA2-light.pdf

International Standards Organization. (2018). *Water and sanitation. ISO-Focus, 126.*

Internet Society. (2017). Paths to Our Digital Future. Retrieved from https://future.internetsociety.org/wp-content/uploads/2017/09/2017-Internet-Society-Global-Internet-Report-Paths-to-Our-Digital-Future.pdf

IPCC. (1996). Climate change impacts on forests. In *Climate Change 1995: Impacts, Adaptations and Mitigation of Climate Change: Scientific-Technical Analyses. Contribution of Working Group II to the Second Assessment Report of the Intergovernmental Panel on Climate Change.* Cambridge University Press.

IPCC. (1997). *Revised 1996 IPCC Guidelines for National Greenhouse Gas Inventories. Intergovernmental Panel on Climate Change.* Bracknell, United Kingdom: Meteorological Office.

Ives, M. (2017). Climate Change Lands at the Airport. *New York Times*. Retrieved from https://www.nytimes.com/2017/09/30/business/airports-climate-change-global-warming.html

Janesick, V. J. (2011). *Stretching Exercises for Qualitative Researchers* (3rd ed.). Thousand Oaks: Sage.

Jansi, M., & Ramadhas, V. (n.d.). Effect of salinity & dissolved nutrient on the occurrence of some seaweeds in Manakkudy Estuary. *Indian Journal of Marine Sciences*, *38*(4).

Jennings, L., Parra-Medina, D., Hilfinger-Messias, D., & McLoughlin, K. (2006). Towards a critical social theory of youth empowerment. Journal of Community Practice, 14(1/2), 31-55.

Jensen, B. B., Simovska, V., Larsen, N., & Holm, L. G. (2005). *Young people want to be part of the answer*. European Network of Health Promoting Schools.

Johnson, R.M., Edwards, E., Gardner, J.S., Diduck, A.P. (2018). Community vulnerability and resilience in disaster risk reduction: An example from Phojal Nalla, Himachal Pradesh, India. doi:10.100710113-018-1326-6

Johnson-Pynn, J. S., & Johnson, L. R. (2010). Exploring Environmental Education for East African Youth: Do Program Contexts Matter? *Children, Youth and Environments*, *20*(1), 123–151.

Jones, C. (2018). Sustainable futures. In W. Leal (Ed.), The Encyclopedia of Sustainability in Higher Education. Springer.

Jones, C. (1998). Envisioning Sustainable Futures. In D. Hicks & R. Slaughter (Eds.), *World Yearbook of Education 1998: Futures Education*. London: Kogan Page.

Jones, P., Selby, D., & Sterling, S. (2010). *Sustainability education: perspectives and practice across higher education*. EarthScan.

Kabendra, E. (2014). Zanzibar's Encroaching Ocean Means Less Water. Retrieved from http://www.ipsnews.net/2013/06/zanzibars-encroaching-ocean-means-less-water

Kais, S. M. and Saidul Islam, (2016). Community Capitals as Community Resilience to Climate Change: Conceptual Connections. *International Journal of Environment, Research and Public Health, 13*(12). Retrieved from https://www.ncbi.nlm.nih.gov/pmc/articles/PMC5201352/

Katz, J. M. (2017) Florida: We Don't Discuss the Cause. *Weather*. Retrieved from https://features.weather.com/us-climate-change/florida/

Kaunda, D. (2013). Warming seas frustrate Zanzibar's seaweed farmers. Thomson Reuters Foundation. http://www.trust.org/item/?map=warming-seas-frustrate-zanzibars-seaweed-farmers

Kawachi, I., Kennedy, B. P., & Wilkinson, R. G. (1999). Crime: Social disorganization and relative deprivation. *Social Science & Medicine, 48*(6), 719–731. doi:10.1016/S0277-9536(98)00400-6 PMID:10190635

Keesstra, S., Nunes, J., Novara, N., Finger, D., Avelar, D., Kalantari, Z., & Cerdà, A. (2018). The superior effect of nature based solutions in land management for enhancing ecosystem services. *The Science of the Total Environment, 610-611*, 997–1009. doi:10.1016/j.scitotenv.2017.08.077 PMID:28838037

Kemp, R. (2017) America's Infrastructure: Is Our Country on the Road to Ruin?! *PA Times*. Retrieved from https://patimes.org/americas-infrastructure-country-road-ruin/

Keshodkar, A. (2013). *Tourism and Social Change in Post-Socialist Zanzibar: Struggles for Identity, Movement, and Civilization*. Lexington Books.

Kezar, A., & Lester, J. (2011). *Enhancing Campus Capacity for Leadership: An Examination of Grassroots Leaders in Higher Education*. Stanford University Press. doi:10.11126tanford/9780804776479.001.0001

Kilian, H. (2012). *Environmental Education Programs: Can Their Effectiveness be Influenced by the Teaching Method of the Visiting School?* IZE Journal.

Killingray, D. (1986). The Maintenance of Law and Order in British Colonial Africa. *African Affairs, 85*(340), 411–413. doi:10.1093/oxfordjournals.afraf.a097799

Kiss, P., & Hunt, A. (2012). *Scoping study: reviewing the coverage of economic impacts in the CCRA*. Paul Watkiss Associates.

Kolbert, E. (2014). *The Sixth Extinction: An Unnatural History*. New York: Henry Holt and Co.

Kopnina, H. (2017). *Teaching Sustainable Development Goals in The Netherlands: a critical approach. Environmental education research*, 1-16. doi:10.1080/13504622.2017.1303819

Kotecha, P. (Ed.). (2010). Climate change, adaptation and higher education: Securing our future. SARUA Leadership Dialogue Series, 2(4).

Krasny, M. E., Lundholm, C., & Plummer, R. (2010) Resilience in social–ecological systems: the roles of learning and education. *Environmental Education Research*, *16*(5-6), 463-474. doi: 10.1080/13504622.2010.505416

Kulig, J. (1999). Sensing collectivity & building skills: Rural communities & community resiliency. In W. Ramp, J. Kulig, I. Townshend et al. (Eds.), Health in rural settings: Contexts for action (pp. 223-244). Lethbridge: University of Lethbridge.

Kulig, J., Reimer, W., Townshend, I., Edge, D., & Lightfoot, N. (2011). Understanding links between wildfires and community resiliency: Lessons learned for disaster preparation and mitigation. Retrieved from http://www.ruralwildfire.ca/files/Final

Kulig, J. C., Edge, D. S., Townshend, I., Lightfoot, N., & Reimer, W. (2013). Community resiliency: Emerging theoretical insights. *Journal of Community Psychology*, *41*(6), 758–775. doi:10.1002/jcop.21569

Kulig, J., Edge, D., & Joyce, B. (2008). Understanding community resiliency in rural communities through multimethod research. *Journal of Rural and Community Development*, *3*(3), 77–94.

Kunstler, J. H. (2005). *The Long Emergency: Surviving the Catastrophes of the Twenty-First Century*. New York: Atlantic Monthly Press.

Lanard, J. (2003). *Scientists And Risk Communication*. GreenFacts.

Lancaster, C. S. (1981). *The Goba of the Zambezi*. Norman, OK: University of Oklahoma Press.

Larsen, K., & Gunnarsson, U. (2009). Climate change scenarios and citizen-participation: Mitigation and adaptation perspectives in constructing sustainable futures. *Habitat International*, *33*(3), 260–266. doi:10.1016/j.habitatint.2008.10.007

Larsen, O. M. (2015). Climate change is here to stay: Reviewing the impact of climate change on airport infrastructure. *Airport Management*, *9*(3), 264–269.

Larsen, P., Goldsmith, S., Smith, O., Wilson, M., Strzepek, K., Chinowsky, P., & Saylor, B. (2008). Estimating future costs for Alaska public infrastructure at risk from climate change. *Global Environmental Change*, *18*(3), 442–457. doi:10.1016/j.gloenvcha.2008.03.005

Lee, J. V. (2010). Policing after 9/11: Community policing in an age of homeland security. *Police Quarterly*, *13*(4), 347–366. doi:10.1177/1098611110384083

Lee, T. D., Jones, M. G., & Chesnutt, K. (2017). Teaching Systems Thinking in the Context of Water Cycle. *Research in Science Education*. doi:10.100711165-017-9613-7

Lemmen, D., Warren, F., & Lacroix, J. (Eds.). (2008). *From impacts to adaptation: Canada in a changing climate 2007*. Ottawa: Government of Canada. doi:10.4095/226455

Leonardi, G. (2010). Adaptation to climate change.

Lerch, D. (2015). *Six foundations for building community resilience. Post Carbon Institute.* Retrieved from http://www.postcarbon.org/wp-content/uploads/2015/11/Six-foundations-for-building-community-resilience.pdf

Levine, A. (2007). Staying Afloat: State Agencies, Local Communities and International Involvement in Marine Protected Area Management in Zanzibar, Tanzania. *Conservation & Society*, *5*(4), 566–567.

Leykin, D., Lahad, M., Cohen, R., Goldberg, A., & Aharonson-Daniel, L. (2016). The dynamics of community resilience between routine and emergency situations. *International Journal of Disaster Risk Reduction*, *15*, 125–131. doi:10.1016/j.ijdrr.2016.01.008

Lichtveld, M. (2018). Disasters through the lens of disparities: Elevate community resilience as an essential public health service. *American Journal of Public Health*, *108*(1), 28–30. doi:10.2105/AJPH.2017.304193 PMID:29211534

Lindsey, R. (2014). Global warming or just weather. *Climate.gov*. Retrieved from https://www.climate.gov/news-features/features/global-warming-or-just-weather

Linnell, M. (2014). Citizen response in crisis: Individual and collective efforts to enhance community resilience. *Human Technology*, *10*(2), 68–94. doi:10.17011/ht/urn.201411203311

Lipscombe, B. P., Burek, C. V., Potter, J. A., Ribchester, C., & Degg, M. R. (2008). An Overview of Extra-Curricular Education for Sustainable Development (ESD) Interventions in UK Universities. *International Journal of Sustainability in Higher Education*, *9*(3), 222–234. doi:10.1108/14676370810885853

Liquete, C., Piroddi, C., Drakou, E. G., Gurney, L., Katsanevakis, S., Charef, A., & Egoh, B. (2013). Current Status and Future Prospects for the Assessment of Marine and Coastal Ecosystem Services: A Systematic Review. *PLoS One*, *8*(7), e67737. doi:10.1371/journal.pone.0067737 PMID:23844080

Lloyd, D., & Wallace, J. (2008). Imaging the Future of Science Education: The Case for Making Futures Studies Explicit in Student Learning. *Studies in Science Education, 40*(1), 139-177. doi:10.1080/03057260408560205

Lofchie, M. (1978). Agrarian Crisis and Economic Liberalisation in Tanzania. *The Journal of Modern African Studies*, *16*(3), 451–475. doi:10.1017/S0022278X00002524

Lotz-Sisitka, H., Wals, A. E. J., Kronlid, D., & McGarry, D. (2015). Transformative, transgressive social learning: Rethinking higher education pedagogy in times of systemic global dysfunction. *Current Opinion in Environmental Sustainability*, *16*, 73–80. doi:10.1016/j.cosust.2015.07.018

Lovelock, J. (2006). *The Revenge of Gaia*. London: Penguin Books.

Lovelock, J. (2009). *The Vanishing Face of Gaia*. New York: Basic Books.

Lovelock, J. (2015). *A Rough Ride to the Future*. New York: The Overlook Press.

Lozano, R., Lukman, R., Lozano, F., Huisingh, D., & Lambrechts, W. (2011). Declarations for sustainability in highereducation: Becoming better leaders, through addressing the university system. *Journal of Cleaner Production*, 1–10.

Lukamba, M. T. (2010). Natural disasters in African countries: What can we learn about them? *TD The Journal for Transdisciplinary Research in Southern Africa*, *6*(2), 478–495.

Lynas, M. (2008). *Six degrees. Our future on a hotter planet*. Washington, D.C.: National Geographic Society.

Macy, J. R. (1983). *Despair and personal power in the nuclear age*. Philadelphia, PA: New Society Publishers.

Madeweya, K., Hiroyasu, O., & Mistsuo, M. (1994). *Proceedings of the International Workshop of Ecotourism and the Environmental Conservation in Zanzibar*. Department of Commercial Crops, Tanzania.

Madzwamuse, M. (2010). *Climate Change Vulnerability and Adaptation Preparedness in South Africa*. Heinrich Böll Stiftung Southern Africa.

Mapolu, H. (1990). Tanzania: Imperialism, the state and the peasantry. In *African agriculture: The critical choices*. Zed Books. Retrieved from http://archive.unu.edu/unupress/unupbooks / uu28ae/uu28ae0h.htm#8

Marcuson, W. F. III. (2008, January). Fixing America's Crumbling Infrastructure, A Call to Action for All. *Public Works Management & Policy*, *12*(3), 473–475. doi:10.1177/1087724X07312145

Margareth, K. (2011). *Institute of Marine Sciences – Zanzibar*. Retrieved From: https://twas.org/sites/default/files/twas_sig_ims.pdf

Marston, A., & Marston, S. (2018). *Type R: Transformative resilience for thriving in a turbulent world*. New York: Public Affairs.

Martine, G., Schensul, D., & Guzmán, J. M. (2013). *The demography of adaptation to climate change*. New York, NY: United Nations Population Fund, International Institute for Environment and Development, and El Colegio de México.

Masih, I., Maskey, S., Mussa, F. E. F., & Trambauer, P. (2014). A review of droughts on the African continent: A geospatial and long-term perspective. *Hydrology and Earth System Sciences*, *18*(9), 3635–3649. doi:10.5194/hess-18-3635-2014

Massoi, L., & Norman, A. S. (2009). *Decentralisation by Devolution in Tanzania: Reflections on community involvement in the planning process in Kizota Ward in Dodoma*. Mzumbe University.

Matamanda, A. R., Dzvimbo, M. A., & Kadebu, R. T. (2017). Climate Change and Infrastructure Planning: Implications for Sustainable Urban Management in Harare, Zimbabwe. *Journal of Public Administration and Development Alternatives*, *2*(11), 95–111.

McCrea, R., Walton, A., & Leonard, R. (2014). A conceptual framework for investigating community wellbeing and resilience. *Rural Society*, *23*(3), 270–282. doi:10.1080/10371656.2014.11082070

McDaniels, T. L., & Gregory, R. (2004). Learning as an objective within a structured risk management decision process. *Environmental Science & Technology*, *38*(7), 1921–1926. doi:10.1021/es0264246 PMID:15112789

McHenry, D. (1979). *Tanzania's Ujamaa Villages: The Implementation of a Rural Development Strategy*. Berkley, CA: Institute of International Studies.

McMahon, E., & Decker, C. (2009). Wives or workers? Negotiating the social contract between female teachers and the colonial state in Zanzibar. *Journal of Women's History*, (21), 2.

McMillin, J., & Dyball, R. (2009). Developing a Whole-of-University Approach to Educating for Sustainability: Linking Curriculum, Research and Sustainable Campus Operations. *Journal of Education for Sustainable Development*, *3*(1), 55–64. doi:10.1177/097340820900300113

McNamara, K. E., & Buggy, L. (2016). Community-based climate change adaptation: A review of academic literature. *Local Environment*, *21*, 1–18. doi:10.1080/13549839.2016.1216954

MDG. (2015). Millenium Development Goals. UNDP. Retrieved from http://www.undp.org/content/undp/en/home/sdgoverview/mdg_goals.html

MEA. (2005). *Ecosystems and Human Well-being: Synthesis*. Washington DC: Island Press. Retrieved from http://millenniumassessment.org/en/Synthesis.html

Meadows, D. H., Randers, J., & Meadows, D. (2004). *Limits to Growth. The 30-year Update*. White River Junction, VT: Chelsea Green Publishing Co.

Mertens, D. M. (2010). *Research and evaluation in education and psychology: Integrating with quantitative, qualitative and mixed methods*. Thousand Oaks, CA: Sage.

Merton, R. K. (1968). Social theory and social structure (Enl. Ed.). Free Press.

Mesfin, M., & Arjen, Y. (2016). Four billion people facing severe water scarcity. *Science Advances*, *2*(2). doi:10.1126ciadv.1500323

Mileti, D. (2006). Promoting Risk Wise Behavior. *Natural Hazards Workshop*.

Modh, S. (2010). *Introduction to disaster management*. New Delhi: Macmillan Publishers.

Mohan, V., & Shellard, T. (2014). Providing family planning services to remote communities in areas of high biodiversity through a population-health-environment programme in Madagascar. *Reprod. Health Matters*, *43*, 93-103. doi:10.1016/S0968-8080(14)43766-2

Mohr, L. B. (1999). The qualitative method of impact analysis. *The American Journal of Evaluation*, *20*(1), 69–84. doi:10.1177/109821409902000106

Montanari, S. (2017). Plastic garbage patch bigger than Mexico found in Pacific. *National Geographic*. Retrieved from https://news.nationalgeographic.com/2017/07/ocean-plastic-patch-south-pacific-spd/

Mooney, C. (2017). Scientists challenge magazine story about 'uninhabitable Earth.' *The Washington Post*. Retrieved from https://www.washingtonpost.com/news/energy-environment/wp/2017/07/12/scientists-challenge-magazine-story-about-uninhabitable-earth/?noredirect=on&utm_term=.019e1445a7d4

Mora & Frazier. (2013). The projected timing of climate departure from recent variability. *Nature, 502*, 183-187.

Morchain, D. (2018). *Technical Expert Meeting on Adaptation 2018: Session #2 Adaptation planning for vulnerable communities* [Webcast]. Retrieved from: http://tep-a.org/technical-expert-meetings-on-adaptation/agenda/

Morrow, B. H. (2009). *Risk Behaviour and Risk Communication: Synthesis and Expert Interviews*. Final Report for the NOAA Coastal Services Center.

Moser, S. (2007). More bad news: The risk of neglecting emotional responses to climate change information. In S.C. Moser & L. Dilling (Eds.), Creating a climate for change: Communicating climate change and facilitating social change. (pp. 64-80). New York, NY: Cambridge University Press.

Moser, S. C., & Luers, A. L. (2008). Managing climate risks in California: The need to engage resource managers for successful adaptation to change. *Climatic Change*, 87.

Mshigeni, K. E. (1973). Exploitation of seaweeds in Tanzania: The Genus *Eucheuma* and algae. *Tanzania Notes and Records*, *72*, 19–36.

Mshigeni, K. E. (1976). Seaweed farming: A possibility for Tanzania's coastal Ujaama Villages. *Tanzania Notes and Records*, *79-80*, 99–105.

Mshigeni, K. E. (1983). Algae resources, exploitation and use in East Africa. In F. E. Round & D. J. Chapman (Eds.), *Progress in Phycological research* (Vol. 2, pp. 387–419). Amsterdam: Elsevier Science.

Mshignei, K. (1996). Zero Emissions Projects in Tanzania. In *Proceedings of the 2nd Second Annual UNU*.

Msuya, F. (2008). Seaweed Farmers use seaweeds directly instead of selling raw product. Retrieved from http://www.wiomariculture.org/forum/?p=114

Msuya, F. (2011b). *Experimental Farming Of The Seaweed Kappaphycus In Floating Rafts In Pemba Island, Zanzibar, Tanzania - BIRR-MACEMP Project Phase II - Final Report*. Institute of Marine Sciences, University of Dar es Salaam.

Msuya, F. E. (1996). Seaweed farming in Lindi and Mtwara regions, Phase-two, Implementation and Expansion. Consultancy report for Rural Integrated Project Support (RIPS) Programme, Mtwara, Tanzania. IMS 1997/01.

Msuya, F. E., & Porter, M. (2009). Impacts of environmental changes on the farmed seaweed and seaweed farmers in Songosongo Island, Tanzania. Report submitted under a Collaborative Project on Sustaining Coastal Fishing Communities, Memorial University of Newfoundland, University of Dar-es-Salaam.

Msuya, F.E., Ngoile, M.A.K., & Shunula, J.P. (1996). *The impact of seaweed farming on the macrophytes and macrobenthos of the East Coast of Unguja Island, Zanzibar, Tanzania.* Report submitted to the Canadian International Development Agency (CIDA), Institute of Marine Sciences, University of Dar es Salaam, Zanzibar, Tanzania, IMS 1997/05.

Msuya, F. E. (2006). The Impact of Seaweed Farming on the Social and Economic Structure of Seaweed Farming Communities in Zanzibar, Tanzania. In A. T. Critchley, M. Ohno, & D. B. Largo (Eds.), *World seaweed resources: An authoritative reference system.* Amsterdam: ETI Bio-Informatics.

Msuya, F. E. (2011a). Environmental changes and their impact on seaweed farming. *World Aquaculture, 42*(4), 34–37, 71. Retrieved from http://www.academia.edu/2043063/Environmental_changes_and_their_impact_on_seaweed_farming_in_Tanzania on 5/28/2014

Msuya, F. E., Buriyo, A., Omar, I., Pascal, B., Narrain, K., Ravina, J. J. M., ... Wakibia, J. G. (2014). 'Cultivation and Utilisation of red seaweeds in the Western Indian Ocean (WIO)Region. *Journal of Applied Phycology, 26*(2), 699–705. Retrieved from https://www.ft.com/content/e1ec5434-50f8-11e5-b029-b9d50a74fd14 doi:10.100710811-013-0086-4

Mukoni, H. (2013). Environmental education in Zimbabwean secondary schools:' greening or transformative social change? *International Journal of Asian Social Science, 3*(4), 971–991.

Munich, R. E. (2018). *Natural disasters: The year in figures.* Retrieved from https://www.munichre.com/topics-online/en/2017/topics-geo/overview-natural-catastrophe-2016

Mustelin, J. (2009). Practical measures to tackle climate change: coastal forest buffer zones and shoreline change in Zanzibar, Tanzania. *UNESCO.* Retrieved from http://www.unesco.org/csi/climate-frontlines/PracticalMeasuresZanzibar_Ebook.pdf

Muttarak, R., & Pothisiri, W. (2013). The role of education on disaster preparedness: Case study of 2012 Indian Ocean earthquakes on Thailand's Andaman Coast. *Ecology and Society, 18*(4), art51. doi:10.5751/ES-06101-180451

National Research Council. (2011). *A framework for K-12 science education: Practices, cross cutting concepts, and core ideas. Committee on a Conceptual Framework for New K-12 Science Education Standards. Board on Science Education, Division of Behavioral and Social Sciences Education.* Washington, DC: National Academies Press.

National Research Council. (2011). *America's climate choices. Committee on America's Climate Choices. Board on Atmospheric Sciences and Climate, Division on Earth and Life Sciences.* Washington, DC: The National Academies Press.

National Research Council. (2012). Climate change education in formal settings, K-14: A workshop summary. A. Beatty, Rapporteur. Steering Committee on Climate Change Education Goals and Objectives. Board on Science Education, Division of Behavioral and Social Sciences and Education. Washington, DC: The National Academies Press.

National Research Council. (2012). Dam and Levee Safety and Community Resilience: A Vision for Future Practice. Washington. D.C., The National Academies Press.

Ngowi, R. (2005). Seaweed Farming helps Women in Tanzania. Retrieved from http://myhero.com/myhero/heroprint.asp?hero=Seaweed_Farming_AP

NOAA. (2013). *State of the Climate: Sea Surface Temperature – 2013 National Oceanic and Atmospheric Administration.* Retrieved from https://www.climate.gov/news-features/understanding-climate/2013-state-climate-sea-surface-temperature

Norris, F. H., Stevens, S. P., Pfefferbaum, B., Wyche, K. F., & Pfefferbaum, R. L. (2008). Community resilience as a metaphor, theory, set of capacities, and strategy for disaster readiness. *American Journal of Community Psychology, 41*(1-2), 127–150. doi:10.100710464-007-9156-6 PMID:18157631

NRC. (1996). *Understanding Risk: Informing Decisions in A Democratic Society.* Washington, DC: National Academy Press.

Odlyzko, A. (2004). *The future of scientific communication.* AT&T Labs - Research.

Ogbone, E. N. (2012). *Water, Sanitation and hygiene Education: Children and adolescents leading the way in Niger Delta.* Academic Press.

Orr, D. (2004). *Earth in Mind: on education, environment and the human prospect* (2nd ed.). Chicago: Island Press.

Orr, H. A. (2005). The Genetic theory of Adaptation: A Brief History. *Nature Reviews. Genetics, 6*(2), 119–127. doi:10.1038/nrg1523 PMID:15716908

Ott, H. J. (2018). Climate change eroding women's status in Zanzibar. *CBS News.* Retrieved from https://www.cbsnews.com/news/climate-change-eroding-womens-status-in-zanzibar/

Owen, R. (2014). Emerging issues in ocean and human health: Managing uncertainty and new knowledge. In R. Bowen, M. Depledge, C. Carlarne, & L. Fleming (Eds.), *Oceans and human health: Implications for society and well-being.* Hoboken, NJ: John Wiley & Sons, Ltd.

Owen, R., Depledge, M., Dr Lurio, J., Bardsley, S., & Palomino, M. (2014). Emerging issues in oceans and human health: Managing uncertainty and new knowledge. In R. E. Bowen, M. H. Depledge, P. Carlarne, & L. E. Fleming (Eds.), *Oceans and human health; implications for society and well-being* (pp. 267–287). Chichester, England: John Wiley & Sons.

Pachauri, R.K. & Meyer L.A. (Eds.). (2014). *Climate Change 2014: Synthesis Report, Contribution of Working Groups I, II and III to the Fifth Assessment Report of the Intergovernmental Panel on Climate Change*. Geneva, Switzerland: IPCC.

Palmer, J. (2002). *Environmental education in the 21st century: Theory, practice, progress and promise*. Routledge.

Parpart, J., & Staudt, K. (1988). *Women and the State in Africa*. Lynne Reiner Publishers.

Parry, M., Lowe, J., & Hanson, C. (2009). Overshoot, adapt and recover. *Nature*, *458*(7242), 1102–1103. doi:10.1038/4581102a PMID:19407775

Patz, J. A., Epstein, P. R., Burke, T. A., & Balbus, J. M. (1996). Global climate change and emerging infectious diseases. *Journal of the American Medical Association*, *275*(3), 217–223. doi:10.1001/jama.1996.03530270057032 PMID:8604175

Paul, D. (2018, February 8). ISO journal highlights need to meet SDG6. *IISD SDG Knowledge Hub*. Retrieved from www.iso.org/files/live/sites/isoorg/files/news/magazine/ISOfocus%20 (2013-NOW)/en/2018/ISOfocus_126/ISOfocus_126_en.pdf

Paulete, E. F., Orr, B., & Samteh, S. (2008). Incorporating EE into grassroots reformation efforts: The Gambia All School Tree Nursery Competetion. In A.B. Temu, S.A.O. Chamshama, J. Kung'u et al. (Eds.), *New perspectives in forestry education* (pp. 319 – 331). Republic of Kenya: Nairobi.

Pielke, R. A., Prins, G., Rayner, S., & Sarewitz, D. (2007). Climate change 2007: Lifting the taboo on adaptation. *Nature*, *445*(7128), 597–598. doi:10.1038/445597a PMID:17287795

Pilke, O., Pilkey-Jarvis, L., & Pilkey, K. (2017). *Retreat from a rising sea: Hard choices in an age of climate change*. New York, NY: Columbia University Press.

Plattner, T. (2005). Modelling Public Risk Evaluation Of Natural Hazards: A Conceptual Approach. *Natural Hazards and Earth System Sciences*, *5*(3), 357–266. doi:10.5194/nhess-5-357-2005

Plodinec, M. J., Edwards, W. C., & White, R. K. (2014). Applications of a "Whole Community" Framework for Enhancing Community or Campus Resilience. *Procedia Economics and Finance*, *18*, 9–16. doi:10.1016/S2212-5671(14)00907-1

Pokrant, R. J., & Stocker, L. (2011). Anthropology, climate change and coastal planning. In H. Kopnina & E. Shoreman-Ouimet (Eds.), *Environmental anthropology today* (pp. 179–194). New York: Routledge.

Polak, F. (1973). *The Image of the Future*. New York: Jossey-Bass.

Popkin, B. M., D'Anci, K. E., & Rosenberg, I. H. (2010). Water, hydration and health. *Nutritional Review*, *68*(8), 439-548. doi:10.1111/j.1753-4887.2010.00304.x

Pratt, W. M., McGuigan, C. C., & Katzev, A. R. (2000). Measuring program outcomes: Retrospective pretest methodology. *The American Journal of Evaluation*, *21*(3), 341–349. doi:10.1177/109821400002100305

Price, D. T., & Apps, M. J. (1996). Boreal forest responses to climate-change scenarios along an ecoclimatic transect in central Canada. *Climatic Change*, *34*(2), 179–190.

Priyanto, Y., Fanani, Z., & Sasmitojati, S. (2013). Environmental awareness as an education paradigm for sustainable development in vocational high schools of Kediri East Java. *International Journal of Academic Research*, *5*(5), 7–12. doi:10.7813/2075-4124.2013/5-5/B.1

Projects Abroad. (2014). Teen volunteer abroad: High School specials. Retrieved from: http://www.projects-abroad.org/volunteer-projects/high-school-specials/

PROVIA. (2013). *PROVIA guidance on assessing vulnerability, impacts and adaptation to climate change. Consultation document.* Nairobi, Kenya: United Nations Environment Programme. Retrieved from http://www.unep.org/provia

Prutsch, A., Grothmann, T., Schauser, I., Otto, S., & McCallum, S. (2010). Guiding principles for adaptation to climate change in Europe (ETC/ACC Technical Paper 2010/6).

Quirog, S., & Suarez, C. (2016). Climate change and drought effects on rural income distribution in the Mediterranean: A case study for Spain. *Natural Hazards and Earth System Sciences*, *16*(6), 1369–1385. doi:10.5194/nhess-16-1369-2016

Radford, T. (2018). Coral reefs face infection risks from plastics. *Climate Change Network.* Retrieved from https://climatenewsnetwork.net/coral-reefs-face-infection-risk-plastic/

Rajsekhar, D., & Gorelick, S. M. (2017). Increasing drought in Jordan: Climate change and cascading Syrian land-use impacts on reducing transboundary flow. *Science Advances*, *3*(8), 1–15. doi:10.1126ciadv.1700581 PMID:28875164

Ramsar Convention Secretariat. (2016). *The Fourth Ramsar Strategic Plan 2016–2024. Ramsar handbooks for the wise use of wetlands* (5th ed., Vol. 2). Gland, Switzerland: Ramsar Convention Secretariat.

Ramsar. (2015). *12th Meeting of the Conference of the Parties to the Convention on Wetlands, Compilation of COP12 Resolutions.* Retrieved from https://www.ramsar.org/sites/default/files/documents/library/cop12_resolutions_pdf_e.pdf

Randall, R. (2009). Loss and climate change: The cost of parallel narratives. *Ecopsychology*, *1*(3), 118–129. doi:10.1089/eco.2009.0034

Rapaport, E., Manuel, P., Krawchenko, T., & Keefe, J. (2015). How Can Aging Communities Adapt to Coastal Climate Change? Planning for Both Social and Place Vulnerability. *Canadian Public Policy*, *41*(2), 166–177. doi:10.3138/cpp.2014-055

Redman, C. L. (2014). Should sustainability and resilience be combined or remain distinct pursuits? *Ecology and Society*, *19*(2), 37. doi:10.5751/ES-06390-190237

Redman, C. L. (2015). Adaptation, transformation and transition. In K. C. Seto, W. D. Solecki, & C. A. Griffith (Eds.), *The Routledge Handbook of Urbanization and Global Environmental Change*. London, UK: Routledge.

Reininger, B. M., Rahbar, M. H., Lee, M., Chen, Z., Alam, S. R., Pope, J., & Adams, B. (2013). Social capital and disaster preparedness among low income Mexican Americans in a disaster prone area. *Social Science & Medicine*, 8350–8360. doi:10.1016/j.socscimed.2013.01.037 PMID:23465204

Repetto, R., & Easton, R. (2010). Changing Climate, More Damaging Weather. *Issues in Science and Technology*, *26*(2).

Report of Intergovernmental Panel on Climate Change. (2007). *Working Group 2: Climate Change Impacts, Adaptation and Vulnerability*. Author.

Resilience. (n.d.). Oxford English dictionary. Retrieved from https://en.oxforddictionaries.com/definition/resilient

Reuter, C., Hughes, A. L., & Kaufhold, M. A. (2018). Social media in crisis management: An evaluation and analysis of crisis informatics research. *International Journal of Human-Computer Interaction*, *34*(4), 280–294. doi:10.1080/10447318.2018.1427832

Rivera, M.A.J., Manning, M.M. & Krupp, D.A. (2013). A Unique Marine and Environmental Science Program for High School Teachers in Hawai'i: Professional Development, Teacher Confidence, and Lessons Learned. *International Journal of Environmental and Science Education*, *8*(2), 217-239.

Rosenzweig, C., Iglesias, A., Yang, X. B., Epstein, P., & Chivian, E. M. (2000). *Climate Change and U.S. Agriculture: The Impacts of Warming and Extreme Weather Events on Productivity Plant Diseases, and Pests, Center for Health and the Global Environment*. Harvard.

Rothstein, B. (2013). Corruption and social trust: Why the fish rots from the head down. *Social Research*, *80*(4), 1009–1032.

Rwebangira, M. (1996). *The Legal Status of Women and Poverty in Tanzania*. Nordiska Afrikainstitutet.

Salick & Byg. (2007). *Indigenous Peoples and Climate Change*. Tyndall Centre for Climate Change Research.

Sammon, A. (2018). Meet the Women Growing the California Seaweed Economy. *Civil Eats*. Retrieved from https://civileats.com/2018/03/23/meet-the-women-behind-californias-first-open-water-seaweed-farm/

Sampson, R. J., & Groves, W. B. (1989). Community structure and crime: Testing social-disorganization theory. *American Journal of Sociology*, *94*(4), 774–802. doi:10.1086/229068

Sander, S. (199). *Hunting for hope: A father's journey*. Boston: Beacon Press.

Sardar, Z. & Sweeney, J. (2015). The Three Tomorrows of Postnormal Times. *Futures*, *75*, 1-13.

Sardar, Z. (2017). *The Postnormal Times Reader*. Great Britain: Centre for Postnormal Politics and Futures Studies.

Sardar, Z., & Sweeney, J. A. (2016). The three tomorrows of postnormal times. *Futures*, *75*, 1–13. doi:10.1016/j.futures.2015.10.004

Sarmiento, J. L., Slater, R. D., Fasham, M. J. R., Ducklow, H. W., Toggweiler, J. R., & Evans, G. T. (1993). A seasonal three-dimensional ecosystem model of nitrogen cycling in the North Atlantic Euphotic Zone. *Global Biogeochemical Cycles*, *7*(2), 417–450. doi:10.1029/93GB00375

Scanlon, B. R., Faunt, C. C., Longuevergne, L., Reedy, R. C., Alley, W. M., McGuire, V. L., & McMahon, P. B. (2012, June 12). Groundwater depletion and sustainability of irrigation in the US High Plains and Central Valley. *Proceedings of the National Academy of Sciences of the United States of America*, *24*(24), 9320–9325. doi:10.1073/pnas.1200311109 PMID:22645352

Scherner, F., Horta, P. A., de Oliveira, E. C., Simonassi, J. C., Hall-Spencer, J. M., Chow, F., ... Pereira, S. M. B. (2013). Coastal urbanization leads to remarkable seaweed species loss and community shifts along the SW Atlantic. *Marine Pollution Bulletin*, *76*(1-2), 106–115. doi:10.1016/j.marpolbul.2013.09.019 PMID:24090881

Schipper, E. L. F., Thomalla, F., Vulturius, G., Davis, M., & Johnson, K. (2016). Linking disaster risk reduction, climate change and development. *International Journal of Disaster Resilience in the Built Environment*, *7*(2), 216–228. doi:10.1108/IJDRBE-03-2015-0014

Schmidt, C. W. (2009). Beyond Mitigation. *Environmental Health Perspectives*, *117*(7), A306–A309. doi:10.1289/ehp.117-a306 PMID:19654902

Scholtz, W. (2011). Climate change. USAID disaster reduction training course for southern Africa (USAID/OFDA. Project ACDS1).

Schreiner, C., Henriksen, E. K., & Kirkeby Hansen, P. J. (2005). *Climate education: Empowering today's youth to meet tomorrow's challenges*. Academic Press.

Schultz, M., Tyrrell, T. D., & Ebenhard, T. (2016). The 2030 Agenda and Ecosystems - A discussion paper on the links between the Aichi Biodiversity Targets and the Sustainable Development Goals. *SwedBio at Stockholm Resilience Centre*, Stockholm, Sweden. Retrieved from http://swed.bio/wp-content/uploads/2016/11/The-2030-Agenda-and-Ecosystems_spread.pdf

Schusler, T., & Krasny, M. (2008). Youth participation in local environmental action: Developing political and scientific literacy. In A. Reid, B. B. Jensen, J. Nikel, & V. Simovska (Eds.), *Participation and Learning: Perspectives on Education and the Environment, Health and Sustainability* (pp. 268–284). Springer. doi:10.1007/978-1-4020-6416-6_17

Schuster, P. F., Schaefer, K. M., Aiken, G. R., Antweiler, R. C., Dewild, J. F., Gryziec, J. D., ... Zhang, T. (2018). Permafrost stores a globally significant amount of mercury. *Geophysical Research Letters*, *45*(3), 1463–1471. doi:10.1002/2017GL075571

Scott, A. (2017). Making governance work for water–energy–food nexus approaches (working paper). *Overseas Development Institute and Climate and Development Knowledge Network*. Retrieved from https://cdkn.org/resource/working-paper-making-governance-work-water-energy-food-nexus-approaches/?loclang=en_gb

SeaTrust Institute. (2014). AWARE. Retrieved from http://www.seatrustinstitute.org/AWARE.html

SeaTrust Institute. (n.d.). Current Projects. Retrieved from www.seatrustinstitute.org/projects

Secretariat. (2016). *Biodiversity and the 2030 Agenda for Sustainable Development [Technical Note]*. Montreal, Quebec, Canada: CBD, FAO, The World Bank, UNEP, UNDP

Serenella, S., Ciuffo, B. I., & Mijkamp, P. (2015). A systemic framework for sustainability assessment. *Ecological Economics*, *119*, 314–325. doi:10.1016/j.ecolecon.2015.09.015

Sheriff, A., & Ferguson, E. (1991). *Zanzibar Under Colonial Rule*. London: Currey.

Simovska, V., & Jensen, B. (2009). Conceptualizing participation – the health of children and young people. *Report Prepared For the European Commission Conference on Youth Health, Brussels, Belgium, July 9-10*. WHO Regional Office for Europe.

Singha, G.G., Cisneros-Montemayora, A.M., Swartz, W., Cheung, W., Guy, J.A., Kenny, T.A. (2018). A rapid assessment of co-benefits and trade-offs among Sustainable Development Goals. *Marine Policy, 93*, 223-231. doi:10.1016/j.marpol.2017.05.030

Slaughter, R. A. (2004). *Futures Beyond Dystopia*. London: Routledge Falmer.

Slaughter, R. A. (2010). *The Biggest Wake up Call in History*. Indooroopilly, Australia: Foresight International.

Smedley, T. (2017, April 12). Is the world running out of fresh water? *BBC*. Retrieved from www.bbc.com/future/story/20170412-is-the-world-running-out-of-fresh-water

Smith, K., & Lawrence, G. (2014). Flooding and food security: A case study of community resilience in Rockhampton. *Rural Society*, *23*(3), 216–228. doi:10.1080/10371656.2014.11082066

Smith, S. (1986). *The Development of Capitalism in Africa*. London: Methuen.

Soekawa, E. (2008). Women Rake it in. Marie Claire Magazine. Retrieved from http://www.accessmylibrary.com/coms2/summary_0286-36094964_ITM

Sovacool, B. K., D'Agostino, A. L., Meenawat, H., & Rawlani, A. (2012). Expert views of climate change adaptation in least developed Asia. *Journal of Environmental Management, Vol.*, *97*, 78–88. doi:10.1016/j.jenvman.2011.11.005 PMID:22325585

St. Lawrence University. (2014). *Aksik: Stories about adaptation and subsistence: Native voices from the frontiers of climate change*. Retrieved from http://aksik.org/

Stake, R. E. (2010). *Qualitative Research: Studying How Things Work*. New York, NY: Guilford. Press.

Stapp, W. B. (1997). *The concept of environmental education*. Academic Press.

Stapp, B. W. (1997). The Concept of Environmental Education. *The Journal of Environmental Education*, *1*(1), 30–31. doi:10.1080/00139254.1969.10801479

Steffen, W., Richardson, K., Rockström, J., Cornell, S. E., Fetzer, I., Bennett, E. M., ... Sorlin, S. (2015). Planetary boundaries: Guiding human development on a changing planet. *Science*, *347*(6223), 1259855. doi:10.1126cience.1259855 PMID:25592418

Stepath, C., & Whitehouse, H. (2006). *Encouraging Proximal Relations*. Queensland.

Stern, N. (2006). *Stern review: the economics of climate change*. Cambridge, UK: Cambridge University Press.

Stive, M., Fresco, L., Kabat, P., Parmet, B., & Veerman, C. (2011). How the Dutch plan to stay dry over the next century *Proceedings of the Institution of Civil Engineers (ICE) –. Civil Engineering (New York, N.Y.)*, *164*(CE3).

Stokols, D., Misra, S., Runnerstrom, M. G., & Hipp, J. A. (2009). Psychology in an age of ecological crisis: From personal angst to collective action. *The American Psychologist*, *64*(3), 181–193. doi:10.1037/a0014717 PMID:19348519

Strazdins, L., & Skeat, H. (2011). Weathering the future: Climate change, children and young people, and decision making: A Report to the Australian Research Alliance for Children and Youth. Canberra, Australia.

Subramanian, S. V. (2004). The relevance of multilevel statistical methods for identifying causal neighborhood effects. *Social Science & Medicine*, *58*(10), 1961–1967. doi:10.1016/S0277-9536(03)00415-5 PMID:15020011

Summary for policymakers. (2014). In Field, C. B., Barros, V. R., Dokken, D. J., Mach, K. J., Mastrandrea, M. D., Bilir, T. E., ... White, L. L. (Eds.), *Climate Change 2014: Impacts,Adaptation, and Vulnerability. Part A: Global and Sectoral Aspects. Contribution of Working Group II to the Fifth Assessment Report of the Intergovernmental Panel on Climate Change. IPCC* (pp. 1–32). Cambridge, UK: Cambridge University Press.

Tagicakibau. (2007). *"Pollution in Paradise": The Impact of Nuclear Testing and Radio-Active Pollution on Indigenous Peoples in the Pacific and Strategies for Resolution*. Pacific Concerns Resource Centre.

Tainter, J. A. (1988). *The Collapse of Complex Societies*. Cambridge, UK: Cambridge University Press.

Taleb, N. (2007). *The black swan*. New York: Random House.

Taleb, N. (2007). *The black swan: The impact of the highly improbable*. New York: Random House.

Taylor, A. H. (2011). *The dance of air and sea: How oceans, weather and life link together*. Oxford, UK: Oxford University Press.

TerraDaily. (2006). Zanzibar Plastic Bag Ban takes Effect as Environment Woes Mount. Retrieved from http://www.terradaily.com/reports/Zanzibar_Plastic_Bag_Ban_Takes_Effect_As_Environment_Woes_Mount_999.html

The Rockefeller Foundation. (2016, March). *Cities resilience index research report (Vol. 4).* Retrieved from www.cityresilienceindex.org/wp-content/uploads/2016/05/Vol4-MeasuringCityResilience.pdf

The Zanzibar Women Corporation. (2010). The International Collaborative for Science Education and the Environment.

Thomalla, F., Johnson, K., Bharwani, S., Johannessen, A., & Butterfield, R. (2015). *Transforming disaster risk reduction for more inclusive, equitable and sustainable development.* Global Sustainable Development Report. *Briefs*. Retrieved from https://sustainabledevelopment.un.org/content/documents/616462-thomalla%20et%20al%20-%20transforming%20disaster%20risk%20reduction%20for%20more%20inclusive%20equitable%20sd.pdf

Tilbury, D., Stevenson, R., Fien, J. & Schreuder, D. (2002). Education and Sustainability. Responding to the Global Challenge. *The World Conservation Union, Commission on Education and Communication CEC.*

Troger, S. (2011). Failing Seasons, ailing societies: Climate change and the meaning of adaptation. *Horn of Africa REC.* Retrieved from www.hoarec.org

U.S. Department of Health and Human Services. (2012). Community resilience fact sheet. Retrieved from http://www.phe.gov/Preparedness/planning/abc/Documents/community-resilience.pdf

U.S. Department of Homeland Security. (2010). Quadrennial homeland security report review: A strategic framework for a secure homeland. Retrieved from http://www.dhs.gov/xlibrary/assets/qhsr_report.pdf

U.S. Department of Homeland Security. (2013). National Response Framework (2nd ed.). Retrieved from http://www.fema.gov/national-response-framework

U.S. Department of Homeland Security. (2014). The 2014 quadrennial homeland security review. Retrieved from http://www.dhs.gov/sites/default/files/publications/qhsr/2014-QHSR.pdf

U.S. Geological Survey. (2016). *The world's waters.* Retrieved from: http://water.usgs.gov/edu/earthwherewater.html

U.S. Global Change Research Program. (2001). *Climate Change Impacts On The United States: The Potential Consequences Of Climate Variability And Change*. Cambridge, UK: Cambridge University Press.

Uitto, J. I., & Shaw, R. (2016). Sustainable Development and Disaster Risk Reduction: Introduction. In J. I. Uitto & R. Shaw (Eds.), *Sustainable Development and Disaster Risk Reduction* (pp. 1–12). Japan: Springer. doi:10.1007/978-4-431-55078-5_1

UN Department of Economic and Social Affairs. (2017, June 28). *'Inclusive, equitable and quality education' at the heart of high-level UN event.* Retrieved from https://www.un.org/development/desa/en/news/sustainable/high-level-event-education.html

UN Fact Sheet. (2009). Women, Gender Equality and Climate Change. Retrieved from www.un.org/womenwatch/feature/climate_change/downloads/Women_and_Climate_Change_Factsheet.pdf

UN General Assembly (2016, December 1). *Report of the open-ended intergovernmental expert working group on indicators and terminology relating to disaster risk reduction* A/71/644. Retrieved from undocs.org/A/71/644

UN General Assembly. (2015, October 21). *Transforming our world: The 2030 Agenda for Sustainable Development*. A/RES/70/1. Retrieved from http://www.refworld.org/docid/57b6e3e44.html

UN General Assembly. (2018). *Global indicator framework for the Sustainable Development Goals and targets of the 2030 Agenda for Sustainable Development* A/RES/71/313 E/CN.3/2018/2. Retrieved from https://unstats.un.org/sdgs/indicators/indicators-list/

UN. (2015). *Water for a sustainable world, The United Nations World Water Development Report 2015*, United Nations. Paris, France: UNESCO on behalf of UN-Water. Retrieved from http://unesdoc.unesco.org/images/0023/002318/231823E.pdf

UNCCD. (2016). A Natural Fix: A Joined-up Approach to Delivering the Global Goals for Sustainable Development. Retrieved from https://www.unccd.int/publications/natural-fix-joined-approach-delivering-global-goals-sustainable-development

UNCCD. (2017). *The future strategic framework of the Convention, The UNCCD 2018–2030 Strategic Framework ICCD/COP(13)/L.18 (14 September 2017)*. Retrieved from https://www.unccd.int/official-documents/cop13-l-documents/iccdcop13l18

UNDESA. (2010). *World Youth Report: Youth and Climate Change*. United Nations Department of Economic and Social Affairs.

UNESCO & UNEP. (2011). *YouthXChange guidebook series: Climate change and lifestyles*. Paris: United Nations Educational, Scientific and Cultural Organization (UNESCO). Retrieved from www.youthxchange.net

UNESCO. (2008). Education and the search for a sustainable future, policy dialogue 1: ESD and Development Policy. Retrieved from http://unesdoc.unesco.org/images/0017/001791/179121e.pdf

UNESCO. (2009). Trends in global higher education: Tracking an academic revolution. Retrieved from http://unesdoc.unesco.org/images/0018/001831/183168e.pdf

UNESCO. (2009). *UNESCO International Seminar on Climate Change Education*. Paris: UNESCO.

UNESCO. (2010). The UNESCO climate change initiative. Climate change education for sustainable development. Paris: UNESCO. Retrieved from http://unesdoc.unesco.org/imageS/0019/001901/190101e.pdf)

UNESCO. (2012). 7th UNESCO Youth Forum Assessment. United Nations Educational, Scientific and Cultural Organization.

UNESCO. (2017). Education for Sustainable Development Goals: Learning objectives. Retrieved from http://unesdoc.unesco.org/images/0024/002474/247444e.pdf

UNESCO. (2018). *Ocean Acidification.* Retrieved from http://www.unesco.org/new/en/natural-sciences/ioc-oceans/focus-areas/rio-20-ocean/blueprint-for-the-future-we-want/ocean-acidification/

UNFCCC. (2006). *United Nations Framework Convention on Climate Change: Handbook. Intergovernmental and Legal Affairs*. Bonn, Germany: Climate Change Secretariat.

UNFCCC. (2015). *Adoption of the Paris Agreement (FCCC/CP/2015/L.9/Rev.1)* United Nations Framework Convention on Climate Change, Retrieved from: https://unfccc.int/resource/docs/2015/cop21/eng/l09r01.pdf

UNFCCC. (2015). *Paris agreement. FCCC/CP/2015/L.9/Rev1*. Retrieved from http://unfccc.int/resource/docs/2015/cop21/eng/l09r01.pdf

UNFCCC. (2017). *Opportunities and options for integrating climate change adaptation with the Sustainable Development Goals and the Sendai Framework for Disaster Risk Reduction 2015–2030* (Report No. FCCC/TP/2017/3). Retrieved from http://unfccc.int/resource/docs/2017/tp/03.pdf

UN-HABITAT. (2012). *Climate Change and Pro-Poor Urban Governance Capacity Development Workshop: Promoting Sustainable Human Development in Asian Cities*. Workshop Summary Report. Bangkok: UNDP.

UNICEF. (2014). Climate change and environmental education: a companion to the child friendly schools. Retrieved from www.unicef.org/publications/CFS-climate-e-pdf

UNISDR. (2015). *Sendai Framework for Disaster Risk Reduction 2015 – 2030*, United Nations International Strategy for Disaster Reduction. Retrieved from http://www.unisdr.org/we/coordinate/sendai-framework

UNISDR. (2017). *2017 Global Platform for Disaster Risk Reduction Proceedings.* United Nations International Strategy for Disaster Reduction. Retrieved from https://www.unisdr.org/we/inform/publications/55465

UNISDR. (2017). *Sendai Framework for Disaster Risk Reduction*. Available from: www.unisdr.org/we/coordinate/sendai-framework

United Nations (2017). Report of the Secretary-General, Progress towards the Sustainable Development Goals, E/2017/66.

United Nations Department of Economic and Social Affairs. (2015). Sustainable Development Goals. Retrieved From https://sustainabledevelopment.un.org/?menu=1300

United Nations Department of Economic and Social Affairs. (2018). Division for Sustainable Development Goals: Transforming our world: The 2030 Agenda for Sustainable Development. Retrieved from https://sustainabledevelopment.un.org/about

United Nations Department of Economic and Social Affairs. (2018). Sustainable Development Goal Four. Retrieved from https://sustainabledevelopment.un.org/sdg4

United Nations Department of Economic and Social Affairs. (n.d.). *United Nations Conference on Sustainable Development, Rio+20*. Retrieved from https://sustainabledevelopment.un.org/rio20.html

United Nations Development Programme. (2009). *Climate Change Perspectives from India*. UNDP.

United Nations Development Programme. (2018). *Foresight manual: Empowered futures for the 2013 Agenda*. Singapore: UNDP Global Centre for Public Service Excellence.

United Nations Division for Sustainable Development Goals. (2018). Sustainable development goals. Retrieved from https://sustainabledevelopment.un.org/sdgs

United Nations Educational, Scientific and Cultural Organization. (2012). Youth and skills: Putting education to work (EFA Global Monitoring Report). France. Retrieved from http://unesdoc.unesco.org/images/0021/002180/218003e.pdf

United Nations Educational, Scientific and Cultural Organization. (2018). World Radio Day 2018. Retrieved from http://www.diamundialradio.org/home

United Nations Environment Assembly of the United Nations Environment Programme. (2014). *Programme of research on climate change vulnerability, impacts and adaptation*. Nairobi: Author.

United Nations Environment Programme. (2014). *Emerging issues for Small Island Developing States: Results of the UNEP Foresight Process*. Nairobi, Kenya: UNEP.

United Nations Environmental Program/ United Nations Environmental Program/Food and Agriculture Organization of the U.N./Priority Action Programs/CDA. (2000). Progress in Integrated Coastal Management (ICAM) for Sustainable Development of Zanzibar's Coast: Unguja Island Coastal Profile and Management Strategy. East African Regional Seas Technical Reports, Split, Croatia.

United Nations Framework Convention on Climate Change. (2015, December 15). *Adoption of the Paris Agreement* (Report No. FCCC/CP/2015/L.9/Rev.1). Retrieved from http://unfccc.int/resource/docs/2015/cop21/eng/l09r01.pdf

United Nations Office for Disaster Risk Reduction (UNISDR). (2015). Chart for the Sendai Framework for Disaster Risk Reduction 2015-2030. Retrieved from https://www.preventionweb.net/publications/view/44983

United Nations. (2015). Transforming our world: The 2030 agenda for sustainable development. Retrieved from https://sustainabledevelopment.un.org/post2015/transformingourworld

United Nations. (2017). International migration report 2017. Retrieved from http://www.un.org/en/development/desa/population/migration/publications/migrationreport/docs/MigrationReport2017.pdf

United Nations. (2017). Progress towards the Sustainable Development Goals. Retrieved from https://sustainabledevelopment.un.org/sdg2

United Nations. (2017, July 6). General Assembly resolution 71/313, Work of the Statistical Commission pertaining to the 2030 Agenda for sustainable development. Retrieved from undocs.org/A/RES/71/313/

United Nations. (n.d.). UNGAR 67/290 High Level Political Forum - General Assembly Resolution 67/290. Retrieved from http://www.un.org/ga/search/view_doc.asp?symbol =A/RES/67/290&Lang=E

United States Environmental Protection Agency. (2014). Adaptation overview. Retrieved from http://epa.gov/climatechange/impacts-adaptation/adapt-overview.html

University of Leicester. (2015, December 1). Failing phytoplankton, failing oxygen: Global warming disaster could suffocate life on planet Earth. *ScienceDaily*. Retrieved from www.sciencedaily.com/releases/2015/12/151201094120.htm

UN-Water. (2013). *Water Security & the Global Water Agenda A UN-Water Analytical Brief.* Canada: United Nations University Institute for Water, Environment & Health. Retrieved from https://collections.unu.edu/view/UNU:2651

UN-Water. (2016). *Water and Sanitation Interlinkages across the 2030 Agenda for Sustainable Development*. Geneva, Switzerland: UN-Water. Retrieved from http://www.unwater.org/publications/water-sanitation-interlinkages-across-2030-agenda-sustainable-development/

UN-Water. (2018). *Executive Summary – SDG 6 Synthesis Report 2018 on Water and Sanitation*. UN-Water. Retrieved from http://www.unwater.org/publications/executive-summary-sdg-6-synthesis-report-2018-on-water-and-sanitation/

UN-Water. (n.d.). *Water scarcity*. Retrieved from http://www.unwater.org/water-facts/scarcity/

UNWWAP. (2018). *2018 UN World Water Development Report, Nature-based Solutions for Water*. World Water Assessment Programme. Retrieved from: http://www.unesco.org/new/en/natural-sciences/environment/water/wwap/wwdr/2018-nature-based-solutions/

Up in Smoke? Latin America and Caribbean: The Threat from Climate Change to the Environment Human Development. (2006). The Third Report for the Working Group on Climate Change and Development.

USAID Office of Population and Reproductive Health. (2014). Marine Ecosystems and Management. Interview with Heather D'Agnes. Seattle: Marine Affairs Research and Education.

USCD. (2018). About (Cal-CAB), California Center for Algae Biotechnology, UC San Diego. Retrieved from https://algae.ucsd.edu/about/index.html

USCD. (2018b). Women in Algae, California Center for Algae Biotechnology, UC San Diego. Retrieved from https://algae.ucsd.edu/events/womeninalgae.html

Uy, N., Delfino, R. J. P., & Shaw, R. (2016) Ecosystem-Based Disaster Risk Reduction: Experiences, Challenges, and Opportunities in the Post-2015 Development Agenda. In J. I. Uitto & R. Shaw (Eds.), *Sustainable Development and Disaster Risk Reduction* (pp. 119–142). Japan: Springer. doi:10.1007/978-4-431-55078-5_8

Vacca, R. (1973). *The Coming Dark Age*. Garden City, N.Y: Doubleday.

Van Dam-Mieras, R. (2006). Learning for sustainable development: Is it possible within the established higher education structures? In J. Holmberg & B. E. Samuelsson (Eds.), *Drivers and barriers for implementing sustainable development in higher education. Education for sustainable development in Action* (pp. 13–18). Paris: UNESCO Education Sector.

Van Poeck, K., Læssøe, J., & Block, T. (2017). An exploration of sustainability change agents as facilitators of nonformal learning: Mapping a moving and intertwined landscape. *Ecology and Society*, *22*(2), 33. doi:10.5751/ES-09308-220233

Vladimirova, K., & Le Blanc, D. (2016). Exploring Links Between Education and Sustainable Development Goals Through the Lens of UN Flagship Reports. *Sustainable Development*, *24*(4), 254–271. doi:10.1002d.1626

VolunteerMatch.org. (2014). Opportunities. Retrieved from https://www.volunteermatch.org/search?l=USA&k=Environment

Vorosmarty, C. J., Green, P., Salisbury, J., & Lammers, R. B. (2000, July 14). Global water resources: Vulnerability from climate change and population growth. *Science*, *289*(5477), 284–288. doi:10.1126cience.289.5477.284 PMID:10894773

Wagner, L., & Pasini, O. (2018). 28th UN-Water Meeting. *UN Water Bulletin*, *82*(32). Retrieved from http://enb.iisd.org/water/un/28/html/enbplus82num32e.html

Walker, B. H., Gunderson, L. H., Kinzig, A. P., Folke, C., Carpenter, S. R., & Schultz, L. (2006). A handful of heuristics and some propositions for understanding resilience in social-ecological systems. *Ecology and Society*, *11*(1), 13. Retrieved from http://www.ecologyandsociety.org/vol11/iss1/art13/ doi:10.5751/ES-01530-110113

Wallace-Wells, D. (2017). The Uninhabitable Earth. *New York Magazine*. Retrieved from http://nymag.com/daily/intelligencer/2017/07/climate-change-earth-too-hot-for-humans.html

Wallace-Wells, D. (2017). The uninhabitable earth. The New Yorker. Retrieved from http://nymag.com/daily/intelligencer/2017/07/climate-change-earth-too-hot-for-humans.html

Wallevik, H., & Jiddawi, N. S. (1999). Impacts of tourisms on woman activities at the South Coast of East of Unguja. In *Proceedings of the 20th Anniversary Conference on Advances in Marine Sciences in Tanzania* (pp. 535-550).

Wals, A. (2010). Mirroring, Gestaltswitching and Transformative Social Learning: Stepping Stones for Developing Sustainability Competence. *International Journal of Sustainability in Higher Education, 11*(4), 380–390. doi:10.1108/14676371011077595

Wals, A. E. J., Mochizuki, Y., & Leicht, A. (2017). Critical case-studies of non-formal and community learning for sustainable development. *International Review of Education, 63*(6), 783–792. doi:10.100711159-017-9691-9

Walton, M. (2008). Algae: The Ultimate in Renewable Energy. *CNN*. Retrieved from http://www.cnn.com/2008/T ECH/science/04/01/algae.oil/index.html

Waters, T. (2018) Seaweed Farming: A Gateway to Conservation and Empowerment, Cool Green Science (Smarter by Nature), *The Nature Conservancy*. Retrieved from https://blog.nature.org/science/2018/02/12/seaweed-farming-a-gateway-to-conservation-and-empowerment/

Webler, T., Tuler, S., Dow, K., Whitehead, J., & Kettle, N. (2014). Design and evaluation of a local analytic-deliberative process for climate adaptation planning. *Local Environment: The International Journal of Justice and Sustainability*. doi:10.1080/13549839.2014.930425

Weiss, H. B., Little, P. M., & Bouffard, S. M. (2005). More than just being there: Balancing the participation equation. *New Directions for Youth Development*, 105. PMID:15943135

WHO/EHA. (2002). *Disasters & Emergencies: Definitions. Training Package*. WHO/EHA Panafrican Emergency Training Center.

Wibek, V. (2014). Enhancing learning, communication and public engagement about climate change – some lessons from recent literature. *Environmental Education Research, 20*(3), 387–411. doi:10.1080/13504622.2013.812720

Wikipedia. (n.d.). Complex adaptive systems. Retrieved from https://en.wikipedia.org/wiki/Complex_adaptive_system

Wilbanks, T. J., Kane, S. M., Leiby, P. N., Perlack, R. D., Settle, C., Shogren, J. F., & Smith, J. B. (2003). Possible Responses to global climate change: Integrating mitigation and adaptation. *Environment, 45*(5), 28–38. doi:10.1080/00139150309604547

Willig, J. (Ed.). (2008). Qualitative Research Methods in Education and Educational Technology. Charlotte: Information Age Publishing.

Wilson, L. (2014, December). *Notes taken at the United Nations Framework Convention on Climate Change, COP20 / CMP10 Special Session*. Intergenerational Inquiry: Youth as Agents of Change, Lima, Peru.

Wilson, L. (2014, October 22). Running dry in a changing world [Commentary]. *Westlaw Journal Thomson Reuters Blog*. Retrieved from http://thomsonreuters.com/index.php/running-dry-in-a-changing-world

Wilson, D. (2018). *Designing the Purposeful World: The Sustainable Development Goals as a Blueprint for Humanity*. New York: Routledge.

Wilson, E. O. (1998). *Consilience: A unity of knowledge*. New York: Vintage Books.

Wilson, L. (2017, November). The AWARE(SM) Pacific Islands/Global Climate Ambassadors Program. Presentation at *United Nations Framework Convention on Climate Change COP23*, Bonn, Germany.

Wilson, L. A. (2013). Every day is day #1: Resilience, climate change and health: How must our own work adapt? Presented at the *United Nations Framework Convention on Climate Change COP19*, Warsaw, Poland.

Wilson, L. A. (2013). *Interview with Dr. Charles Chiyama, Chiawa region, Zambia, July 13, 2013*. Lower Zambezi, Zambia: Chiawa Water and Beds Project.

Women Action Committee (WAC). (2010). *Using indigenous method to fight environmental disaster in schools*. Paper presented to secondary and primary schools in Aghamelum local Government area.

Wood, S.L.R., Jones, SK., Johnson J.A., Brauman, K.A., Chaplin-Kamer, R., Fremier, A., … DeClerk, F.A. (2018) Distilling the role of ecosystem services in the Sustainable Development Goals, *Ecosystem Services*, *29*(Part A), 70-82. doi:10.1016/j.ecoser.2017.10.010

Woodruff, P. (2001). *Reverence: Renewing a forgotten virtue*. Oxford, UK: Oxford University Press.

World Congress on Zero Emissions, United Nations University Zero Emissions Initiative. (n.d.). Retrieved from http://www.unu.edu/zef/publications _e/96_congress_proc.pdf

World Economic Forum. (2015). *WEF Outlook 2015: Outlook on the global agenda*. Retrieved from http://reports.weforum.org/outlook-global-agenda-2015/top-10-trends-of-2015/3-lack-of-leadership/

World Health Organization. (2014). *Malaria Fact Sheet*. Retrieved from http://www.who.int/mediacentre/factsheets/fs094/en/

World Health Organization. (n.d.). *Synergies for beating NCDs and promoting mental health and well-being*. Retrieved from http://www.who.int/ncds/governance/high-level-commission/synergies-beating-ncds/en/

WWAP (United Nations World Water Assessment Programme). (2015). *Facing the challenges; Case studies and indicators*. Paris, France: UNESCO.

Xue, Y., Hu, Z., Kumar, A., Banzon, V., Smith, T. M., & Rayner, N. A. (2014). Sea surface temperatures. *Bulletin of the American Meteorological Society*, S51–S54.

Yin, R. K. (2010). *Qualitative Research from Start to Finish*. New York, NY: Guilford Press.

Yohe, G. W., Lasco, R. D., Ahmad, Q. K., Arnell, N. W., Cohen, S. J., Hope, C., . . . Perez, R. T. (2007). Perspectives on climate change and sustainability. In *Climate Change 2007: Impacts, Adaptation and Vulnerability*. Contribution of Working Group II to the Fourth Assessment Report of the Intergovernmental Panel on Climate Change. Cambridge University Press.

Zarcadoolas, Boyer, Krishnaswami, & Rothenberg. (2007). GIS Maps To Communicate Emergency Preparedness: How Useable Are They For Inner City Residents? *Journal of Homeland Security and Emergency Management*, 4.

Zembroski, D. (2011). Sociological theories of crime and delinquency. *Journal of Human Behavior in the Social Environment*, *21*(3), 240–254. doi:10.1080/10911359.2011.564553

Zimmerman, R. (2009). Making Infrastructure Competitive in an Urban World. *The Annals of the American Academy of Political and Social Science*, *226*.

Zolli, A. (2012). *Resilience: Why things bounce back*. London: Headline Publishing Group.

About the Contributors

Lynn Wilson is Executive Director for SeaTrust Institute, a scientific and educational nonprofit organization in the Pacific Northwest. She is also senior analyst for the consulting firm OSSIA through which she has been offering research, conflict management, strategic and policy guidance for over 25 years. Her work appears in books, lectures, workshops, global and regional conferences and in academic, technical and trade publications. Recent publications include Environmental Issues, Policy and Management Collaboration Toolbox (2011), book chapters in E-Research Collaboration: Frameworks, Tools and Techniques, (2010), and Handbook of Research on Electronic Collaboration and Organizational Synergy (2009) which she co-edited. Current research on climate and health collected as Co-Chair of the United Nations Framework Convention on Climate Change (UNFCCC) Coalition on Health and the Environment: Climate Change Initiative, in collaboration with global climate and health professionals, is being continuously presented through the United Nations climate change negotiations process throughout 2011-2013 and is accompanied by a further partnership with African NGO Coalition Co-Chair to deliver local training on climate adaptation and disease surveillance throughout the globe. Research from the Coalition and the local projects inform the book Climate and Health Literacy: A Guide to Understanding and Action" scheduled for publication in 2012. Dr. Wilson develops and delivers face-to-face and online courses related to science and policy, sustainability and collaboration and is a graduate instructor for Environmental Policy in the School of Legal Studies at Kaplan University. As Director of the Consortium on Climate Change and Population Health, she leads a project for innovative research and publications on climate and human health concerns. Lynn serves on advisory and review boards for policy, education, science and the environment and is Head of Delegation, SeaTrust Institute, NGO Focal Point to Conference of Parties/ UNFCCC, United Nations Climate Change Conference, Copenhagen, DK, 2009, Cancun, MX 2010, and Durban SA, 2011.

Carolyn N. Stevenson is a Full-Time Faculty member at Purdue University Global. She has over 18 years teaching and administrative experience in higher education. She holds a Master of Arts degree in Communication, Master of Business Administration, and Doctor of Education with an emphasis in Higher Education. Prior to pursuing a career in higher education, she worked in the publishing field and served as a technical writing consultant. She currently serves as Associate Editor for the International Journal of Technologies and Educational Marketing (IJTEM), published by IGI-Global; Editorial Board Member and Reviewer for the Journal of Education and Learning published by the Canadian Center of Science and Education. Recent publications include a chapter entitled: "Leading across Generations: Issues for Higher Education Administrators" published in the Handbook of Research on Transnational Higher Education Management, IGI Global; Technical Writing: A Comprehensive Resource for Technical Writers at all Levels, (Martinez, Hannigan, Wells, Peterson and Stevenson) Revised and Updated Edition, Kaplan Publishing, Building Online Communities in Higher Education Institutions: Creating Collaborative Experience (with Joanna Bauer), published by IGI Global, and Promoting Climate Change Awareness through Environmental Education (with Lynn Wilson), published by IGI Global.

* * *

Innocent Chirisa is a Full Professor in the Department of Rural and Urban Planning at the University of Zimbabwe. He holds a DPhil in Social Sciences with an inclination towards regional and urban planning. He is also the current Deputy Dean of the Faculty of Social Studies and the Chairman of the Department of Rural and Urban Planning. He joined the University in 2007 having taught, for two years in the Department of Civil Engineering at Harare Polytechnic. He is also a Research Fellow in the Department of Urban and Regional Planning, University of Free State, Bloemfontein, South Africa. A prolific writer and author of a combination of over 100 refereed journal articles, book chapters and books achieved in 10 years, peri-urban housing is the chief focus of his research. He is responsible for teaching Environmental Planning and Management and has a keen interest in urban and peri-urban dynamics. He is currently focusing on environmental systems dynamics with respect to land-use, ecology, water and energy. Prof Chirisa strongly believes in diligence as his anchoring philosophy of which Proverbs 22:29 is his fuel in achieving his ambition: "Seest thou a man diligent in his business? He shall stand before kings; he shall not stand before mean men." (The Bible, King James Version).

LaMesha Craft received her PhD in Public Policy and Administration from Walden University in 2017. In 2018, she was hired by Coastal Carolina University as a Lecturer of Intelligence and National Security Studies. Before full-time teaching, she served 20 years of active duty in the U.S. Army as an All-Source Intelligence Warrant Officer. She has provided strategic and operational intelligence analysis of nation-state and non-state threats to U.S. interests, policy, data, and networks in/around Asia, Europe, the Horn of Africa, the Middle East, Southwest Asia. Dr. Craft writes and presents on the threats of agro-terrorism and the potential for anomie; setting the conditions for community development and competence during crises; and emerging threats during post-normal times.

Janos Csala is a research assistant at SeaTrust Institute supporting education, and policy-oriented research. He is driven to contribute to transforming challenges to opportunities for regenerative sustainable development. His work supported the development and presentation of materials at the 23rd meeting of the UNFCCC's Conference of the Parties and the 48th meeting of the Subsidiary Body for Implementation. He is currently pursuing his MSc degree in Industrial Ecology at TU Delft University and Leiden University.

Nadra Hashim works as reviewer for the Journal of Agriculture, Food Systems, and Community Development. She has also assisted served as grant writer for Hunger Reduction, International, an East African food security NGO. In 2011, Dr. Hashim was appointed visiting/ founding assistant dean and associate professor at Jindal Global University School of International Affairs in Sonipat, India. Since returning from India, Ms. Hashim has published a book entitled - Hemp and the Global Economy The Rise of Labor, Innovation, and Trade. Dr. Hashim was an inaugural Ford Fellow at Amherst College in 2001 and received her PhD from the University of Virginia in 2006.

Christopher Burr Jones is a faculty member of the School of Public Policy and Administration, Walden University, where he has worked for 9 years. He holds a PhD in Political Science and MA in Alternative Futures from the University of Hawaii at Manoa and has taught in higher education systems in Hawaii, Oregon, Texas, and Colorado. He is engaged in research in accelerating climate change, futures studies, and postnormal theory and politics. He is a Fellow in the World Futures Studies Federation, Senior Fellow in the Centre for Postnormal Policy and Futures Studies, and member of the Association of Professional Futurists.

Mphemelang Ketlhoilwe is a lecturer of Environmental Education at the University of Botswana. He teaches both post graduate and undergraduate courses. His research interest are gender based natural resources management, education for sustainable development, teacher education, and project based learning.

Abraham R. Matamanda is a PhD student in the department of Urban and Regional Planning, University of Free State. His PhD seeks to explore the emerging human settlement forms and urban dilemmas nexus focusing on citizens' access to basic services such as water, sanitation, public transport and security issues. Abraham is a keen researcher and emerging academic who has shown his interest in research work through publications in peer-reviewed local and international journals such as City, Territory and Architecture; Regional Development Dialogue and Zambezia among others. His recent work has been published in IGI Global and Journal of Public Administration and Development Alternatives. Specifically Abraham's research focuses on urbanisation, sustainability, environmental planning, rural development and the planning of cities and towns.

Jennifer Wanjiku Mwangi is a Program Manager at SeaTrust Institute where she develops AWARE(SM) curriculum and enjoys exploring ways individuals and communities can discover the value of their environment and become better stewards of it. Jennifer has contributed to material that was presented at several UNFCCC Conference of the Parties meetings. She also works with interns on developing their research skills. Jennifer holds a Master of Science degree in Environmental Engineering from Michigan Technological University.

Ediola Pashollari is of Albanian nationality and holds a Masters in Business Administration (UK), a Masters of Arts in Political Science (UK), a Masters in Entreprenology (USA) and also a Bachelors Degree in LLB (UK). She is presently the Secretary General of the World Assembly of Youth (WAY) which is headquartered in Melaka, Malaysia. Through her capacity as the Secretary General of WAY and the Director of the World Youth Institute (WYI), she is responsible for the administration of the Secretariat and coordination of regional and international programmes and events. She is also the liaison for the membership and affiliate organizations, representative of the organization at the United Nations, intergovernmental and non-governmental platforms, and ministries responsible for youth. She coordinates humanitarian and relief programmes around the world and reports to the President, Executive Committee, and General Assembly.

Mona Zoghbi is a sustainable development consultant. She has worked with international organizations such as UNESCO and UNDP, as well as various universities, research institutes, and NGOs including in Lebanon, Arab region, Europe and Africa on climate change, sustainable development, public health, and youth environmental education and leadership. She also worked as national coordinator of youth projects at UNESCO Office in Beirut/Regional Bureau for Education in the Arab States and as a University Instructor on sustainability. Holder of a Master in Environmental Sciences and PhD in Sustainable Development.

Index

A

B

C

M

N

O

P

Q

R

S

Ensure Quality Research is Introduced to the Academic Community

Become an IGI Global Reviewer for Authored Book Projects

The overall success of an authored book project is dependent on quality and timely reviews.

In this competitive age of scholarly publishing, constructive and timely feedback significantly expedites the turnaround time of manuscripts from submission to acceptance, allowing the publication and discovery of forward-thinking research at a much more expeditious rate. Several IGI Global authored book projects are currently seeking highly qualified experts in the field to fill vacancies on their respective editorial review boards:

Applications may be sent to:
development@igi-global.com

Applicants must have a doctorate (or an equivalent degree) as well as publishing and reviewing experience. Reviewers are asked to write reviews in a timely, collegial, and constructive manner. All reviewers will begin their role on an ad-hoc basis for a period of one year, and upon successful completion of this term can be considered for full editorial review board status, with the potential for a subsequent promotion to Associate Editor.

If you have a colleague that may be interested in this opportunity, we encourage you to share this information with them.

www.ingramcontent.com/pod-product-compliance
Lightning Source LLC
LaVergne TN
LVHW081344300926
849161LV00008B/2061